Household and Living Arrangement Projections

THE SPRINGER SERIES ON

DEMOGRAPHIC METHODS AND POPULATION ANALYSIS

Series Editor

KENNETH C. LAND

Duke University

In recent decades, there has been a rapid development of demographic models and methods and an explosive growth in the range of applications of population analysis. This series seeks to provide a publication outlet both for high-quality textual and expository books on modern techniques of demographic analysis and for works that present exemplary applications of such techniques to various aspects of population analysis.

Topics appropriate for the series include:

- General demographic methods
- Techniques of standardization
- Life table models and methods
- Multistate and multiregional life tables, analyses and projections
- Demographic aspects of biostatistics and epidemiology
- Stable population theory and its extensions
- Methods of indirect estimation
- Stochastic population models
- Event history analysis, duration analysis, and hazard regression models
- Demographic projection methods and population forecasts
- Techniques of applied demographic analysis, regional and local population estimates and projections
- Methods of estimation and projection for business and health care applications
- Methods and estimates for unique populations such as schools and students

Volumes in the series are of interest to researchers, professionals, and students in demography, sociology, economics, statistics, geography and regional science, public health and health care management, epidemiology, biostatistics, actuarial science, business, and related fields.

For further volumes:
http://www.springer.com/series/6449

Yi Zeng • Kenneth C. Land • Danan Gu
Zhenglian Wang

Household and Living Arrangement Projections

The Extended Cohort-Component Method and Applications to the U.S. and China

Springer

Yi Zeng
Center for Study of Aging and Human Development Medical School
Duke University
Durham, NC, USA

National School of Development
Center for Healthy Aging and Development Studies
Peking University
Beijing, China

Danan Gu
Population Division
United Nations
New York, NY, USA

Kenneth C. Land
Department of Sociology and Center for Population Health and Aging
Population Research Institute
Duke University
Durham, NC, USA

Zhenglian Wang
Center for Population Health and Aging
Population Research Institute
Duke University
Durham, NC, USA

Household and Consumption Forecasting, Inc.
Chapel Hill, NC, USA

ISSN 1389-6784
ISBN 978-94-024-0496-8 ISBN 978-90-481-8906-9 (eBook)
DOI 10.1007/978-90-481-8906-9
Springer Dordrecht Heidelberg New York London

Printed on acid-free paper

Springer is part of Springer Science+Business Media (www.springer.com)

Preface

This book presents an innovative demographic toolkit known as the ProFamy extended cohort-component method for the projection of household structures and living arrangements with empirical applications to the United States, the largest developed country, and China, the largest developing country. The ProFamy method uses demographic rates as inputs to project detailed distributions of household types and sizes, living arrangements of all household members, and population by age, sex, race/ethnicity, and urban/rural residence at national, sub-national, or small area levels. It can also project elderly care needs and costs, pension deficits, and household consumption. The book consists of four parts. The first part presents the methodology, data, estimation issues, and empirical assessments. The next two parts present applications in the United States (Part II) and China (Part III), concerning demographic, social, economic, and business research; policy analysis, including forecasting future trends of household type/size, elderly living arrangements, disability, and home-based care costs; and household consumptions, including housing and vehicles. The fourth part includes a user's guide for the ProFamy software to project households, living arrangements, and home-based consumptions.

The very initial idea of the research presented in this book began when I was a Ph.D. student at Brussels Free University and attended the International Union for Scientific Studies of Population (IUSSP) 1983 family demography seminar. At that seminar, I was especially interested in a paper presented by Professor John Bongaarts on "The projection of family composition over the life course with the family status life table." With strong support from my supervisors, Professors Frans Willekens and Ron Lesthaeghe, and stimulated by Professor John Bongaarts' initial nuclear family status life table model, I conducted my Ph.D. thesis research at the Netherlands Interdisciplinary Demographic Institute (NIDI) in 1984–1986 to develop a general family status life table model including both nuclear and three-generation families, with an empirical application to China. During my study at NIDI, I learned a great deal of multistate demography from Professor Frans Willeken, who not only supervised my Ph.D study but also helped my long-term professional career development including work on this book. As a Frank Notestein

Post-doctoral Fellow in 1986–1987, I further studied this demographic topic at the Office of Population Research at Princeton University, under the supervision of Professors Ansley Coale and Jane Menken. My research at Brussels Free University, NIDI, and Princeton University enabled me to win the Population Association of America 1987 Dorothy Thomas Award, and my paper on "Changes in Family Structure in China: A Simulation Study" was published in *Population and Development Review* (Zeng 1986). I greatly appreciate what I learned from Professors Willekens, Lesthaeghe, Coale, Menken, and Bongaarts during my Ph.D. and post-doctoral studies, knowledge which led to the new research reported in this book.

After I became a faculty member at the Institute of Population Research at Peking University in August 1987, I continued my research in family demography and tried to expand it to healthy aging and population policy analysis. I particularly enjoyed and learned a lot from long-term productive collaborations with Professor James Vaupel, who I first got to know and started to collaborate with at the International Institute of Applied Systems Analysis in 1985, a summer young scientists' program which was supervised by Professors Nancy Keyfitz and James Vaupel. Among many collaborative research projects and co-authored peer-reviewed published papers with James since 1985, our joint articles (Zeng, Vaupel, and Wang 1997, 1998) are most relevant to this book. In our 1997–1998 papers, we initially developed the household and living arrangement projection model known as ProFamy, which includes two sexes and time-varying demographic rates as input, and applied it to China, based on the one-sex family status life table models (with constant demographic rates) of Bongaarts (1987) and Zeng (1986) mentioned above. In a 1992 interview entitled "Talk to Demographer: Chinese demography starts to go to the world" published in the journal *Population* by Fudan University in Shanghai, I told the interviewer who was one of the journal's editors: "I learned a lot from my collaborator James Vaupel; his mind is like a computer. When I visit him every year, we start to discuss research immediately after saying hello and shaking hands. Sometimes he would be driving but continuing to talk with me about the mathematical demographic formulas and computer programs of our joint papers (which was one of the prior research bases of this book), and I had to remind him to be careful to avoid a car accident. I am sure that many Chinese people were deeply impressed from Professor James Vaupel's style of scientific research by reading this published interview report.

Since late 1998, I have been a faculty member at Duke University, while I keep my faculty position at Peking University and divide my work time between these two universities across the ocean. Professor Ken Land, who is the second author of this book, picked me up at the airport when I first came to Duke, an occasion that I will remember for life. Since then we began to frequently discuss our work and closely collaborate on research. Our joint work includes substantially extending and developing the ProFamy extended cohort-component model, software, and applications to the United States and China. Ken's extremely broad mind and strong mathematical, statistical, and demographic skills have helped me a lot. For example, he innovatively developed and summarized the basic procedures of the ProFamy model into four "core ideas" (refer to Sect. 2.2 of Chap. 2 of this book), which are remarkably helpful for readers to easily understand the relatively complicated structure of the new method. He presented these four "core ideas" at the American Statistical Association 2006 Annual Meeting, a presentation that was well received by statistical and demographic colleagues at the meeting. During the rather long period of preparing the manuscript of this book, Ken repeatedly reminded me that "We need to take our time to work this manuscript over and over again, so that it is a product of the high quality that we want." Clearly, Ken's solid scientific research style and knowledge significantly contributed to the merit of this book.

I am of course very grateful for the outstanding contributions of the other two co-authors of this book, Dr. Danan Gu and Dr. Zhenglian Wang; this book would never have been produced without their hard work and close collaborations. We would like to sincerely thank Dr. Jessica Sautter, who very carefully edited, questioned, and commented on the entire manuscript to help us to significantly improve the quality of this book. We also would like to thank the supports from Duke University, Peking University, and the Max Plank Institute for Demographic Research, and our colleagues at these institutions listed in the Acknowledgments.

Professor of Duke University and Peking University

Yi Zeng

Acknowledgments

The research reported in this book was mainly supported by NIA/NIH SBIR Phase I and Phase II project grants, which received the best possible evaluation score (zero percentile rank), and China National Natural Science Foundation grants (71110107025 and 71233001). We also thank the Population Division of U.S. Census Bureau, NIA (grant 1R03AG18647-1A1), NICHD (grant 5 R01 HD41042-03), Duke University, Peking University, and the Max Planck Institute for Demographic Research for supporting related basic and applied research.

We sincerely thank our colleagues who contributed to our previous collaborative research or supported our research related to this book, including the following researchers (listed alphabetically): John Bongaarts, H. Booth, O'Neil C. Brain, Shuai Cao, Huaming Chai, Huashuai Chen, Lingguo Cheng, Simon Choi, Ansley Coale, Harvey J. Cohen, Ted Chu, Paul Demeny, Zhiheng Ding, Qiushi Feng, Changqing Gan, Ning Gao, Linda K. George, Daniel Goodkind, Todd Graham, Zhigang Guo, Helin Huang, Angang Hu, Nicole Keilman, Nancy Keyfitz, Cheng Jiang, Leiwen Jiang, Weiping Jiang, Ron Lee, Ron Lesthaeghe, Chunhua Li, Faju Li, Jianxin Li, Lan Li, Ling Li, Qiang Li, Shuang Liang, Zhiwu Liang, Nan Lin, Justin Yifu Lin, Yuzhi Liu, Jiehua Lu, Wolfgang Lutz, Andrew Mason, Jane Menken, Philip S. Morgan, Jessie Norris, Diane Parham, Xiezhi Peng, Alexia Prskawetz, Jessica Sautter, Melania Sereny, Ke Shen, Weiheng Shi, Wenzhao Shi, Stanley K. Smith, Eric Stallard, Paul Shultz, Becky Tesh, Feng Wang, James W. Vaupel, Frans Willekens, John Wilmoth, Deqing Wu, Qiong Wu, Zhenyu Xiao, Yu Xie, Changli Yang, Anatoli Yashin, Xuejun Yu, Chunyuan Zhang, Zhen Zhang, Zhongwei Zhao, Zhenzhen Zheng, Zhiheng Ding, Liquan Zhou, Yun Zhou, and Hania Zlotnik. Thanks also go to many scholars who participated in the software usability tests and provided helpful comments to make ProFamy software a user-friendly package.

Note that most of the contents of this book are based on our updated, extended, and improved previous research related to the ProFamy extended cohort-component method and its applications to the United States and China, published since 1997 in *Demography, Population and Development Review, Population Research and Policy Review, Demographic Research, Journal of Aging and Health,*

GENUS, *Mathematical Population Studies*, and *International Journal of Market Research.* While we have cited our relevant publications in various chapters of this book, we would like to thank the editors and the anonymous reviewers of these journals who provided invaluable comments that were substantially helpful for the improvement of our research reported in this book.

Disclaimer: Dr. Danan Gu's work was mainly completed when he was at Duke University. He now works at the United Nations Population Division. The views expressed in this book are solely those of the authors and do not necessarily reflect those of the United Nations, Duke University, Peking University, Max Planck Institute, or the funding agencies.

Contents

List of Figures

List of Tables

Chapter 1
Introduction

Demographers have developed likely scenarios of changes in family households for many national and sub-national populations during the twenty-first century. These anticipated demographic changes will alter the number and proportion of different kinds of households, producing important questions for the future. How many elderly persons will live alone, with spouse only, with children or other relatives, or be institutionalized? How many elderly persons will need assistance in daily activities, but will not have children and/or spouse to provide help? How many middle-age persons will have responsibilities to care for both elderly parents and young children? How many children will live in a single-parent household? How many teenage and adult single mothers will have to care for their children with no spouse or partner present? What are the implications of these changing scenarios for family caregiving and the health service system? The new method and user-friendly software for family household and living arrangement projection presented in this book can be used to answer these and other important questions. The method of projecting and evaluating the consequences of demographic changes on future family household dynamics and living arrangements is clearly useful in empirical studies, development of theories, policy analyses, and business management.

Subsequent chapters are grouped into four parts. The first part establishes basic concepts, presents the ProFamy extended cohort-component household projection methodology, data and estimation issues, and empirical assessments, as well as a few illustrative applications directly connected to the general methods. The second and third parts deal with applications in the United States (the largest developed country) and China (the largest the developing country); within each part, chapters consist of applications to academic and policy analysis and business/market research, including forecasting future trends of household composition, elderly living arrangements, disability and home-based care needs and costs, and household consumption (housing and vehicles). The fourth part includes the User's Guide of the ProFamy software for household and consumption forecasting. The last chapter concludes the book by briefly summarizing the major findings and discussing future perspectives in the field of household and consumption

Y. Zeng et al., *Household and Living Arrangement Projections*,
The Springer Series on Demographic Methods and Population Analysis 36,
DOI 10.1007/978-90-481-8906-9_1, © Springer Science+Business Media Dordrecht 2014

projections. This book is practically useful for faculty, analysts, and students in academics, public, and private business whose work is related to households and populations.

1.1 Definitions of Family and Household

As defined by the United Nations, a *household* consists of one or more persons living together who make of his/her own or a common provision for food and other essentials for living (U.N. 2008). A distinction is made between private (or domestic) households and institutional households. Institutional households are comprised of persons living in dormitories of schools, universities and other units, religious institutions, hospitals, military installations, correctional and penal institutions, nursing homes, centers for hospice care or rehabilitation, and so forth. In most cases, we abbreviate private (or domestic) households as "households". The general term 'households' refers to either private households or private and institutional households as a whole; readers may distinguish them by the context of the discussion.

The term *family* usually remains a less precise word for kinship groups, construed either narrowly as the "nuclear family" consisting of parents and unmarried children only or broadly as encompassing all a person's kin (Wachter 1987: 216). Ryder (1987: 117) proposed defining a family as two or more persons, each of which is either married to, a parent of, or a child of, another member of the co-residence group.

In this volume, we adopt Ryder's definition with some elaborations. We define a *family household* as a group of co-residing persons related through marriage or consensual union, blood, or adoption; Furthermore, the family households may also contain co-residing non-relatives (e.g., a housekeeper or care provider). In sum, while this volume deals with both families and households as units of analysis, we use "family households" or simply "households" as a short term.

Another terminological distinction: "forecast" usually refers to short-term projections for business and socioeconomic planning, while "projection" usually includes both short- and long-term simulations for policy analysis or academic investigation. We use these two terms interchangeably in this book because the methodology and data issues we discuss are applicable to both short- term forecast and long-term simulations.

1.2 Why Project Household and Living Arrangements?

1.2.1 Household Projections, Socioeconomic Planning and Policy Analysis

Family household projections are useful in socio-economic planning and policy analysis, especially when the number and composition of family households are changing in response to demographic and socio-economic changes. For example, several welfare programs in the United States restrict eligibility to single-parent families (Yelowitz 1998). As a result, projecting the costs of such programs depends heavily upon projections of the numbers, types, and sizes of future single-parent family households. Moffitt (2000) argued that demographic changes, including increased numbers of female-headed families, were the primary factors leading to increased Aid to Families with Dependent Children (AFDC) expenditures. This is in contrast to many contemporary observers who interprete the increased expenditures as evidence of increased "take up" rates among those eligible or increased expenditures per recipient. Moffitt argued that these latter explanations were based on misunderstandings of the forces causing the increased expenditures and led to retrenchment reforms. Moffitt further argued that "better forecasting of demographic trends of households" may reduce the "amplitude of the surprise-reaction-retrenchment cycle in welfare policy"; thus, better more accurate forecasting of households with dependent children should be a major policy goal (Moffitt 2000).

Past research has also established that households and living arrangements are the major determinants of the amount and type of long-term care for the elderly (e.g., Chappell 1991; FIFARS 2010; Morris et al. 1998; Soldo et al. 1990). In particular, the use of long-term care varies by family household status (Freedman 1996). Clearly, demographic projections of family households that include elderly living arrangements are directly relevant for urgently needed academic and policy research to face the challenges of rapid population aging in most countries all over the world. Long-term care issues are especially important in China, where the one-child policy has important implications for future Chinese family household structure in the context of rapid population aging. More specifically, policy makers, researchers, and the public need to understand the consequences of alternative policy options in the coming decades. For example, what would happen to Chinese family household structure and family support for the elderly if the current one-child policy is unchanged? What would happen if the one-child policy is smoothly changed into a two-child plus later childbearing policy? The new demographic method for family household projections presented in this book is highly responsive to this type of policy analysis.

Finally, family household projections are important for governmental models of population, environment, and development. For example, private energy consumption patterns are largely defined in terms of numbers of households rather than individuals (Lutz and Prinz 1994: 225). Creating a new household, by divorce or union dissolution, generates an immediate increase in energy consumption. A divorce, by creating a new household, may cause more additional CO_2 emission than an additional birth Mackellar et al. (1995). Two articles published in *Nature* show that a rapid increase in households of smaller size, which results in higher per capita energy consumption, implies a larger demand for resources (Keilman 2003) and poses serious challenges to biodiversity conservation (Liu et al. 2003).

1.2.2 Household Projections and Consumption Market Analysis

For a variety of goods and services in both the public and private sectors, households are more relevant units of demand than individuals because households are the basic units of consumption into which people are organised as units for living. The obvious example is that planning for housing needs and successful housing policy depend on projected numbers, types, and sizes of households in the future years. The demand for consumer durables such as appliances, furniture, and vehicles; the development of residential utilities such as gas, water, and electricity; and the construction of local community service facilities are also determined by anticipated changes in household size and structure (e.g., Dalton et al. 2008; Davis 2003, 2004; Myers et al. 2002; Prskawetz et al. 2004). Consequently, market analyses of demand for housing, automobiles, and many other products and services consumed by households rather than individuals depend heavily on household projections. As a specific example, with rapid population aging in the North and South Americas, Europe, Asia and some countries in Africa, the market for elder-care services is growing with extraordinary speed, which creates a strong demand for projections of household and elderly living arrangements for more efficient and profitable elder-care businesses (Goldscheider 1990; Himes 1992). Because of these and other potential applications, there is a growing demand for projections of households in terms of their size and structure.[1]

Note that the national projections are not applicable to smaller regions due to the regional differentials. Since the late 1990s, more researchers and policymakers have demanded household projections at sub-national levels such as provinces (or states),[2] sub-regional areas such as counties and cities, and other small areas[3] (Crowley 2004; Ip and McRae 1999; Rao 2003; Treadway 1997). Sub-national household projections are useful for distributing government funds, allocating various types of resources, planning the development of infrastructure and public facilities, manufacturers' market research, production planning for household-related goods and services, and decisions on the expansion or reduction of local businesses (Smith et al. 2001; Swanson and Pol 2009).

1.2.3 A Tool for Development of Demographic Theories

A theory is a coherent set of ideas that represents some aspects of part of the real world in such a way that it can explain a phenomenon or class of phenomena (Boland 1989; Burch 1999). Many social scientists develop theoretical concepts or

[1] Household projection reports have been among Statistics Canada's best sellers George (1999: 8–9).

[2] Note that a state in the United States is equivalent to a province or other kind of administrative region immediately underneath the nation in other countries.

[3] The "small-area" term refers to small towns and places, possibly even tracts or block groups which have a small population size.

propositions from field observations and/or limited empirical data analysis. With revolutionary progress in information and computer sciences, the practice of computer simulation to model theoretical ideas is commonplace in the physical and biological sciences. Demographic computer modelling, such as the household projections/simulations discussed in this book, can similarly serve as a tool for demographic theory development.

Family household demography is perhaps the most complicated sub-field in demography because it deals simultaneously with almost all of the main demographic processes, such as fertility, marriage/union formation and dissolution, mortality, migration, leaving the parental home, etc. These complex model manipulations will often exceed the capacity of analytic mathematics, straining the analyst's ability to derive logical implications from theoretical assumptions. Therefore, user-friendly computer software and numerical simulations based on demographic models are practical tools for the more rigorous statement and manipulation of theoretical ideas. Such tools also go beyond analytic mathematics in power and flexibility, and are accessible to social scientists who do not have mathematical and computer programming expertise (Burch 1999). Such tools have in effect extended our power to deduce theoretical propositions, which are much more complicated than can be dealt with using logic only or even analytic mathematics. This enables us to match the breadth and depth of our insights with expanded power of theoretical development, leading to a true marriage of theory and empirical research (Burch 1999).

The usefulness of the ProFamy household projection model and software can potentially serve for theory development as well. For example, the second demographic transition theory (van de Kaa 2008) may be further developed through projecting or simulating detailed implications for household structure if the current trends of low fertility, low and late marriage, more cohabitation, and high divorce rates remain stable, increase, or reverse direction. Another example is that based on mathematical modeling and empirical data analysis on gender differentials of survival, Lakdawalla and Philipson (2002) proposed a theory stating that the faster increase of male elders will lead to a relative reduction in the need for institutional elder care because the supply of spousal care will also increase since old men are less likely to be widowed than old women. These effects could be offset by the large increase in divorce rates and decrease in marriage and remarriage rates since the 1960s, implying that the future elderly may less likely be married (Goldscheider 1990) and therefore less likely to receive spousal care. On the other hand, cohabitation has become more and more popular, which increases the supply of family care. These issues can be integrated and investigated, and better theoretical ideas can thus be gained through demographic simulations and projections of elderly family households which simultaneously deal with mortality, marriage and union formation, and dissolution.

Another example is that, application of the ProFamy extended cohort-component model and software for household projection presented in this book has shown that, in scenarios of U.S. household projections with constant demographic rates, the distributions of households and elderly living arrangements will change considerably in the next a couple of decades. Such scenario simulations using the ProFamy tool have led to a new theory of “family household momentum” (Zeng et al. 2006; see also Sect. 8.3 of Chap. 8 of this book), which is similar to the well-known population momentum theory developed by Keyfitz (1971) over four decades ago. The new theory is based on the ground that older cohorts, who had more traditional family patterns, will be replaced by the younger cohorts with modern family patterns even if current demographic rates remain unchanged. Obviously, the ProFamy extended cohort-component model and software tool have played an important role in finding empirical evidence to formulate the new theory of family household momentum.

Clearly, household projections and simulations may extend the reach of theory development from summarizing past observed data only to much more sophisticated empirical databases that combine the observed past and anticipated future.

1.3 Why Do Household Projections Need to Use Demographic Rates as Input?

1.3.1 Demographic Change and Family Household Dynamics

Changes in demographic processes and rates of fertility, marriage/union formation and dissolution, and mortality etc., in various combinations and strengths in different countries, are yielding new patterns and distributions of family household structure and living arrangements all over the world. In his influential Population Association of America (PAA) presidential address, for example, Samuel Preston labeled marriage change “the earthquake (that) shuddered through the American families” (Preston 1984: 451). In the United States, divorce rates doubled by the mid-1970s as compared to the 1950s and have remained at a high plateau (or perhaps in a slight decline) since about 1980 (Bramlett and Mosher 2002; Goldstein 1999; Strow and Strow 2006). Roughly 40–50 % of all marriages end in divorce (see Cherlin 1992, 1999: 421; Schoen and Standish 2001). Cohabitation (unmarried heterosexual cohabiting unions) has increased dramatically over the last few decades (Bumpass and Sweet 1995; Casper and Cohen 2000; Zeng et al. 2012b), and has become a widespread and acceptable living arrangement (Bumpass and Lu 2000; Smock 2000; Thornton et al. 2007). In the United States, about half of young adults cohabit with a partner before marrying (Bumpass and Lu 2000), first marriage rates have declined, and the likelihood that divorced women will remarry has also declined (Zeng et al. 2012b). After increasing sharply in the late 1980s, birth rates declined for American teenagers from 1991 to 1997 (U.S. National Centre for Health Statistics 1998) and continue to decline, reaching historical lows in 2009–2010 (Hamilton and Ventura 2012). Between 1970 and 2000, the median age at first marriage for women increased by 4.3 years to 25.1; for men, the increase was 3.6 years, reaching 26.8 in 2000 (U.S. Census Bureau 2001). In the last decade, the median age at first marriage continued to climb to 26.9 for women and 28.9 for men in 2011 (U.S. Census Bureau

2011). Age patterns of childbearing also changed, with more people having their first child at older ages (U.S. Census Bureau 1998).

Large changes in demographic regimes in European countries over the last few decades have led scholars to call this most recent period an era of Second Demographic Transition (Lesthaeghe and Van de Kaa 1986; Van de Kaa 2008). The Second Demographic Transition is characterised by postponement of marriage and increasing unmarried cohabitation, a decline of fertility to well below replacement level, a rising proportion of non-marital births, a declining average size of households, and a growing variety of household types (Van de Kaa 2008).

Fertility in China has declined dramatically from more than six children per woman in the 1950s and 1960s to about 1.6 children per woman today, significantly lower than that in the U.S. The propensity of divorce in China increased 42 % between 1982 and 1990 (Zeng and Wu 2000), and has gradually increased since then. Average life expectancy at birth for both sexes combined in China has increased from about 43.3 years in 1950 to 69.5 years in 1990, 72.1 years in 2000, and 74.1 years in 2005 and will continue to increase in the future (U.N. 2013). The large cohorts of baby boomers, who were born in the 1950s and 1960s and subjected to dramatic fertility decline since 1970, are rapidly approaching the elder ages. The family household structures of the forthcoming Chinese elderly will differ tremendously from those of the elderly today, mainly due to the dramatic demographic changes in fertility and mortality. Generations born after the 1970s have fewer siblings than the proceeding cohorts. Thus, they will have a smaller chance of moving out of the parental home to formulate an independent nuclear family household if the Chinese tradition that most parents live with one married child remains more or less stable (Zeng 1986, 1991a). The dramatically changing demographic regimes have determined that the population of China is aging at an extremely rapid speed and to a large scale, with significant implications for family households.

It is important to note that China is not alone in this trend, as many other developing countries are also experiencing significant changes in family households and rapid population aging induced by remarkable demographic changes (Zeng and George 2010). Remarkable demographic changes have been causing, and will continue to cause, substantial dynamic changes in family households; thus, household projection must use demographic rates as input.

1.3.2 Demographic Change, Elderly Living Arrangements, and Care Needs and Costs

The elderly population will increase tremendously in the next few decades in all developed and developing countries (U.N. 2013), with implications for change in family household structure (see, e.g., Wolf 1994). Because people in almost all countries are living longer (U.N. 2013), an increasing number of middle-age persons have living children, parents, and even grandparents, and thus they may need to care for the previous and subsequent generation simultaneously (Watkins et al. 1987). The substantial disadvantage in life expectancy of males has left more female elderly widows, although there is evidence that the gender gap is declining in most countries (Clark and Peck 2012). Increased mobility is leading children to move to areas distant from their parents.

There are important interactions among changes in demographic rates, household structure, living arrangements and health status of the elderly, and their caregiving needs and costs. Living alone due to absence of close-by children and other relatives can cause or worsen ill health and disability, because close-by family members often supply support to ill or disabled older adults. Even in developed countries, elderly persons depend upon spouses and children for emotional and psychological support, and occasionally financial aid as well. In the developing world, where pension and social security systems are not widely available, the elderly heavily depend on family support (National Research Council 2001). In the absence of family support, needs and costs for nursing homes, social services, and health-care services will increase. Costs of health-care and social services provided to the elderly now account for over 10 % of GDP in many developed countries, including the United States (OECD 2013). As the proportion of elderly grows and demographic changes decrease the availability of family support, these costs will grow quickly. Long term care costs in the U.S. have doubled each decade since 1970, reaching an annual level of $106.5 billion in 1995 and up to $243 billion in 2009 (Frank 2012). Home-based care costs grew 90.7 % from 1990 to 1995 and 39 % from 1999 to 2004, in contrast to a 33.4 % and 24 % increase for institutional care costs in the corresponding periods (Hartman et al. 2008; Stallard 2000). Clearly, needs and costs for home-based care have been increasing much faster than for institutional care, especially for the oldest-old (Cutler and Meara 1999; Hartman et al. 2008).

Changes in family household structure and living arrangements due to demographic changes strongly affect caregiving needs and costs, the long-term care health service system, and health-related policy-making (Doty 1986; Himes 1992). The ProFamy extended cohort-component method and software presented in this book can be used to investigate how demographic changes in fertility, mortality, marriage/union formation and dissolution, and migration would affect family household structure, living arrangements of the elderly, and care needs and costs (see Chaps. 5 and 10 for the relevant methodology and illustrative applications). Addressing these issues is important to improve the old age care system and health-related policy-making.

1.3.3 Demographic Changes and Living Arrangements of Children

As a result of demographic changes, including high divorce rates, about 30 % and 40 % of American children were born to unmarried mothers in the 1990s and early 2000s, respectively (NCHS 2012); the proportion of children who spend time in a single-parent family increased substantially, from 30 % in 2000 to 35 % in 2011 (The Annie E. Casey Foundation 2012). The proportion of single-parent households has increased from 24 % in 1990 to nearly 30 % in 2008 (U.S. Census Bureau 2012). Survey data have repeatedly shown that children who grow up in single-parent families have relatively lower economic well-being, school performance, and mental health compared to children who grow up in two-parent families (Cherlin 1999). Demographers can document the changing family household structure and living arrangements of children and identify the effects of proximate demographic determinants (i.e., delayed marriage, cohabitation, divorce, remarriage, non-marital fertility, etc.) on children's living arrangements. Previous demographic research on children's family life course focused on either cross-sectional counts of children living in single-parent families or cohort evidence on 'life time experience' in single parent families. Several researchers studied children's family life course by means of multi-state life table techniques, with fixed demographic rates and limited statuses identified (Dykstra et al. 2006; Heuveline and Timberlake 2003; Hofferth 1987; van Gaalen and van Poppel 2009). These cross-sectional, retrospective life-time experiences and life table approaches are useful, but they cannot address the following important questions: With anticipated future time-varying demographic rates, how many children will live with two parents and how many will live with a single parent? How long will these different statuses last in children's life course? The ProFamy extended cohort-component method and software presented in this book can be used to answer these questions by projecting children's family household structure and living arrangements. Note that the ProFamy projections are based on anticipated time-varying demographic changes and follow an integrated multidimensional model. This is particularly useful, since children's family structure and living arrangements are affected by several demographic factors such as marriage/union formation and dissolution, marital and non-marital fertility, etc.; treating these factors in isolation does not allow for their offsetting, additive, or synergetic cumulative effects to be estimated.

1.3.4 Demographic Changes and Family Life Course of Adults, Especially Single Mothers

Demographic changes affect the living arrangements and life course not only of elders and children, but also young and middle-aged adults. In adult family life course studies, special attention must be given to women, since women are disproportionately disadvantaged when their marriage or consensual union is dissolved. For example, mothers are more likely than fathers to care for children after divorce or separation; men are more likely than women to work and are likely to earn more income than women both before and after union dissolution. Thus, union disruption poses a greater social and economic threat to women than to men (e.g., Bianchi et al. 1999). Teenage childbearing has been identified as a major social problem and the key factor in the intergenerational transmission of poverty in many countries. The ProFamy extended cohort-component method and software presented in this book can be used to perform simulations with changing marriage/union formation and dissolution rates and teenage fertility rates and assess how the changes affect the family and life course of young- and middle-age adults, especially women and their children. For example, how would a reduction or elimination of teenage childbearing affect the number of single mothers? Consequently, how much money would government programs subsiding single-mother families save?

1.4 A Brief Review of Methodological Approaches for Household Projections and Simulations

Demographers use models for the projection or simulation of households based primarily on three types of methodological approaches: headship rate, micro-simulation, and macro-simulation. We briefly describe these three major models with a few examples of studies following the micro- and macro- approaches in this section.

1.4.1 Headship Rate

The headship rate method is a classic approach that has long been used by demographers to project households. In a census or a survey, a "head" of household (or householder) is identified for each household. Age-sex-specific headship rates are computed by dividing the number of persons who are head of a household by the total number of persons of the same age and sex. Households in future years are projected by extrapolating the headship rates. Despite the widespread use of this approach, the headship rate method suffers several serious shortcomings and has been widely criticized by demographers for about three decades. The head is often an arbitrary choice, a vague one status which largely depends on who answers or fills the census or survey questionnaire (Murphy 1991). Another major disadvantage is no clear linkage between headship rates and underlying demographic rates. This creates great difficulties for projections on how changes in demographic rates may affect households (Mason and Racelis 1992: 510; Spicer et al. 1992: 530). Furthermore, the headship rate method lumps all household members other than heads into the very heterogeneous category "non-heads" (Burch 1999). This feature makes it impossible to study the family life courses of elders, children, and others who are "non-heads" of the households. Further discussion and a detailed comparison between the headship rate method and the new ProFamy extended cohort-component method for household projection will be presented in Sect. 4.2 of Chap. 4.

1.4.2 Micro-Simulation Models

Micro-simulation models simulate life course events and keep detailed records of demographic status transitions for each individual of a sample population. Several models following a micro-simulation approach have made important contributions to the study of kinship patterns and family support for older adults: the SOCSIM model developed by Hammel, Wachter, and other colleagues (Hammel et al. 1981; Hammel 2005; Hammel et al.1991; Murphy 2004, 2011; Wachter 1987; Zagheni 2011), the KINSIM model developed by Wolf (1988, 1990, 1994), the MOMSIM model developed by Ruggles (1987, 1993), the CAMSIM developed by Laslett (1986, 1988, 1994), and the APPSIM (Australian Population Policy Simulation Model) developed by Bacon and Pennec (2007).

As compared with the macro-simulation approach, micro-simulation offers three major advantages: (1) it can handle a large state space with many covariates, (2) the relation of individuals can be explicitly retained in the modelling process, and (3) it provides richer output including the probabilistic (and stochastic) distribution of outcomes. It is particularly powerful for complex kinship simulations and projections, which the macro-simulation and headship rate methods cannot do.

As always, however, these advantages have a cost. Three kinds of random variations in micro-simulation have been discussed in detail in the literature (see, for example, Van Imhoff and Post 1998; Van Imhoff 1999). The first is due to the nature of Monte Carlo random experiments: different runs of the same model and input produce different sets of outcomes. This inherent stochasticity can be reduced by increasing the sample size or by taking an average over a large number of runs, but it cannot be wholly eliminated.

In the second random variation, the starting population of the micro model is a sample from a given total population. Thus, it is subject to classic sampling errors, especially for small sub-groups such as oldest- old persons and very young teenage mothers. At least for simple projections, this type of sampling error may be reduced by the use of ex-post weights, which cause the weighted sum of persons with stated characteristics to agree with the population total at baseline. Accurately estimating the ex-post weights and the weighted sum for future years, however, is problematic. In a large population in which households are classified by a large number of characteristics, the size of the representative sample to be used as the starting point of a projection should also be large. For example, a sample of 1 % of the populations of China and the United States consists of about 13.5 and 3.2 million persons, respectively. To simulate so many persons one by one and repeat the entire simulation many times to obtain stable average results would take very substantial computing power and time, which could last for days, depending on the sample size and complexity of the models.

Another problem is that a census usually asks simple questions that cannot provide enough data for the micro-simulation to model detailed characteristics of individuals. Hammel, Wachter, and their colleagues handled this problem by initiating their simulations with a pre-simulation for a few decades before the beginning year of their projections. Using a manageable sample for this pre-simulation, they were able to approximate the family, household, and kinship distribution of the beginning year of the projection; they then simulated it forward. The results made very good sense in policy analysis. From a practical point of view for planning purposes, however, such an approach is not as accurate and practical as directly using the 100 % population and household distributions obtained from a census as the starting point of the projection. This is because the simulated family and population distributions at the beginning year of the projection may not be precisely the same as that of the census enumeration. A related problem is that the procedure demands additional detailed data for a few decades before the beginning year of the projection, which may not be available, especially for many developing countries.

The third random variation, micro-simulation, which includes many explanatory variables and complex relationships among individuals, increases the stochasticity and measurement error to which the model outcomes are subject. Some scholars call this kind of bias "specification randomness" (Pudney and Sutherland 1994). A correctly specified micro-simulation model, with many explanatory variables and complex relationships, may significantly increase the specification randomness (Van Imhoff and Post 1998: 111).

1.4.3 Macro-Simulation Models

The unit of macro-simulation models is a group, for instance, a group of persons of the same age, who are members of a cohort, with the same parity and/or marital status. For the specified groups, the calculations proceed iteratively, group/cohort by group/cohort and time period by time period, generally by means of status transition probabilities. The macro-simulation approach does not suffer the shortcomings inherent in headship rate methods. Although not as flexible as micro-simulation models in analysing variability and probability distributions and simulating relationships among individuals, macro-simulation models do not have the problems of the inherent random variations of Monte Carlo experiments and sampling errors in the starting population. The macro approach can fully and effectively use data from a census or population registers as a starting point.

The problem of specification randomness is also present in a macro model, but is likely to be less serious than that in a micro-simulation model, since it is much less tempting to introduce additional complications into a macro model (Van Imhoff and Post 1998: 111). However, the debate continues: some scholars suggest that adding more covariates to a projection model may improve forecasts, while others think that a very complicated projection model may not perform as well as a simpler one does (Ahlburg 1995; Smith 1997). Furthermore, it is much easier to develop a macro-simulation model into a user-friendly demographic computer software tool for use by typical researchers and policy analysts who lack sophisticated mathematical and programming expertise. Planners and policy analysts can conduct macro-simulation projections relatively easily on a personal computer obtaining the projection and simulation results in less than 1 min if user-friendly software and a clear manual are provided.

The choice between a micro- or macro-model depends on the complexity of the user's task. For detailed analyses of behavioral patterns and complex family kinship relationships, a micro-simulation approach may be preferable. For relatively straightforward demographic and household consumption projections for the purposes of policy analyses, market trend studies, and socioeconomic planning, especially by non-experts, a macro-simulation approach may be satisfactory.

Keilman (1988) and Van Imhoff and Keilman (1992) reviewed dynamic household models based on the macro-simulation approach. Most of these models require data on transition probabilities among various statuses of household types, data that have to be collected in a special survey because they are not available in conventional demographic data sources – vital statistics, censuses, and ordinary surveys.

A typical example of macro-simulation model is LIPRO, a well-known model developed by Van Imhoff and Keilman (1992). In a typical application of the LIPRO model, 11 household statuses are distinguished: single (one-person household), married living with spouse but without children, married living with spouse and one or more children, cohabiting without children present, cohabiting with one or more children, head of one-parent family, child in family with married parents, child in family with cohabiting parents, child in family with one parent, non-family related adult, and other position in private household. In such a model, a household status transition probability matrix of size 11 $\times$ 11 (=121 elements) is needed for each age for males and females respectively (Van Imhoff and Keilman 1992).

As stated by Van Imhoff and Keilman (1992), such stringent data demands, especially for data not commonly available on transition probabilities of household type statuses, is an important factor in the slow development and infrequent application of macro-simulation models. Due to the aggregated nature of the household statuses identified in the model, there is no projection information about groups other than those based on the pre-defined state space (e.g., the 11 statuses defined in the typical application of LIPRO, as listed above). For instance, a frequency distribution of households by size is impossible to construct unless the size of the household is explicitly incorporated in the state space (Van Imhoff et al. 1995: 348). Incorporation of household size in the state space would largely increase the number of household statuses distinguished (e.g., from 11 to 20) and enlarge the size (e.g., from 11 $\times$ 11 to 20 $\times$ 20) of the transition probability matrix for which estimates are needed at each age for males and females, which is likely to a size that is not feasible in practical applications. Furthermore, the household type status-transition-based model cannot directly link changes in household structure with demographic rates. For example, the probability of transition from the household status of married living with spouse and one or more children to cohabiting without children jointly depends on changes in divorce rates, death rates of spouse, cohabiting rates, fertility rates, and rates of children's leaving the parental home. How these demographic rates are linked to the transition probability is not clear. It is thus difficult for such models to identify the impacts of demographic factors on changes in family household structure because the input for these models is transition probabilities among household types, which have no clear linkage with demographic rates.

This implies that it is important to develop a dynamic household projection/simulation model that requires as input only conventional demographic rates which can be obtained from vital statistics, censuses, and ordinary surveys. This is one of the key features of the family status life table models initially developed by Bongaarts (1987) and extended by Zeng (1986, 1988, 1991a), and the household projection model developed by Zeng et al. (1997, 1998), extended by Zeng et al. (2006, 2013a), and to be presented and elaborated in this book.

Benefiting from methodological advances in multidimensional demography (Rogers 1975; Willekens et al. 1982; Land and Rogers 1982; Schoen 1988), and especially the multi-state marital status life table model (Willekens et al. 1982; Schoen and Standish 2001), Bongaarts (1987) developed a nuclear family status life table model. He presented illustrative applications to three hypothetical life table populations at three different points in the demographic transition (pre-transitional, transitional, and post-transitional). Watkins et al. (1987) applied Bongaarts' model to the U.S. population, using U.S. demographic data in 1800, 1900, 1960 and 1980 to estimate the length of life spent in various family statuses. One of the many interesting findings derived from this study is that not only have the years spent with at least one parent over age 65 risen (from less than 10 years under the 1900 regime of demographic rates to nearly 20 years under the 1980 regime) but, as a result, the proportion of adult lifetime spent with at least one parent over age 65 has also increased, from 15 % to 29 %. Applying the Bongaarts model, Lee and Palloni (1992) estimated cohort family status life tables for women born in 1890–1894, 1910–1914, 1930–1934, 1950–1954, and 1970–1974, and conducted cohort and cross-sectional analyses of changes in the family status of elderly women in Korea.

Zeng (1986, 1988, 1991a) extended the Bongaarts model into a general family status life table model that includes both nuclear and three-generation households, and developed associated software called FAMY (Zeng 1990). The life table models by Bongaarts and Zeng are female-dominant one-sex models and assume that age-specific demographic rates are constant. Building on the family status life table models, Zeng et al. (1997, 1998) developed a two-sex dynamic projection model that permits demographic schedules to change over time. This model was extended by Zeng et al. (2006, 2013a). The new model is an extended cohort-component approach and its associated software is named ProFamy. The basic mechanism of this model is that projections of changes in demographic components (marriage/union formation and dissolution, fertility, leaving parental home, mortality, and migration) are made for each of the cohorts to produce household type and size distributions for future years. This is analogous to, and a substantive extension of, the conventional cohort-component population projection model. The ProFamy extended cohort-component model uses groups of individuals as the basic unit of projection and thus requires only data that are available from conventional demographic data sources. It can be used to identify the effects of changes in demographic rates on household structure. This is substantially important progress compared with most other macro-simulation models (such as LIPRO) which use groups of households as the basic unit of projection and thus require household status transition probability matrices as input, data which are not available from conventional demographic data sources. As pointed out by Lutz and Prinz (1994: 225), the previous population and household models provided no feasible way to convert information based on individuals directly into information on households. Even if these two different aspects could be matched for the starting year, there exists no way of guaranteeing consistent changes in both patterns when they are projected into the future. As is shown in Zeng et al. (1997, 1998), Zeng et al. (2006, 2013a), and detailed in the next chapter, the ProFamy extended cohort-component model projects households and individuals simultaneously and consistently based on demographic rates.

1.5 Concluding Remarks

Household projection is useful in socioeconomic, actuarial, and welfare planning, in policy analysis, and in market trend studies. Demographers use models for the projection of households based primarily on three types of methods: headship rates, micro-simulation, and macro-simulation. The classic headship rate method for household projection, which is still widely used despite about three decades of criticism by demographers, is not linked to demographic rates, projects a few household types or size, and does not deal with household members other than heads. The choice between a micro-simulation or macro-simulation model depends on the complexity of the user's task. For detailed analyses of behavioural patterns and complex family kinship relationships, a micro-simulation approach may be preferable. For relatively straightforward demographic and household consumption projections based on commonly available data for the purposes of policy analyses, market trends studies, and socioeconomic planning, especially projections used by non-experts, a macro-simulation approach may be satisfactory. The ProFamy extended cohort-component approach, initially proposed by Zeng et al. (1997, 1998) and substantially extended in Zeng et al. (2006, 2013a), is a typical macro-simulation approach. In contrast to the widely-criticized headshiprate method, the ProFamy extended cohort-component approach uses demographic rates as input and projects much more detailed household types, sizes, and living arrangements for all members of the population under study. The methodology, data needed, and empirical assessments of the ProFamy extended cohort-component approach, including a detailed comparison between the headship rate method and the ProFamy approach for household projection, will be presented in the next three chapters.

Part I
Methodology, Data, and Assessments

Chapter 2
ProFamy: The Extended Cohort-Component Method for Household and Living Arrangement Projections

2.1 Units of Analysis and Demographic Status Identification

In the ProFamy model, all individuals of a population are grouped and projected forward by age; sex; marital/union status; parity; number of co-residing children; co-residence with two, one, or no parents; rural or urban status (optional); and race (optional) (see Table 2.1). We also distinguish individuals' status of living in a private household or institutional household. In other words, the ProFamy model uses *groups of individuals* as the unit of analysis; only conventional and normally available data from censuses, surveys, and vital statistics are required as inputs Zeng et al. (2006). As reviewed in Chap. 1, this unit of analysis is in contrast to the *groups of households* used as the unit of analysis in most other macro-simulation models for household projections; these other models require data for the estimation of transition probabilities among household-type statuses – data that have to be collected in special surveys as they are not available in vital statistics, censuses, or ordinary surveys (Van Imhoff and Keilman 1992; Keilman 1988). This strong data requirement is an important factor in the slow development and infrequent application of these models (Van Imhoff et al. 1995). Furthermore, household-status-transition-based models cannot directly link changes in household structure to demographic rates. Thus, it is difficult for such a model to identify the impacts of demographic factors on changes in household structure.

Users can choose either the classic four marital status model or a seven marital status model in applications of the ProFamy extended cohort-component approach and its associated software. The classic model includes four marital statuses: (1) Never-married; (2) Married; (3) Widowed; (4) Divorced (Willekens et al. 1982) (see Fig. 2.1). The four marital status model requires the least data but it does not include cohabiting, which is increasingly prevalent in many societies. Accordingly, we have extended the classic four marital status model to include seven marital/union statuses: (1) Never-married and not-cohabiting; (2) Married; (3) Widowed and not-cohabiting; (4) Divorced and not cohabiting;

Y. Zeng et al., *Household and Living Arrangement Projections*,
The Springer Series on Demographic Methods and Population Analysis 36,
DOI 10.1007/978-90-481-8906-9_2,

Table 2.1 Demographic statuses of individuals in the ProFamy model

Status	Symbol	Definition	e.g. U.S. application
Age	x	0,1,2,3,. . .,W; W is chosen by user	x = 0,1,2,3,. . .,100
Sex	s	1. Female; 2. Male	s = 1,2
Race (optional)	r	Determined by user	r = 1,2,3,4
Marital/union status	m	4 or 7 marital/union status chosen by user	m = 1,2,3,4,5,6,7
Co-residence with parent(s)	k	1. Living with two parents; 2. Living with one parent only; 3. Not living with parents.	k = 1,2,3
Parity	p	p = 0,1,2,. . ., H; H is chosen by user	p = 0,12,3,4,5+
# co-residing children	c	c = 0,1,2,. . ., H (c ≤ p)	c = 0,1,2,3,4,5+
Residence (optional)	u	1. Rural; 2. Urban	Not considered
Projection year	T	Single year from t_1 to t_2, chosen by user	t_1 = 2,000; t_2 = 2,050

Notes: (1) Status k can also be defined in the ProFamy model as having 0, 1, or 2 surviving parents, disregarding co-residence. Status c can also be defined as having 1,2,. . .,P surviving children, disregarding co-residence. With this option, one can project the future availability of surviving parents and children. (2) Parity is defined in the ProFamy model as the total number of children ever-born and ever-adopted

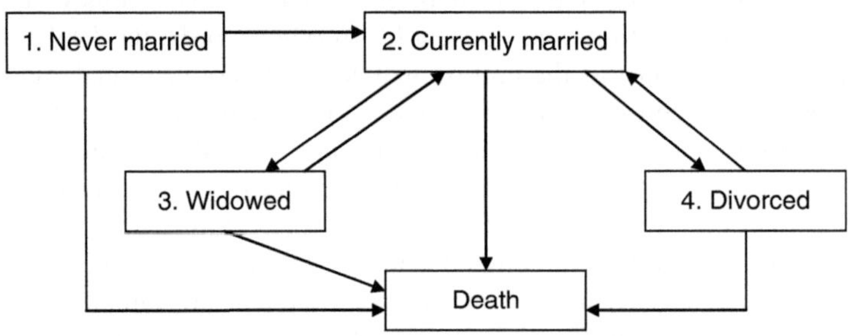

Fig. 2.1 A model of four marital statuses

(5) Never-married and cohabiting; (6) Widowed and cohabiting; (7) Divorced and cohabiting (see Fig. 2.2).[1]

In the United States and other countries with substantial racial differentials in demographic rates, it may be necessary to classify the population by race for accurate projections. For example, there are large differences in demographic rates for all races

[1] We do not include "married and legally separated but cohabiting with a partner" because of unavailability of data. "Legal separation" is combined with "Divorced" to simplify the model. One may consider lumping never-married and cohabiting, widowed and cohabiting, and divorced and cohabiting into one status of "cohabiting", which leads to a simpler model that contains five statuses only. In a five-status model, however, the three kinds of cohabiting people with different legal marital statuses are not distinguishable and they all become "single" once their union is broken, which is not appropriate.

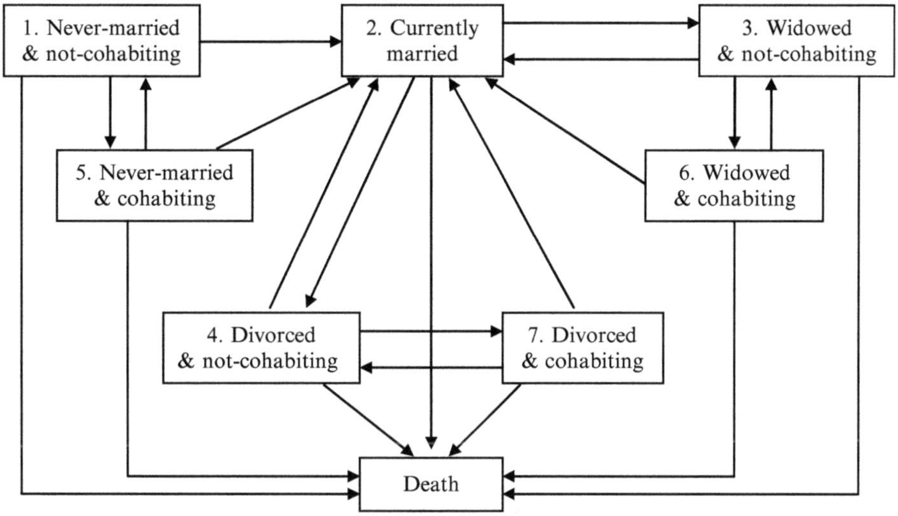

Fig. 2.2 A model of seven marital/union statuses

combined between Minnesota and Florida in the U.S. But a simple standardization of age and race makes these differences mostly disappear, which demonstrates that most of the differences in demographic rates between Minnesota and Florida are due to racial composition rather than different race-specific demographic rates (Morgan 2004). Thus, inclusion of race is an efficient way of capturing most demographic variations in the U.S. and other countries with major racial groups and varying demographic rates. The model cannot simply be run separately for each racial group because of the need to allow interracial marriages and ensure two-sex consistency. For example, the total number of male marriages of all races should be equal to female marriages of all races. The ProFamy model allows interracial marriages/unions and ensures two-sex consistency, which has been tested and evaluated through applications to U.S. household projections by race groups, as will be described later. Users can choose the number of race groups, or no race classification, based on the degree of racial differentials and availability of race specific data.

If the rural/urban differentials are substantial and data classified by rural and urban sectors are available, such as is the case for China and many other developing countries, one may include rural/urban dimensions in the applications of the ProFamy model and software. If the demographic differences between rural and urban sectors are small, as is the case in many developed countries, or if the data classified by rural/urban sectors are not available, the rural/urban dimension can be omitted.

2.2 Core Ideas of the ProFamy Extended Cohort-Component Model

The ProFamy model is built on four core ideas (Zeng et al. 2013a), which are described below.

2.2.1 Core Idea 1: A Multi-State Accounting Model

The innermost core of the ProFamy model is a multi-state accounting model for transforming the marital/union statuses and co-residence with children and parents statuses of members of a population in year t into their corresponding statuses in year $t + 1$. We follow Brass' marker approach to identify households based on individuals' characteristics. Brass (1983) calls the reference person a household "marker". In Brass's original work, the nuclear family-status life table models developed by Bongaarts (1987), and the general family-status life table model including nuclear and three-generation families of Zeng (1986, 1988, 1991a), only female adults are chosen as markers, which implies a female-dominant one-sex model. In the ProFamy model developed in this book, both sexes are included; a female adult, or a male adult when a female adult is not available, is identified as the reference person (or "marker") of the household.

In addition to identifying the individual members of a population by single years of age, sex, optional race and rural–urban residence, the ProFamy model keeps track of changes in individuals' living arrangements including marital/union status, statuses of co-residence with one or two parent(s), and number of co-residing children in each year of the projection. To derive the distributions of household types and sizes, we follow Brass's (1983) basic concept of using a marker or reference person to identify and classify households based on the individuals' marital/union and co-residence statuses with parents/children. For example, a married or cohabiting woman who is not co-residing with parents ($k = 0$) and whose number of co-residing children is c ($c = 0,1,2,3,4,5+$ in this example), is a reference person representing a two-generation and couple household of $2 + c$ family members. If this married or cohabiting women's status of co-residence with parents is 1 ($k = 1$, living with one parent) or 2 ($k = 2$, living with two parents), she represents a three-generation household with a size of $2 + c + k$. If the reference person is not married and not cohabiting (either a man[2] or a woman), he or she is the reference person for a single-parent household of $1 + c$ family members. The formulas for computing the number of households of various types and sizes are presented in Table 2.2. Note that, as indicated in the formulas in Table 2.2, the households of one-woman or one-man without children and the households of a lone-mother with children or a lone-father with children are classified by age and marital/union status of the woman or man. The households of a couple without children and the households of a couple with children are classified by age and married or cohabiting status of the woman. The three-generation households with one or two grandparents are classified by age and marital/union status of the middle-generation. The categories of three-generation households are applicable to Asian and many other developing countries where three-generation households are common household types, but they may be omitted

[2] Note that a married or cohabiting man cannot be a reference person because we already chose the married or cohabiting woman as the reference person and one household cannot have two reference persons.

Table 2.2 Basic formulas for computing number of households by types and sizes identified in the model

Household types	Reference person's status s	k	m	P	C	Household size	Number of households in the year t
One-generation							
One woman	1	3	1, 3, 4	≥ 0	0	1	$G_a^1(t) = \sum_{x=\alpha}^{W} \sum_{m=1,3,4} \sum_{p=0}^{H} N_{3,m,p,0}(x,t,1)$
One man	2	3	1, 3, 4	≥ 0	0	1	$G_b^1(t) = \sum_{x=\alpha}^{W} \sum_{m=1,3,4} \sum_{p=0}^{H} N_{3,m,p,0}(x,t,2)$
One couple	1	3	2, 5,6,7	≥ 0	0	2	$G_c^1(t) = \sum_{x=\alpha}^{W} \sum_{m\neq 1,3,4} \sum_{p=0}^{H} N_{3,m,p,0}(x,t,1)$
Two-generation							
A couple and children	1	3	2, 5,6,7	>0	>0	2 + c	$G_a^2(t) = \sum_{x=\alpha}^{W} \sum_{m\neq 1,3,4} \sum_{p=1}^{H} N_{3,m,p,c}(x,t,1) - (G_a^3 + G_c^3 + G_e^3)$
Lone-mother and children	1	3	1, 3, 4	>0	>0	1 + c	$G_b^2(t) = \sum_{x=\alpha}^{W} \sum_{m=1,3,4} \sum_{p=1}^{H} N_{3,m,p,c}(x,t,1) - \left(G_b^3 + G_d^3 + G_f^3\right)R$
Lon-father and children	2	3	1, 3, 4	>0	>0	1 + c	$G_c^2(t) = \sum_{x=\alpha}^{W} \sum_{m=1,3,4} \sum_{p=1}^{H} N_{3,m,p,c}(x,t,1) - \left(G_b^3 + G_d^3 + G_f^3\right)(1-R)$
Three-generation							
Two grandparents and a couple and children	1	1	2, 5,6,7	>0	>0	2 + 2 + c	$G_a^3(t) - \sum_{x=\alpha}^{W} \sum_{m\neq 1,3,4} \sum_{p=1}^{H} N_{1,m,p,c}(x,t,1)$
One grandparent and a couple and children	1	2	2, 5,6,7	>0	>0	1 + 2 + c	$G_b^3(t) = \sum_{x=\alpha}^{W} \sum_{m\neq 1,3,4} \sum_{p=1}^{H} N_{2,m,p,c}(x,t,1)$
Two grandparents and lone-mother and children	1	1	1, 3, 4	>0	>0	2 + 1 + c	$G_c^3(t) = \sum_{x=\alpha}^{W} \sum_{m=1,3,4} \sum_{p=1}^{H} N_{1,m,p,c}(x,t,1)$

(continued)

Table 2.2 (continued)

Household types	Reference person's status s	k	m	P	C	Household size	Number of households in the year t
One grand-parent and lone-mother and children	1	2	1, 3, 4	> 0	>0	1 + 1 + c	$G_d^3(t) = \sum_{x=\alpha}^{W} \sum_{m=1,3,4} \sum_{p=1}^{H} N_{2,m,p,c}(x,t,1)$
Two grandparents and lone-father and children	2	1	1, 3, 4	> 0	>0	2 + 1 + c	$G_e^3(t) = \sum_{x=\alpha}^{W} \sum_{m=1,3,4} \sum_{p=1}^{H} N_{1,m,p,c}(x,t,2)$
One grand-parent and children	2	2	1, 3, 4	> 0	>0	1 + 1 + c	$G_f^3(t) = \sum_{x=\alpha}^{W} \sum_{m=1,3,4} \sum_{p=1}^{H} N_{2,m,p,c}(x,t,2)$

Notes: (1) $x, s, m, k, p, c, t, W, H$ are defined in Table 2.1. $N_{k,m,p,c}(x,t,s)$ is the population of age x, sex s, and k, m, p, c status in year t. (2) Those elderly couples who live together with an ever-married child (and the child's spouse if the child is currently married) and grandchildren, are not reference persons of a family household, since the ever-married child with whom they live has already taken the position of a reference person. A household cannot have two reference persons. Therefore, the number of those elderly couples that is equal to ($G_a^3(t) + G_c^3(t) + G_e^3(t)$) should be subtracted when we compute $G_a^2(t)$; (3) Similarly, ($G_b^3(t) + G_d^3(t) + G_f^3(t)$) and ($G_b^3(t) + G_d^3(t) + G_f^3(t)) \times (1 - R)$ should be subtracted when we compute $G_b^2(t)$ and $G_c^2(t)$, respectively. R is equal to the number of non-married women over age 49 not living with parents but living with child and grandchild divided by the total number of non-married women over age 49 and non-married men over age 51 not living with parents but living with child and grandchild

in household projections for Western countries where the number of three-generation households is very small.

How accurate is our accounting model, as described above, in modeling the real world? We have tested it using the real data sets. First, we identified each individual code for sex, marital/union status, maternal status, and status of co-residence with parents and children. According to these codes, we identified the reference person. Based on the characteristics of the reference person and following the accounting model described above and the formulas listed in Table 2.2, we derived the distribution of households by types and sizes. The household distribution derived in this way may be called "model-count". Second, we followed the standard census tabulation approach and derived the household distribution directly using the codes that record household membership and relationship to the household reference person. This kind of census tabulation may be called "direct-count". Both model-count and direct-count distributions exclude relatives other than parents and children and exclude non-relatives, creating comparability for our testing purpose.

Comparisons of household type distributions derived by the "model-count" and the "direct-count" methods using the U.S. 2000 census datasets at national and sub-national levels as well as the Chinese 2000 census dataset are shown in panel (A) of Table 2.3. The comparisons show that the relative differences are very small – all are less than 1 %.

Table 2.3 Comparisons of distributions of household types derived by the "model-counts" and "direct-counts" using the U.S. 2000 census datasets at national and sub-national levels as well as the Chinese 2000 census dataset

(A) Household types				(B) Household sizes			
	Direct count	Model count	Diff. %		Direct count	Model count	Diff. %
U.S. whole country 2000							
One-couple only	0.2574	0.2571	−0.11	1 person	0.2908	0.2909	0.04
1 generation	0.5482	0.5480	−0.03	2–3 person	0.4843	0.4843	0.01
2 generations	0.4158	0.4162	0.11	4–5 person	0.1992	0.1993	0.05
3 generations	0.0361	0.0358	−0.78	6+ person	0.0258	0.0255	−1.04
U.S. Minnesota state 2000							
One-couple only	0.2937	0.2936	−0.03	1 person	0.2853	0.2852	−0.02
1 generation	0.5790	0.5788	−0.03	2–3 person	0.4812	0.4812	0.01
2 generations	0.4070	0.4073	0.06	4–5 person	0.2056	0.2057	0.05
3 generations	0.0140	0.0139	−0.63	6+ person	0.0280	0.0279	−0.36
U.S. Los Angeles County 2000							
One-couple only	0.1803	0.1793	−0.52	1 person	0.2979	0.2983	0.15
1 generation	0.4781	0.4777	−0.10	2–3 person	0.4127	0.4128	0.03
2 generations	0.4521	0.4531	0.22	4–5 person	0.2398	0.2401	0.10
3 generations	0.0697	0.0692	−0.70	6+ person	0.0496	0.0488	−1.65
China 2000							
One-couple only	0.1313	0.1308	−0.38	1 person	0.0957	0.0956	−0.06
1 generation	0.2271	0.2264	−0.29	2–3 person	0.4896	0.4891	−0.12
2 generations	0.5934	0.5942	0.13	4–5 person	0.3602	0.3648	1.27
3 generations	0.1795	0.1794	−0.06	6+ person	0.0546	0.0506	−7.25

Notes: Diff. % = 100 × (Model count − Direct count)/Direct count

Comparisons of household size distributions derived by the "model-count" and the "direct-count" methods using the U.S. 2000 census datasets at national and sub-national levels as well as the Chinese 2000 census dataset are shown in panel (B) of Table 2.3. The relative differences in frequency distributions of household sizes between model-count and direct-count methods are also small – almost all are below 1 %, with a few exceptions. The largest difference is that the model count underestimates the Chinese large households with six or more persons by 7.25 %. This is likely due to two factors. First, the ProFamy model accounts for up to three generation family households but omits a small number of four or more generation family households and joint family households in which married siblings and their children live together. Second, we limit the highest parity to five in this application, which may underestimate the size of the small number of large family households that have more than five children in the model count. Such underestimation may be reduced in the future because joint family households, households with four or more generations, and party higher than five will be substantially reduced in future years.

Note that this multi-state accounting model includes the reference person's co-residence status with spouse, children, and parent(s), but does not identify the reference person's co-residence status with other relative(s) or non-relative(s), which can create an inaccurate accounting of the distribution of the population by household size in the starting and future projection years. This inaccuracy can be reasonably corrected by a procedure described in Appendix 1. Moreover, a reference person's other relatives and non-relatives who are also part of our projected population accounts cannot be chosen as reference person of the household because one household cannot have more than one reference person. A simple procedure to meet the requirement that other relatives and non-relatives cannot be reference persons of the household is presented in Appendix 2.

2.2.2 *Core Idea 2: Distinguishing Continuously Occurring from Periodic Demographic Accounting Processes*

With the model design and individual statuses identified in our ProFamy model, a conventional multistate computation strategy would require estimation of very high dimensional matrices of cross-status transition probabilities. For example, if seven marital/union statuses (see Fig. 2.2), three statuses of co-residence with parents (i.e., $k = 0, 1, 2$), six parity (i.e., parity 0, 1, 2, 3, 4, 5+) and six co-residence statuses with children are distinguished as in the U.S. household and living arrangement projections of Zeng et al. (2006), one would have to estimate a cross-status transition probabilities matrix with 194,481 $\left(= 441 \times 441; \text{where } 441 = 7 \times 3 \times \sum_{p=0}^{5} (p+1)\right)$ elements[3] at each age of each sex for each race group. This certainly is not practical, as it would be impossible to have a sufficiently large dataset with appropriate sub-sample sizes to reasonably estimate so many elements of the cross-status transition probabilities matrix at each age of each sex for each race group, although there are considerable numbers of structural zero elements such as transitions to lower parity. Thus, we adopt a computational strategy of calculating individual group marital/union, co-residence (with parents/children), migration, and survival status changes by assuming: (a) births occur throughout the first and second half of the single-year age interval, and (b) marital/union status changes, leaving parental home, migration, and death occur in the middle of the age interval (see Fig. 2.3). This strategy, which was originally proposed by Bongaarts (1987) and further justified mathematically and numerically by Zeng (1991a: 61–63 and 80–84), circumvents the problems of estimating huge matrices of cross-status transition probabilities.

---

[3] Because the number of co-residing children is less than or equal to parity, the number of composite statuses of parity and co-residing children is $\sum_{p=0}^{5} (p+1)$ rather than (6×6).

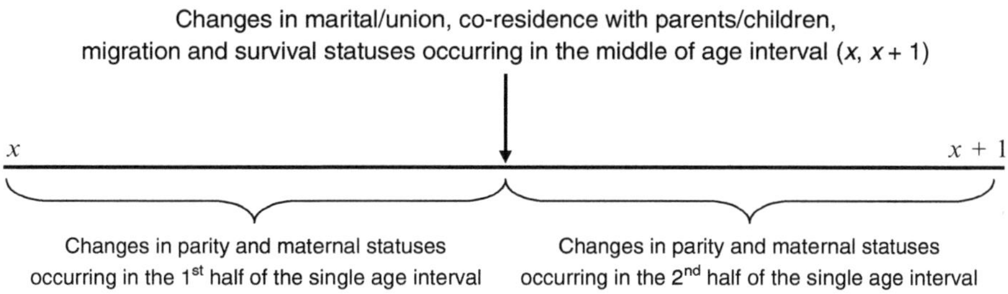

Fig. 2.3 Computational strategy to calculate changes in marital/union, co-residence with parents/children, migration and survival statuses

2.2.3 *Core Idea 3: A Judicious Use of Independence Assumptions*

Coupled with Core Idea 2, the third core idea of the ProFamy model greatly simplifies the estimation of the multi-state transition probabilities. This idea, also originally suggested by Bongaarts (1987) and adapted and generalized by Zeng et al. (1997, 1998), is that not all of the elements of the transition probability matrix depend on many of the other elements; indeed, some of their real-world dependencies are sufficiently small that they can be reasonably assumed to be independent. In other cases, the reality of limited data sources available for estimation of transition probabilities that depend on many other covariates forces the application of an independence assumption. In either case, the consequences of the independence assumption are that either (a) some statuses do not affect or condition on the risks of transition between other statuses, or (b) marginally or partially conditioned on estimates of risk for each of two or more statuses can be multiplied to estimate the corresponding transition probabilities. More specifically, in the ProFamy extended cohort-component model, marital/union status transitions depend on age, sex, and race, but are assumed to be independent of parity and co-residence status with parents and children.[4] Fertility rates depend on age, race, parity, and marital/union status, but are assumed to be independent of co-residence status with parents and children. Mortality rates are age, sex, race, and marital/union status specific, but are assumed to be independent of parity, co-residence status with parents, and children. The probability of two parents dying in the same year is estimated by multiplying the corresponding probabilities of death of the mother and father; the probability of more than one child leaving home in the same year is estimated by multiplying the corresponding probabilities of leaving home of each of the children.

[4] Ideally, one may wish to differentiate the marital/union status transition probabilities by parity and co-residence status with children. Such differentiation is, however, not feasible because it would require a dataset with a very large sample size (not available to us currently but not theoretically impossible at some future time for some specific populations) to estimate the parity-co-residence-marital/union-status-specific transition probabilities at each single age for men and women of each race group with reasonable accuracy.

2.2.4 *Core Idea 4: Employing the National Age-Sex-Specific Model Standard Schedules and the Summary Parameters at the Sub-National Level*

Note that data for estimating race-sex-age-specific standard schedules of the demographic rates of fertility, mortality, marriage/union formation and dissolution, and leaving the parental home may not be available at the sub-national level. However, once the age-race-sex-specific standard schedules at the national level are prepared, they can be employed as model standard schedules for projections at the sub-national level. This is similar to the widely practiced application of model life tables (e.g., Coale et al. 1983; U. N. 1982), the Brass logit relational life table model (e.g., Murray et al. 2003), the Brass relational Gompertz fertility model (Brass 1974), and other parameterized models (e.g., Coale and Trussell 1974; Rogers 1986) in population projections and estimations. Numerous studies have demonstrated that relational parameterized models consisting of a model standard schedule and a few summary parameters offer an efficient and realistic way to project or estimate demographic age-specific rates Brass (1978; Booth 1984; Paget and Timaeus 1994; Zeng et al. 1994).

The theoretical foundation of applications of model life tables and other model standard schedules associated with our core idea 4 is that the demographic summary parameters are crucial for determining changes in level and age pattern of the age-specific rates which affect the projections or estimations. At the same time, the projection and estimation results typically are not highly sensitive to the race-sex-age-specific model standard schedules, as long as the possible changes in the general shape of the standard schedules and timing of the demographic events are properly modeled by relevant summary parameters (e.g., mean or median age, interquartile range). Three kinds of tests in our previous publications Zeng et al. (2000, 2006, 2013a) have further corroborated this theoretical foundation. These three tests are summarized here.

The most recent tests are the applications of the ProFamy extended cohort-component model using the national age-sex-specific demographic model standard schedules and sub-national demographic summary measures to simultaneously project household composition, living arrangements and population age structure and sizes from an earlier census year to a later census year of 10 years apart, at the sub-national level for the 50 states and Washington DC in the U.S. and for the regions and province in China. Using this approach, comparisons of projections from 1990 to 2000 with census counts in 2000 for each of the U.S. 50 states and DC show that 68.0 %, 17.0 %, 11.2 %, and 3.8 % of the absolute percent errors are <3.0 %, 3.0–4.99 %, 5.0–9.99 % and 10.0 %, respectively; similarly small forecasting errors were also fund in comparisons of the projected main indices of households and population from 1990 to 2000 or from 2000 to 2010 with census counts in 2000 or 2010 for the Chinese sub-national regions and province (see Sect. 4.1 of Chap. 4 for details).

Zeng et al. (2006) performed two scenarios of U.S. national projections by race from 1990 to 2020 with all race-specific demographic summary parameters identical

Table 2.4 Comparing the main indices of ProFamy forecasts of household numbers, types/sizes between using the standard schedules observed in the 1980s and using the standard schedules observed in the 1990s, while the projected input summary measures are identical

	2000			2010			2020		
	1980s Stand	1990s Stand	Diff %	1980s Stand	1990s Stand	Diff %	1980s Stand	1990s Stand	Diff %
Total number of household	105,779,128	105,901,696	0.1	120,269,184	120,454,376	0.2	134,421,136	134,627,744	0.2
Average household size	2.53	2.53	−0.1	2.47	2.47	−0.2	2.44	2.44	−0.2
Percentage of									
1-person household	25.77	25.19	−2.3	26.85	26.09	−2.9	27.36	26.52	−3.1
2-person household	32.95	33.81	2.6	34.68	35.77	3.2	35.72	36.95	3.4
3-person household	18.02	18.09	0.4	17.22	17.33	0.7	16.36	16.48	0.8
4-person household	13.98	13.8	−1.3	12.38	12.17	−1.7	11.74	11.48	−2.2
5 + -person household	9.28	9.11	−1.9	8.86	8.64	−2.6	8.83	8.58	−2.9
Married couple family	53.85	53.69	−0.3	47.36	47.20	−0.3	42.38	42.23	−0.3

to each other, but with one scenario using the race-sex-age-specific rates observed in the 1990s and the other scenario using the race-sex-age-specific rates observed in the 1980s. The 17 main indices of the household, living arrangement, and population projections from each scenario were compared for the years 2000, 2010, and 2020. The results showed that, when the projected input summary parameters are identical, standard schedules observed in the 1980s and standard schedules observed in the 1990s produced almost the same projections. About two-thirds of the relative discrepancy rates of the projected main indices between these two scenarios are less than 1 % and the other one-third are 1.0–3.4 % (see Tables 2.4 and 2.5).

In another empirical evaluation, Zeng et al. (2000) applied the extended Brass relational Gompertz model, observed summary parameters (total rate, median age, and interquartile range) in one population, and a model standard schedule derived from another demographically similar population to estimate future age-specific demographic rates in the population under study and compare the estimated and observed age-specific rates. In addition to successfully estimating the demographic rates and comparing them with the observed rates, using the observed rates of the same population under study in the earlier period as the standard schedules (see Figs. 1a–e and Tables A-1–4 in Zeng et al. 2000), the following specific estimates and comparisons were made:

- Estimates of female age-specific divorce rates in Australia in 1975 based on observed Australian summary measures and age-specific divorce schedule in 1970 as a standard schedule, and then compared with the observed divorce rates in Australia in 1975;

Table 2.5 Comparing the main indices of ProFamy forecasts of population between using the standard schedules observed in the 1980s and using the standard schedules observed in the 1990s, while the projected input summary measures are identical

	2000			2010			2020		
	1980s Stand	1990s Stand	Diff %	1980s Stand	1990s Stand	Diff %	1980s Stand	1990s Stand	Diff %
Total population	276,483,456	276,417,600	0.0	306,898,912	306,896,384	0.0	339,270,464	339,189,344	0.0
Percentage of children age < 18	25.35	25.33	−0.1	24.30	24.31	0.0	23.92	23.91	0.0
Percentage of 60+	16.92	16.92	0.0	19.00	19	0.0	22.93	22.93	0.0
Percentage of 65+	13.03	13.04	0.1	13.60	13.6	0.0	16.74	16.75	0.1
Percentage of 80+	3.61	3.61	0.0	4.15	4.15	0.0	4.27	4.27	0.0
Percentage group quarters	2.73	2.73	0.0	2.84	2.84	−0.1	2.85	2.86	0.0
Dependent ratio of									
Children	0.41	0.41	0.0	0.39	0.39	0.0	0.40	0.40	0.0
Old	0.21	0.21	0.0	0.22	0.22	0.0	0.28	0.28	0.0
Children and old	0.62	0.62	0.0	0.61	0.61	0.00	0.69	0.69	0.0

- Estimates of female age-specific remarriage rates of widows in the Netherlands in 1978 based on observed Dutch summary measures and the a model standard Australian age-specific remarriage schedule for Australia in 1975 as a standard schedule, and then compared with the observed remarriage rates of widows in the Netherlands in 1978.
- Estimates of age-parity-specific fertility rates in China in 1981–1989 based on observed Chinese summary measures and U.S. average fertility schedule in 1970–1980 as a standard schedule, and then compared with the observed Chinese fertility rates in 1981–1989;
- Estimates of marriage-duration-specific rates of leaving the parental home for females in the 1950–1969 and 1970–1979 marriage cohorts in each of the 12 provinces of China based on observed provincial summary measures and the average of the rates in the other 11 provinces as a standard schedule, and then compared with the observed rates of leaving home in each of the 12 provinces.

For these various comparisons, it was found that almost all estimated schedules based on the observed summary measures in the population under study and a standard schedule derived from another demographically similar population are fairly close to the observed ones and almost all of the Index of Goodness of Fit (IGF) values are above 0.95 (see Fig. 1 and Tables A-4–6 in Zeng et al. 2000), indicating remarkably accurate estimates. These results are based on the fact that model standard schedules adopted from another population can reflect reasonably well the general age pattern of the schedules in the population being studied. For example, the general age patterns of divorce and remarriage in Australia, the Netherlands, and the United States in the 1970s did not differ substantially. The relevant estimates are, therefore, good. In the 1980s, Chinese fertility had already begun to roughly conform to a general age pattern of relatively low fertility comparable to the fertility pattern in the U.S. in the 1970s. Consequently, the average of the age-parity-specific fertility rates in the United States in 1970–1980 could be used as a model standard schedule to estimate Chinese rates in the 1980s. When observed parity-specific total fertility rates, median ages, and interquartile ranges in China in the 1980s were used as input for the estimates, they adequately measured the differences in fertility quantum, timing, and the shape of the curves between China and the U.S.

The successful empirical assessments described above show that it is theoretically and empirically justifiable to use model standard schedules at the national level and to use the estimated or projected demographic summary parameters from a sub-national region under study to reasonably estimate the needed sex-age-specific demographic schedules for household and living arrangement projections. This will be further assessed in Chap. 4 of this book with projections from a census year to the next census year (10 years apart) and comparisons between projections and census-observations in the later census for each of the 50 U.S. states and DC, and for different regions/provinces in China, while using the corresponding national age-sex-specific rates as the model standard schedules.

With model standard schedules in hand, analysts can concentrate on projecting future demographic summary parameters. This can be done using conventional time series analysis with statistical software (e.g., SAS, SPSS, or STATA) or expert opinion approach.[5] Time series data on other related socioeconomic covariates (e.g., average income, education, urbanization, etc.) also can be used in projecting the demographic summary parameters.

2.3 The Demographic Accounting Equations

Demographic accounting equations are used to compute the number of female and male persons and changes in their marital status, parity, co-residence with parents and children, deaths, etc. in each projection year. The basic structure of all accounting equations is as follows:

Number of persons aged $x + 1$ with status i at time $t + 1$ = (number of persons aged x with status i at time t) + (number of new entries into status i which occur in the year ($t,t + 1$) among persons aged $x + 1$ at time $t + 1$) – (number of exits out of status i which occur in the year ($t, t + 1$) among persons aged x at time t).

The events that cause change in status include births, deaths, migration, marriage/union formation and dissolution, leaving and returning to parental home, and so on between age x and $x + 1$ (and between time t and $t + 1$). The number of events is calculated as the number of persons aged x at risk of the occurrence of the event in

[5] According to the Wiki answers, the term of "expert opinion" is defined as "in the opinion of an expert or someone who knows a lot about said topic." (http://wiki.answers.com/Q/Define_expert_opinion; accessed June 23, 2013).

the year multiplied by the probability of occurrence of the event between age x and $x + 1$ (and between time t and $t + 1$).

Based on the principles stated above and the analytical framework presented in the previous sections, we derive accounting equations to calculate changes in statuses between age x and $x + 1$ and between time t and $t + 1$ for all individuals in the population. Calculation of the changes in k, m, p, and c statuses (see Table 2.1 for definitions) between age x and $x + 1$ and between years t and $t + 1$ consists of three steps as described below. Those who are not interested in mathematical formulas can proceed directly to Sect. 2.4.

Note that in the following discussion, dimension "t" refers to the year, and dimension "s" refers to sex ($s = 1$ for females and $s = 2$ for males). If rural–urban or race classification is identified, all of the variables should add additional rural–urban or race dimension r.

Step 1. Updating p and c statuses due to births occurring in the first half of the year $(t, t + 0.5)$

Let $N_{k,m,p,c}(x,t,s)$ and $N_{k,m,p,c}(x + 0.5,\ t + 0.5,\ s)$ denote the population of age x, sex s, and k, m, p, c status at the beginning and middle of year t after updating p and c status, respectively; ${}_{½}b_{p,m}(x,t,s)$, the probability of having a birth of order $p + 1$ in the first half of year t by persons of age x, sex s, parity p, and marital status m; $p(0,t)$, the average probability of the children (both sexes combined) born in year t surviving from birth to the end of the year t; $p(0,t)$ is estimated based on the life table probability of surviving from age 0 to 1 in year t.

When ($p = 0$, $c = 0$) (note: c cannot be greater than p),

$$N_{k,m,0,0}(x+0.5,t+0.5,s) = N_{k,m,0,0}(x,t,s)\ (1-{}_{½}b_{0,m}(x,t,s)) \tag{2.1}$$

When ($p > 0$ and $c = 0$),

$$\begin{aligned} N_{k,m,p,0}(x+0.5,t+0.5,s) = {} & N_{k,m,p,0}(x,t,s)\ (1-{}_{½}b_{p,m}(x,t,s)) \\ & + N_{k,m,p-1,0}(x,t,s)\,{}_{½}b_{p-1,m}(x,t,s)\ (1-p(0,t)) \end{aligned} \tag{2.2}$$

When ($p > 0$, $c > 0$ and $p > c$),

$$\begin{aligned} N_{k,m,p,c}(x+0.5,t+0.5,s) = {} & N_{k,m,p,c}(x,t,s)\ (1-{}_{½}b_{p,m}(x,t,s)) \\ & + N_{k,m,p-1,c-1}(x,t,s)\,{}_{½}b_{p-1,m}(x,t,s)\ p(0,t) \\ & + N_{k,m,p-1,c}(x,t,s)\,{}_{½}b_{p-1,m}(x,t,s)\ (1-p(0,t)) \end{aligned} \tag{2.3}$$

When ($p > 0$, $c > 0$ and $p = c$)

$$\begin{aligned} N_{k,m,p,c}(x+0.5,t+0.5,s) = {} & N_{k,m,p,c}(x,t,s)\ (1-{}_{½}b_{p,m}(x,t,s)) \\ & + N_{k,m,p-1,c-1}(x,t,s)\,{}_{½}b_{p-1,m}(x,t,s)\ p(0,t) \end{aligned} \tag{2.4}$$

The derivation and justification of the formulas for estimating half-year age and parity-specific birth probabilities based on the whole-year age and parity specific data are presented in Appendix 3.

Step 2. Calculating marital status transitions; surviving or dying; in-migrations, out-migrations, and rural–urban migrations (if included); and changes in co-residence with parents and children at the middle of the year

Methods for calculating marital status changes and numbers surviving or dying are well documented in the literature (e.g., Bongaarts 1987; Schoen 1988; Willekens et al. 1982), and therefore are not detailed here.

In the ProFamy model and software, we allow users to choose whether they compute net-migration or in-migration and out-migration with various options as follows:

If the projection is for a country, choose one of the following two options:

- International in-migration and international out-migration;
- International net migration;

(Note: Net migration is defined as in-migration minus out-migration.

If the projection is for a sub-national region, choose one of the following options:

- International net migration, domestic in-migration and out-migration;
- Combined international and domestic in-migration and combined international and domestic out-migration;
- Combined international and domestic net migration;
- International in-migration and out-migration, domestic net migration;
- International net migration, domestic net migration;
- International in-migration and out-migration, domestic in-migration and out-migration.

For example, if a user chooses "Combined international and domestic in-migration and combined international and domestic out-migration" for sub-national region projections, the age-sex-specific numbers of in-migrants to the region under study from anywhere else in the world are calculated by multiplying the projected number of female and male immigrants by standard age-sex-specific frequency distributions of in-migrants (also marital/union status-specific and number of co-residing children-specific, if data are available). The in-migrants are then added to the population. The age-sex-specific numbers of out-migrants from the region under study to anywhere else in the world are calculated by multiplying the age-sex-specific number of female and male persons in the region by the corresponding standard age-sex-specific probability of out-migration, to get rough age-sex-status-specific estimates of out-migrants (again, marital/union status-specific and number of co-residing children-specific data can be included if available). We proportionally adjust these estimated numbers of out-migrants to make sure that the sum of their adjusted values is equal to the projected (or assumed) total number of out-migrants. We then subtract the age-sex-status-specific out-migrants from the population.

If a distinction is made between rural and urban populations, the rural to urban net-migrants in each year are calculated based on the projected proportion of urban population and age-sex-specific standard schedules of rural–urban net-migration (and also marital status and number of co-residing children status specific if the data are available). For each projection year t, we get a projected total number of rural-to-urban net-migrants by multiplying the total population in year t by the difference of proportions of urban population between year t and year $t-1$. We then multiply the projected total number of rural-to-urban net-migrants by the standard age-sex-status-specific frequencies of rural-to-urban net-migrants (estimated from census data) to estimate the age-sex-status-specific numbers of rural-to-urban net-migrants, and subtract them from the rural population and add them to the urban population. Note that the sum of the standard age-sex-status-specific frequencies of rural-to-urban net-migrants is equal to one.

Let $w_{ij}(x,t,s,m)$ denote the probability of transition from co-residence status i at age x in year t to co-residence status j at age $x + 1$ in year $t + 1$, for persons of sex s and marital/union status m. The events that cause transitions in statuses of co-residence with parents are the death of one or two parents, divorce of parents, remarriage of the non-married parent, leaving parental home, and returning to parental home. We assume that these events are independent, so the transition probabilities $w_{ij}(x,t,s,m)$ can be estimated based on the age-specific probabilities of death, divorce, remarriage, leaving and returning home (see Appendix 4);

$N'_{i,m,p,c}(x + 0.5,\ t + 0.5,\ s)$, the population at the middle of year t after computing deaths, immigration and emigration, rural–urban migration (if it is included in the application), and marital/union status transitions;

$N''_{j,m,p,c}(x + 0.5,\ t + 0.5,\ s)$, the population at the middle of year t after updating the k status.

Hence,

$$N''_{j,m,p,c}(x+0.5,t+0.5,s) = \sum_{i=1}^{3} N'_{i,m,p,c}(x+0.5,t+0.5,s)\ w_{ij}(x,t,s,m) \tag{2.5}$$

To illustrate the accounting equations for calculating the co-residence with children status due to children's leaving or returning home or death, and to simplify the presentation, we assume here that the highest parity is 3 (the calculation method will be basically the same but presentation will be more complicated when the highest parity is larger than 3). In year t, let $s_1(t)$, $s_2(t)$, and $s_3(t)$ be the probabilities that the one child, two children, or three children who were living at home at the beginning of the year will survive and live at home at the end of the year; let $d_1(t)$, $d_2(t)$, and $d_3(t)$ denote the probabilities that the one child, both of the two children, and all of the three children will die or leave home during the year. Let $d_{12}(t)$, $d_{13}(t)$, and $d_{23}(t)$ be the probabilities that one of the two children, one of the three children, and two of the three children will die or leave home at the end of the year. Assuming that the events of leaving home and death are independent, we can easily estimate $s_1(t)$, $s_2(t)$, $s_3(t)$, $d_1(t)$, $d_2(t)$, $d_3(t)$, $d_{12}(t)$, $d_{23}(t)$ and $d_{13}(t)$ (See Appendix 5).

Let $N'''_{i,m,p,c}(x + 0.5, t + 0.5, s)$ denote the population at the middle of year t after updating c status due to children's leaving/returning home or deaths.

$$N'''_{\mathrm{k,m,0,0}}(x+0.5,t+0.5,s) = N''_{\mathrm{k,m,0,0}}(x+0.5,t+0.5,s) \tag{2.6}$$

when $p > 0$ and $p \geq c$,

$$\begin{aligned} N'''_{k,m,p,0}(x+0.5,t+0.5,s) = {} & N''_{k,m,p,0}(x+0.5,t+0.5,s) \\ & + N''_{k,m,p,1}(x+0.5,t+0.5,s)\, d_1(t) \\ & + N_{k,m,p,2}(x+0.5,t+0.5,s)\, d_2(t) \\ & + N''_{k,m,p,3}(x+0.5,t+0.5,s)\, d_3(t) \end{aligned} \tag{2.7}$$

$$\begin{aligned} N'''_{k,m,p,1}(x+0.5,t+0.5,s) = {} & N''_{k,m,p,1}(x+0.5,t+0.5,s)\, s_1(t) \\ & + N''_{k,m,p,2}(x+0.5,t+0.5,s)\, d_{12}(t) \\ & + N''_{k,m,p,3}(x+0.5,t+0.5,s)\, d_{23}(t) \end{aligned} \tag{2.8}$$

$$\begin{aligned} N'''_{k,m,p,2}(x+0.5,t+0.5,s) = {} & N''_{k,m,p,2}(x+0.5,t+0.5,s)\, s_2(t) \\ & + N''_{k,m,p,3}(x+0.5,t+0.5,s)\, d_{13}(t) \end{aligned} \tag{2.9}$$

$$N'''_{k,m,p,3}(x+0.5,t+0.5,s) = N''_{k,m,p,3}(x+0.5,t+0.5,s)\, s_3(t) \tag{2.10}$$

Step 3. Updating p and c status due to births occurring in the second half of the year

Note that we assume that a person does not have a birth in the second half of the year if she or he has a birth in the first half of the year. So, only those who do not have a birth in the first half of the year may have a birth in the second half of the year. Therefore, we need to calculate the proportion of those who do not give birth in the first half of the year among those who are of status k, m, p, c, sex s, and age $x + 0.5$ at the middle of the year $t + 0.5$. The proportion is denoted as $B0_{k,m,p,c}(x,t,s)$. The persons who are of status k, m, p, c, sex s, and age $x + 0.5$ at the middle of the year $t + 0.5$ (denoted as $N_{k,m,p,c}(x + 0.5,t + 0.5,s)$) include those whose parity p is the same at age x and $x + 0.5$ (i.e., does not have a birth in the first half of the year) and those whose parity is $p-1$ at age x, but p at age $x + 0.5$ (i.e., have a birth in the first half of the year). The estimator of

$$\begin{aligned} B0_{k,m,p,c}(x,t,s) \text{ is : } B0_{k,m,p,c}(x,t,s) \\ = N_{k,m,p,c}(x,t,s)\, \left(1-\tfrac{1}{2}b_{p,m}(x,t,s)\right)/N_{k,m,p,c}(x+0.5,t+0.5,s) \end{aligned}$$

when ($p = 0$, $c = 0$) (note: c cannot be greater than p)

$$N_{k,m,0,0}(x+1,t+1,s) = N'''_{k,m,0,0}(x+0.5,t+0.5,s)\,(1-{}_{1/2}b_{0,m}(x+0.5,t+0.5,s)) \tag{2.11}$$

when ($p > 0$ and $c = 0$)

$$\begin{aligned} N_{k,m,p,0}(x+1,t+1,s) &= N'''_{k,m,p,0}(x+0.5,t+0.5,s) \\ &- N'''_{k,m,p,0}(x+0.5,t+0.5,s)BO_{k,m,p,0}(x,t,s)_{1/2}b_{p,m}(x+0.5,t+0.5,s) \\ &+ N'''_{k,m,p-1,0}(x+0.5,t+0.5,s)BO_{k,m,p,0}(x,t,s)_{1/2}b_{p-1,m}(x+0.5,t+0.5,s)\,(1-p(0,t)) \end{aligned} \tag{2.12}$$

when ($p > 0$ and $c > 0$ and $p > c$)

$$\begin{aligned} N_{k,m,p,c}(x+1,t+1,s) &= N'''_{k,m,p,c}(x+0.5,t+0.5,s) \\ &- N'''_{k,m,p,c}(x+0.5,t+0.5,s)BO_{k,m,p,c}(x,t,s)_{1/2}b_{p,m}(x+0.5,t+0.5,s) \\ &+ N'''_{k,m,p-1,c-1}(x+0.5,t+0.5,s)BO_{k,m,p-1,c-1}(x,t,s)_{1/2}b_{p-1,m}(x+0.5,t+0.5,s)p(0,t) \\ &+ N'''_{k,m,p-1,c}(x+0.5,t+0.5,s)BO_{k,m,p-1,c}(x,t,s)_{1/2}b_{p-1,m}(x+0.5,t+0.5,s)\,(1-p(0,t)) \end{aligned}$$

when ($p > 0$ and $c > 0$ and $p = c$)

$$\begin{aligned} N_{k,m,p,c}(x+1,t+1,s) &= N'''_{k,m,p,c}(x+0.5,t+0.5,s) \\ &- N'''_{k,m,p,c}(x+0.5,t+0.5,s)BO_{k,m,p,c}(x,t,s)_{1/2}b_{p,m}(x+0.5,t+0.5,s) \\ &+ N'''_{k,m,p-1,c-1}(x+0.5,t+0.5,s)BO_{k,m,p-1,c-1}(x,t,s)_{1/2}b_{p-1,m}(x+0.5,t+0.5,s)p(0,t) \end{aligned} \tag{2.13}$$

2.4 Projecting Households and All Individuals of the Entire Population Simultaneously

The accounting equations discussed above include all individuals of the population at the starting and future projection years. The distribution of household size and structure are derived from the characteristics of the reference persons. The family household statuses, including marital/union status, parity, and number of co-residing (or surviving disregarding co-residence) parents and children, are projected for all individuals of reference and non-reference persons in the entire population. Tabulations of population size, age/sex distributions, and other demographic indices, such as numbers/proportions of the elderly, middle-age, youth and children by various statuses, dependency ratios, labour force size and age structure, are derived from all individuals including reference and non-reference persons. This projection model projects households and all individuals of the entire population simultaneously, and thus consistencies between changes in population size/structure and households are guaranteed. This is not the case in other models for projecting households or population, as they provide no feasible way to convert information based on individuals directly into information on households. Even if these two different aspects could be matched for the starting year, it is very difficult for other models to guarantee consistent changes in both patterns when they are projected into the future (Lutz and Prinz 1994: 225). As shown above, the ProFamy extended cohort-component model projects households and individuals simultaneously and consistently, which is advantageous.

2.5 Consistencies in the Two-Sex and Multi-Generation Model

Because our model deals with both sexes and both children and parents, the following procedures are adopted to ensure the necessary consistencies.

2.5.1 Consistency Between Males and Females

Consistency between the male and female populations is a basic requirement in any two-sex model (including ProFamy) dealing with marriage/union statuses. In any year, the number of male marriages is equal to the number of female marriages,[6] the number of male divorces is equal to the number of female divorces, and the number of newly widowed females (males) is equal to the number of new deaths among currently married men (women). When cohabiting status is distinguished, the number of cohabiting males is equal to the number of cohabiting females, and the number of males (females) who exit from cohabiting status either due to union dissolution or death of partner is equal to the number of female (male) counterparts.

The two-sex consistency requirement described above is applicable to both closed marriage market populations with negligible inter-marriage with outsiders and to open marriage market populations with a sizable number of grooms and brides from outside the region or country under study. In the case of open marriage market populations, we still need to calculate the two-sex consistency because the sizable number of grooms and brides from outside of the region or country become residents of this region or country through immigration. In any case, in the same year in the same region or country, the number of newly married (or newly cohabiting) men is equal to the number of newly married (or newly cohabiting) women. We use the harmonic-mean procedure to ensure two-sex consistency. Mathematical formulas for the harmonic mean used in our two-sex model to ensure the consistency can be found elsewhere (see, e.g., Keilman 1985: 216–221). It has been shown that the harmonic mean satisfies most of the theoretical requirements and practical considerations for handling consistency problems in a two-sex model (Pollard 1977; Schoen 1981; Keilman 1985; Van Imholf and Keilman 1992).

---

[6] The ProFamy two-sex model does not account for same-sex marriages.

2.5.2 Consistency Between Children and Parents

We define three quantities from the children's perspective[7]:

C_1 – number of status transitions from living with two parents or one parent to not living with parents due to leaving home.
C_2 – number of status transitions from living with two parents or one parent to not living with parents due to death of one parent.
C_3 – number of deaths of persons who lived with parents at time of their death;
Let $S_1 = C_1 + C_2 + C_3$. Note that S_1, C_1, C_2, C_3 and all other variables defined in this sub-section are year-specific, but we omit the time dimension for simplicity.

We then define three quantities from the perspective of parents:

P_1 – number of reductions in number of co-residing children[8] for married or cohabiting couples (represented by the wife or female partner) and single parents due to children's leaving home;
P_2 – number of reductions in number of co-residing children for married or cohabiting couples and a single parents due to children's deaths;
P_3 – number of death events of both parents die in the same year multiplied by number of their co-residing children plus number of deaths of single parent multiplied by number of their co-residing children (note: if both parents die in the same year or a simple parent die, their children's status of co-residing with parent(s) will change).

Let $S_2 = P_1 + P_2 + P_3$
In any year, S_1 should be equal to S_2. In the numerical calculation, however, S_1 and S_2 may not be exactly the same due to differences in estimation procedures. Therefore, an adjustment is needed to ensure that S_1 and S_2 are equal to each other.
Following the harmonic mean approach, two equations must be satisfied:

$$C_1a_1 + C_2a_2 + C_3a_3 = 2S_1S_2/(S_1 + S_2) \tag{2.14}$$

$$P_1b_1 + P_2b_2 + P_3b_3 = 2S_1S_2/(S_1 + S_2) \tag{2.15}$$

Where a_1, a_2, a_3, b_1, b_2, b_3 are adjustment factors to be estimated.
Note that the following relations should be satisfied: $C_2a_2 = P_3b_3$ and $C_3a_3 = P_2b_2$. We have used the exact age-specific number of persons at risk and the age-specific death probabilities to compute C_3 and P_3. Only the age of each person is traced, however; no detailed age information for his or her parents and children is distinguished in our macro model. Consequently, we have to estimate the ages of his or her parents when we compute C_2, and estimate his or her children's ages

[7] The definition of children here is relative to parents. For example, a person aged 60 and older is still a child if he or she lives with parent(s).

[8] When the status of number of children living together is reduced by i, i events are accounted.

when we compute P_2.[9] Obviously, C_3 and P_3 are much more accurate than C_2 and P_2. Therefore, a_1, a_2, a_3, b_1, b_2, and b_3 are estimated as follows:

$$a_3 = 1.0 \tag{2.16}$$

$$b_3 = 1.0 \tag{2.17}$$

$$a_2 = P_3/C_2 \tag{2.18}$$

$$b_2 = C_3/P_2 \tag{2.19}$$

$$a_1 = [(2\, S_1 S_2/(S_1 + S_2)) - C_2 a_2 - C_3 a_3]/C_1 \tag{2.20}$$

$$b_1 = [(2\, S_1 S_2/(S_1 + S_2)) - P_2 b_2 + P_3 b_3]/P_1 \tag{2.21}$$

2.5.3 Consistency Between Births Calculated for the Female and Male Populations

Changes in parity and the status of co-residence with children (c) are calculated for both female and male populations in our two-sex model. The total number of births calculated based on the female population should be equal to the total number of births calculated based on the p and c status changes of the male population. Single-year age and parity-specific fertility rates for the male population are rarely available. Therefore, we estimate male birth rates based on female birth rates and the average age difference between the male and female partners. Clearly, the calculation of births for the female population is more accurate than that for the males. Therefore, the number of births produced by the male population is adjusted (by raising or lowering the male age-specific birth rates) to be equal to the number of births produced by the female population. In the ProFamy model, the assumption about how fertility would depend on marital status for females is consistent with that for males.

2.5.4 Consistency Between Females' and Males' Status of Co-residence with Children Before and After Divorce (or Dissolution of Cohabitation) or Remarriage

Children stay with either the mother or father after their parents' divorce or dissolution of cohabitation. Therefore, the number of children living with mother or father immediately after the parents' divorce should be equal to the number before their parental divorce. The living arrangement of the children of divorced couples is a complicated social phenomenon, and data are commonly available on this issue.

[9] One example may help to clarify this. Suppose that there are 1,000 women aged 30 living with one parent and two children, and whose k status is 2 and c status is 2. Although the parents and children of these 1,000 women are also in the pool of individual members of the population, it is impossible to individually link them one-by-one with these 1,000 women in our macro model. Therefore, the model knows that these women have one parent and two children living together, but does not keep track of the exact age of their parents and children. The ages of the parents and children of these 1,000 30-year-old women are estimated based on weighted averages using proper frequency distributions of fertility rates as the weights (see Appendices 4 and 5).

In most contemporary societies, young children are more likely to stay with their mother than with their father after the parents' divorce. For societies where divorced couples do not wish their children to be separated from each other, the ProFamy model and software provide an option for users to choose to assume that all children stay with their mother after their parents' divorce. Another available option is to assume that if a couple has an odd number of children living together before divorce, the mother has one more child than the father has after divorce. If a couple has an even number of children living together before their divorce, each party would have an equal number (half) of their children after divorce.

Children living with a single mother or a single father would join a new household after their parents remarried. A newly remarried couple's number of children living together should equal to the sum of children living with either of the parties before remarriage. For simplicity, we assume that the probability that a remarried woman or man will have additional children from the new partner's previous union depends on the frequency distribution of the status of co-residence with children of newly married men or women in the year.[10]

To conclude the present chapter, it is worthwhile to list the major assumptions discussed and justified in previous sections to provide a clearer picture of the nature of the ProFamy extended cohort-component model and to aid the interpretation of the projection output.

As discussed in Sect. 2.2.2 and shown in Fig. 2.3, we assume that births occur throughout the first half and the second half of the year while other status transitions and deaths occur at the middle of the year. We adopted a judicious use of independence assumptions, as discussed in Sect. 2.2.3. More specifically, some of the events are assumed to be locally independent. The child death events and child leaving home events are independent. Deaths and birth events are independent. Deaths and marital/union status changes are independent of parity and the number of children living at home. Death of one or two parents, divorce of parents, remarriage of the non-married parent, and leaving the parental home, as well as returning home, are independent. We adopted some reasonable assumptions about children's living arrangement after their parents' divorce and remarriage, as discussed in Sect. 2.5.4. In addition, we also adopted the Markovian assumption: status transitions depend on age and the status at the beginning of the single-year age interval, but are independent of duration in the status. The related homogeneity assumption is that people with the same characteristics have the same status transition probabilities.[11] We assume that parents may or may not live with one

[10] We exclude persons who are newly married for the first time with no pre-marital births from the frequency distribution for maternal status of newly remarried persons, since those young people are much less likely to choose a partner whose previous marriage was dissolved.

[11] The homogeneity assumption can be relaxed by introducing more characteristics. For instance, the homogeneity assumption is much less strong for a fertility model that considers age, parity, and marital/union status than for a fertility model that takes account of age only in the classic cohort-component population projection model. Since our family household projection model accounts for more characteristics of the population under study than most other demographic projection models, the Markovian and homogeneity assumptions in our model are less restrictive than those other models.

married child and his (or her) spouse and their unmarried children, while the rare cases of married brothers or sisters living together in modern societies are ignored.

We discuss data and estimation procedures for household and living arrangement projections using the ProFamy extended cohort-component approach in the next chapter. The fourth chapter presents empirical validation tests of projections from an earlier census year to a later census year and comparing projections with census counts in the later census year, using the ProFamy model for the United States and China at national and sub-national levels. Chapter 4 also presents a relatively detailed comparison of the ProFamy model and the classic headship-rate method which is still widely used.

Appendix 1: Procedure to Correct the Inaccurate Accounting of Household Size Distribution Due to the Lack of Capacity to Identify the Reference Person's Co-residence Status with Other Relatives or Non-relatives

Based on the census data set, we can derive $h(i,j,t)$, the proportion of households with i direct family members and j other relatives or non-relatives among the total number of households with i direct family members in year t. The term "direct family members" here refers to spouse (or cohabiting partner), children, and parents of the reference person. $\sum_{j=0}^{M} \mathrm{h(i,j,t)} = 1.0$, for all i. The maximum value of i in our model is $2 + 2 + P$; i.e., the largest three-generation household has two grandparents, two parents, and P (highest parity distinguished) children. $j = 0, 1, 2, 3, \ldots, M$, where M is the largest number of other relatives or non-relatives living in a household. We chose M as 5 in our current version of the ProFamy software since the number of single households with more than five other relatives or non-relatives in modern societies is negligible.

Denote by $H(i,t)$ the number of households of size i accounted for by our model before the adjustment. Denote $N(i,j,t)$ as the number of households with i direct family members and j other relatives or non-relatives in year t. $N(i,j,t) = H(i,t)h(i,j,t)$. The actual household size of $H(i,j,t)$ is $i + j$. Regrouping $H(i,j,t)$ with the sum of i and j as z, we obtain the adjusted number of households with size z in year t, which is denoted as $H(z,t)$, where $z = 1, 2, 3, \ldots, 2 + 2 + P + M$ (i.e., the largest household size is $2 + 2 + P + M$).

The average number of other relatives or non-relatives among all households with i direct family members is $a(i,t) = \sum_{j=0}^{M} \mathrm{h(i,j,t)\ j}$. We can allow $a(i,t)$ to change over time during the projection period. We may assume that the relative changes in $h(i,j,t)$ for all $j > 0$ in year t as compared with year t-1 is the same as the relative changes of $a(i,t)$ as compared with $a(i,t-1)$; more specifically, assuming $h(i,j,$

$t) = h(i, j, t - 1)a(i,t)/a(i, t - 1)$ for all $j > 0$. If the sum of $h(i,j,t)$ $(j > 0)$ over j is greater than one, which will usually never happen in the real world, we will have to standardize $h(i,j,t)$ $(j > 0)$ to ensure their sum is not greater than one. We then estimate $h(i,0,t)$ as $h(i, 0, t) = 1.0 - \sum_{j=1}^{M} \mathrm{h(i,j,t)}$

To help readers to understand how this procedure works, we present a numerical example as follows. Based on the U.S. 1990 census data set, we know that the proportion of American households with four direct family members and 0, 1, 2, 3, 4, 5 other relatives or non-relatives were 0.9320 0.0516 0.0102 0.0040 0.0012, 0.0011, respectively; the average number of other relatives or non-relatives among the households of four direct family members was 0.09 in 1990. If we assume that this average will become 0.11 in year 2000, we then estimate:

$$h(4,1,2000) = h(4,1,1990) \times 0.11/0.09 = 0.0516 \times 1.222 = 0.0631;$$
$$h(4,2,2000) = h(4,2,1990) \times 0.11/0.09 = 0.0102 \times 1.222 = 0.0125;$$
$$h(4,3,2000) = h(4,3,1990) \times 0.11/0.09 = 0.0040 \times 1.222 = 0.0049;$$
$$h(4,4,2000) = h(4,4,1990) \times 0.11/0.09 = 0.0012 \times 1.222 = 0.0015;$$
$$h(4,5,2000) = h(4,5,1990) \times 0.11/0.09 = 0.0011 \times 1.222 = 0.0013;$$
$$h(4,0,2000) = 1.0 - (0.0631 + 0.0125 + 0.0049 + 0.0015 + 0.0013) = 0.9167.$$

Appendix 2: A procedure to Meet the Requirement that Other Relatives and Non-relatives Cannot Be Reference Persons of the Household

The k status is equal to 3 (not living with parents) for those who are relatives other than parents and children or non-relatives of the reference person, since they do not live with parents. The c status of these persons is equal to 0, since they do not live with children. If no adjustment is made, these people would be counted as a one-person household if they are not married and not cohabiting (most likely), or counted as a one-couple household if they are married or cohabiting (less likely). Therefore, adjustment of the number of persons whose $k = 3$ and $c = 0$ must be done in order to derive a correct account of one-person and one-couple households.

Based on the census sample data set of the starting year, we calculate the 5-year-age-specific and marital-status-specific proportions of those, who are relatives (other than parents and children) or non-relatives in reference to the household reference person, among all persons with the same age and marital status not living with parents and children. We may assume that these proportions either remain constant or change over time. Multiplying these projected (or assumed) proportions by the corresponding number of persons whose k status is equal to 3 and c status is equal to 0, we estimate the number of other relatives and non relatives in the future years. Subtracting them from the number of persons whose k status is equal to 3 and c status is equal to 0, we have met the requirement that other relatives and non-relatives cannot be reference persons of the households.

Appendix 3: Parity Transition Probabilities in the 1st and 2nd Half of the Year

The age-specific probabilities of parity status change occurring throughout the first and second halves of the interval are "gross" probabilities, in the absence of the mother's mortality since it had already been taken into account in the middle of the single-year age interval.

Let f_p (x, m) denote the occurrence/exposure rates of the pth birth by age x and marital status m of the mother, which is defined as the number of pth births by women aged x to $x + 1$, with marital status m divided by the person-years lived in parity p-1 and marital status m of women aged x to $x + 1$. The probability that a woman of parity p-1 and marital status m at exact age x will be in parity p at exact age $x + 1$ in the absence of mortality and marital status change, b_p (x, m), can be estimated in a familiar manner with the assumption of a uniform distribution of births between ages x and $x + 1$ (analogous to the estimation of death probabilities from death rates):

$$b_p(x,m) = \frac{f_p(x,m)}{1 + \frac{1}{2} f_p(x,m)} \qquad (p = 1, 2, 3...., N). \tag{2.22}$$

As stated earlier, we will calculate the parity status change in the first half and in the second half of the age interval, respectively, so the corresponding formulas are needed. It should be stated that the following derivation is based on the assumption that no multiple parity transitions take place within a single age interval. There are at least two reasons for making this assumption. First, the multiple parity transitions are very rare. Second, birth rates are usually defined as the number of births divided by the number of women at risk. Multiple births and multiple deliveries in a single year have already been counted in the number of births, which is the numerator of the birth rates to be used.

Define ${}_{1/2}b_p(x,m)$ and ${}_{1/2}b_p(x + 0.5, m)$ as the probabilities of giving a pth birth between exact ages x and $x + \frac{1}{2}$ and between exact ages $x + \frac{1}{2}$ and $x + 1$, respectively, in the absence of mortality. Define W as the number of women of parity p-1 at exact age x. Assuming the uniform distribution of births in a year, the number of pth births to these W women in the first half of the year is equal to those occur in the second half of the year; both are $\frac{1}{2}Wb_p(x,m)$. Therefore, the probability of giving a pth birth in the first half of the year is:

$${}_{1/2}b_p(x,m) = \frac{1}{2}Wb_p(x,m)/W = b_p(x,m)/2. \tag{2.23}$$

There are $W - \frac{1}{2}Wb_p(x,m)$ women of parity p-1 in the middle of the year at risk of giving a pth birth. Since we assume that no multiple births occur in a single age interval, we must assume that the women who were of parity p-2 at the beginning of the age interval but who give a $(p\text{-}1)^{\text{th}}$ birth in the first half of the interval are not at risk of giving a pth birth in the second half of the interval. The probability of giving a pth birth in the second half of the year is

$$ {}_{1/2}b_p(x+0.5,m) = \frac{1}{2}Wb_p(x,m)/\left[W - 0.5Wb_p(x,m)\right] = b_p(x,m)/\left[2 - b_p(x,m)\right]. \tag{2.24} $$

Note that the data $f_p(x,m)$ are for an 1-year age interval, but the calculation of parity transitions between exact age x and $x + 1$ is divided into two steps by formulas (2.23) and (2.24). Fortunately, however, the parity distribution at the end of the age interval calculated by ${}_{1/2}b_p(x,m)$ and ${}_{1/2}b_p(x + 0.5, m)$ with two steps is the same as the parity distribution calculated by one step only, using $b_p(x,m)$ estimated by formula (2.22). This equivalence can be demonstrated as follows: first, combining two steps, the probability of parity progression is

$$ {}_{1/2}b_p(x,m) + \left[1 - {}_{1/2}b_p(x,m)\right] {}_{1/2}b_p(x+0.5,m) $$
$$ = {}_{1/2}b_p(x,m) + \frac{\left[1 - 0.5b_p(x,m)\right]b_p(x,m)}{2 - b_p(x,m)} = b_p(x,m) $$

Second, the probability of no parity progression is

$$ \left[1 - {}_{1/2}b_p(x,m)\right]\left[1 - {}_{1/2}b_p(x+0.5,m)\right] = \left[1 - \frac{b_p(x,m)}{2}\right]\left[1 - \frac{b_p(x,m)}{2 - b_p(x,m)}\right] $$
$$ = 1 - b_p(x,m) $$

This supports our two-step approach for calculating parity transitions (Zeng 1991a: 61–63).

Appendix 4: Procedures for Estimating Transition Probabilities of Status of Co-residence with Parents

Let $w_{ij}(x,t,s,m)$ denote the probability of transition from co-residence status i at age x in year t to j at age $x + 1$ in year $t + 1$ for persons of sex s and marital status m, where i (=1,2,3) and j (=1,2,3);

$q_m(x,t)$ and $q_f(x,t)$, probabilities of death of an x-year-old person's mother and father;

$d_m(x,t)$ and $d_f(x,t)$, probabilities of divorce of an x-year-old person's mother and father in year t;

$q_1(x,t)$ and $q_2(x,t)$, female and male death probabilities in year t;

$d_1(x,t)$ and $d_2(x,t)$, female and male divorce probabilities in year t;

z, the average age difference between the male and female partners;

$$q_m(x,t) = \sum_{i=15}^{49} q_1(x+i,t)f_1(i);\ \ q_f(x,t) = \sum_{i=15}^{49} q_2(x+z+i,t)f_2(i);$$

$$d_m(x,t) = \sum_{i=15}^{49} d_1(x+i,t)f_1(i);\ \ d_f(x,t) = \sum_{i=15}^{49} d_2(x+z+i,t)f_2(i);$$

Note that $f_1(i)$ and $f_2(i)$ are the frequency distributions of a product of age-specific fertility rates and conditional survival probability.

$$f_1(i) = (b(i)l_1(x+i)/l_1(i))/\sum_{i=15}^{49} (b(i)l_1(x+i)/l_1(i)$$

$$f_2(i) = (b(i)l_2(x+i)/l_2(i))/\sum_{i=15}^{49} (b(i)l_2(x+i)/l_2(i)$$

where $b(i)$ are age-specific fertility rates, $l_1(x)$ and $l_2(x)$ are female and male survival probabilities from age 0 to x. It is ideal that $b(i)$,$l_1(x)$ and $l_2(x)$ are cohort data, but it would be a good approximation if one employs the period data since the frequency distribution rather than the fertility and mortality level is used.

The events that cause transitions of the co-residence status from 1 to 2 are death of one of the parents or divorces of the parents. If the death of one parent occurs first, divorce cannot occur. Divorce, however, may precede death. Therefore,

$$\begin{aligned} w_{12}(x,t,s,m) \ &= q_m(x,t) + q_f(x,t) + d(x,t) - q_m(x,t)\, q_f(x,t) \\ &- q_m(x,t)d(x,t)/2 - q_f(x,t)d(x,t)/2 \end{aligned} \tag{2.25}$$

where $d(x,t) = (d_m(x,t) + d_f(x,t))/2$.

The events that cause transitions of the co-residence status from 1 to 3 are an x-year-old person leaving the parental home or numbers of death of both parents. If the deaths of both parents occur first, the event of leaving parental home cannot occur. A person can leave home, however, before either parent dies or after one of them dies. Therefore,

$$\begin{aligned} w_{13}(x,t,s,m) = l(x,t,s,m) &+ q_m(x,t)\, q_f(x,t) \\ &- q_m(x,t)\, q_f(x,t)\, l(x,t,s,m)(2/3) \end{aligned} \tag{2.26}$$

where $l(x,t,s,m)$ is the probability of leaving the parental home at age x in year t for persons of sex s and marital status m.

The events that cause transitions of the co-residence status from 2 to 3 are death of the non-married parent or numbers of an x-year-old person leaving the parental home. If the death of the lone parent occurs first, the event of leaving the parental home cannot occur. Therefore,

$$w_{23}(x,t,s,m) = l(x,t,s,m) + q(x,t) - (l(x,t,s,m)\, q(x,t))/2 \tag{2.27}$$

where $q(x,t) = (q_m(x,t) + q_f(x,t))/2$.

The events that cause transitions of the co-residence status from 2 to 1 are remarriage of the non-married parent, who may be widowed or divorced. Denote $r_{d1}(x,t)$ and $r_{d2}(x,t)$ as divorced female and male remarriage probabilities in year t; $r_{w1}(x,t)$ and $r_{w2}(x,t)$ as widowed female and male remarriage probabilities in year t.

$$\begin{aligned} w_{21}(x,t,s,m) &= \Big(\sum_{i=15}^{49} r_{d1}(x+i,t) f_1(i)\Big) g_{d1}(x) + \Big(\sum_{i=15}^{49} r_{d2}(x+z+i,t) f_2(i)\Big) g_{d2}(x+z) \\ &+ \Big(\sum_{i=15}^{49} r_{w1}(x+i,t) f_1(i)\Big) g_{w1}(x) + \Big(\sum_{i=15}^{49} r_{w2}(x+z+i,t) f_2(i)\Big) g_{w2}(x+z) \end{aligned} \tag{2.28}$$

where,

$$g_{d1}(x) = \sum_{i=15}^{49} N_{d1}(x+i) \Big/ \sum_{i=15}^{49} [N_{d1}(x+i) + N_{d2}(x+z+i) + N_{w1}(x+i) + N_{w2}(x+z+i)]$$

$$g_{d2}(x) = \sum_{i=15}^{49} N_{d2}(x+i) \Big/ \sum_{i=15}^{49} [N_{d1}(x+i) + N_{d2}(x+z+i) + N_{w1}(x+i) + N_{w2}(x+z+i)]$$

$$g_{w1}(x) = \sum_{i=15}^{49} N_{w1}(x+i) \Big/ \sum_{i=15}^{49} [N_{d1}(x+i) + N_{d2}(x+z+i) + N_{w1}(x+i) + N_{w2}(x+z+i)]$$

$$g_{w2}(x) = \sum_{i=15}^{49} N_{w2}(x+i) \Big/ \sum_{i=15}^{49} [N_{d1}(x+i) + N_{d2}(x+z+i) + N_{w1}(x+i) + N_{w2}(x+z+i)]$$

and $N_{d1}(x + i)$, $N_{d2}(x + z + i)$, $N_{w1}(x + i)$, and $N_{w2}(x + z + i)$ are the number of divorced females, divorced males, widowed females, widowed males, age $x + i$ or $x + z + i$, all living with at least one child in year t.

The event that causes a transition of the co-residence status from 3 to 1 is an x-year-old person returning home to join her or his two parents; the event that causes a transition of the co-residence status from 3 to 2 is an x-year-old person returning home to join her or his one non-married parent, so that

$$w_{31}(x,t,s,m) = h(x,t,s,m)\{N_{k1}(x,t,s,m)/[N_{k1}(x,t,s,m) + N_{k2}(x,t,s,m)]\} \tag{2.29}$$

$$w_{32}(x,t,s,m) = h(x,t,s,m)\{N_{k2}(x,t,s,m)/[N_{k1}(x,t,s,m) + N_{k2}(x,t,s,m)]\} \tag{2.30}$$

where $h(x,t,s,m)$ is the probability of returning home between age x and $x + 1$ to join the parental home in year t, for persons of sex s and marital status m. $N_{k1}(x,t,s,m)$

and $N_{k2}(x,t,s,m)$ are numbers of x-year-old persons of sex s and marital status m who are living with two parents and one parent, respectively.

In addition,

$$w_{11}(x,t,s,m) = 1 - w_{12}(x,t,s,m) - w_{13}(x,t,s,m) \quad (2.31)$$
$$w_{22}(x,t,s,m) = 1 - w_{21}(x,t,s,m) - w_{23}(x,t,s,m) \quad (2.32)$$
$$w_{33}(x,t,s,m) = 1 - w_{31}(x,t,s,m) - w_{32}(x,t,s,m) \quad (2.33)$$

Appendix 5: Procedures for Estimation of Probabilities of Change in Number of Children Living Together

Denote by q_1 the average probability of dying for the children of an x-year-old mother or father;

q_2, average probability of leaving the parental home for the children of an x-year-old mother or father;

$q(x-i)$, age-specific average death rates for male and female children;

$h(x-i)$, age-specific average rates of leaving the parental home for male and female children;

$f(i)$, the frequency distribution of fertility rates from age α to age x; $\sum_{i=\alpha}^{x} f(i) = 1.0$, i as the age at birth of the mother or father; α, the lowest age at birth.

From the model, we know an x-year-old person has c ($c = 0, 1, 2, \ldots$) children living together, but the ages of these c children are not kept track of to make the model manageable. The chance that the x-year-old person gave a birth at age i and the child is x−i years old is $f(i)$. The weighted average of the probability of dying of a child of an x-year-old person can be estimated as

$$q_1 = \sum_{i=\alpha}^{x} q(x-i) * f(i) \quad (2.34)$$

The weighted average of the probability of leaving the parental home of a child of an x-year-old person can be estimated as

1. If three-generation households are considered,

$$q_2 = \sum_{i=\alpha}^{x} h(x-i) * f(i) \quad (2.35)$$

2. If three-generation households are negligible, such as in the Western countries, we assume all children who have not left the parental home before marriage (or cohabitation) will do so in the same year of their marriage (or cohabitation). In other words, children who remain single until the end of the year have a leaving home probability of $h(x-i)$; children who newly marry or enter a union in the year have a leaving home probability of 1.0.

$$q_2 = \sum_{i=\alpha}^{x} [h(x-i)(1-m(x-i)) + 1.0m(x-i)] * f(i) \tag{2.36}$$

where $m(x-i)$ is the age-specific average probability of first marriage/union formation for male and female children.

The probability that a child will survive and continue to live at home is $p = (1-q_1)(1-q_2)$; the probability that a child will either leave home or die is $1-p$.

Assuming the events of death and leaving the parental home are locally independent, we could easily estimate the probability of changes in the c status of the number of co-residing children. To simplify the presentation, we assume that the highest parity is 3 here (the calculation method will be the same when the highest parity is larger than 3).

let $s_1(t)$, $s_2(t)$, and $s_3(t)$ denote the probabilities that the one child, two children, or three children who were living at home at the beginning of the year will survive and live at home at the end of the year t;

$d_1(t)$, $d_2(t)$, and $d_3(t)$, the probabilities that the one child, both of the two children, and all of the three children will die or leave home during the year t.

$d_{12}(t)$, $d_{13}(t)$, and $d_{23}(t)$, the probabilities that one of the two children, one of the three children, and two of the three children will die or leave home at the end of the year t.

The estimators of $s_1(t)$, $s_2(t)$, $s_3(t)$, $d_1(t)$, $d_2(t)$, $d_3(t)$, $d_{12}(t)$, $d_{23}(t)$ and $d_{13}(t)$ are as follows:

$$s_1(t) = p \tag{2.37}$$
$$s_2(t) = p \times p \tag{2.38}$$
$$s_3(t) = p \times p \times p \tag{2.39}$$
$$d_1(t) - 1 - p \tag{2.40}$$
$$d_2(t) = (1-p) \times (1-p) \tag{2.41}$$
$$d_3(t) = (1-p) \times (1-p) \times (1-p) \tag{2.42}$$
$$d_{12}(t) = 2 \times p(1-p) \tag{2.43}$$
$$d_{23}(t) = 3 \times p \times (1-p) \times (1-p) \tag{2.44}$$
$$d_{13}(t) = 3 \times p \times p \times (1-p) \tag{2.45}$$

In the case in which two children leave home or die in the year t among the three co-residing children ($d_{23}(t)$), for example, there are three combinations of one of the three children leaving home or dying while the other two survive and continue to stay at home. Therefore, we multiply '$p \times (1-p) \times (1-p)$' by '3'.

Chapter 3
Data Needs and Estimation Procedures

3.1 Data Needed

The following data are needed for household and living arrangements projections at the national or sub-national level employing the ProFamy model:

1. Base population at the national or sub-national level: A sample dataset, based on a census micro data file or population register, needs to contain variables of sex, age, marital/union status, relationship to the household head or householder, parity (optional), and whether live in private household or institutional household (see (1) in Table 3.1). Based on this sample dataset, the ProFamy software will derive the base population of the starting year of the projection, classified by age, sex, marital/union status, parity (optional), number of co-residing children and co-residing status with none, one or two parents, and whether living in a private household or group quarter. Parity (number of children ever born) information is optional and not crucial, given that the role it plays in the model is to express parity-specific (also age- and marital status-specific) fertility rates only. If no parity information is available, we assume that birth probabilities depend on age, marital status, and number of children living at home. If a sample dataset is used, 100 % tabulations of age-sex distributions of the entire population and those living in group quarters, as well as the total number of households, must be derived from the entire census or population register data. This is to ensure accurate total population size and age/sex distributions and total number of households in the starting year of the projection (see Appendix 1 for more details), while the sample dataset provides more detailed information of the status distributions.
2. Model standard schedules at the national level, which also may be used for projections at the sub-national level (see justification in Sect. 2.4 of Chap. 2).
   (a) Age-sex- (and marital-status, if possible) specific probabilities of surviving, derived from recent life tables.

Y. Zeng et al., *Household and Living Arrangement Projections*,
The Springer Series on Demographic Methods and Population Analysis 36,
DOI 10.1007/978-90-481-8906-9_3,

Table 3.1 Data needs to project households and living arrangements using the ProFamy extended cohort-component method, with data sources for U.S. applications as an illustration

Contents of the data	e.g. data source for U.S. application
(1) Base Population at national and sub-national levels	
(a) A census micro sample or population register or an exceptionally large survey data file with a few needed variables, including sex, age, marital/union status, relationship to the householder, and whether living in a private or institutional household	(a) Census 5 % micro data or more recent and cumulative American Community Survey (ACS) data files (b) the published online 100 % census or ACS cross-tabulations.
(b) If a sample data set is used for the status distributions, published 100 % census tabulations of age-sex-specific (and marital status-specific if possible) distributions of the entire population, including those living in group quarters, as well as the aggregated numbers of households will be needed	
(2) Model standard schedules at the national level (may be used for projections at sub-national level)	
(a) Age-, sex-, (and marital-status if possible) specific probabilities of surviving	(a): Census Bureau's estimates, and Schoen and Standish (2001)
(b) Age-sex-specific o/e rates of marriage/union formation and dissolution	(b),(c): Pooled NSFH, NSFG, CPS, SIPP data sets, see Zeng et al. (2006) and Zeng et al. (2012b)
(c) Age-parity-specific o/e rates of marital and non-marital fertility	
(d) Age-sex-specific net rates of leaving the parental home, estimated based on two adjacent census micro data files and the intra-cohort iterative method Coale (1984, 1985; Stupp 1988; Zeng et al. 1994), using the ProFamy software	(d): The 1990 and 2000 censuses micro data files
(e) Age-sex-specific rates of international immigration and emigration, or Age-sex-specific rates of international net-migration	(e): Census 5 % micro data files.
(f) Age-sex-specific rates of domestic in-migration and out-migration for the sub-national regions (may be estimated for the nation and sub-national regions based on the census micro data files)	(f): Census 5 % micro data, ACS data files
(3) Demographic summary parameters for the national and sub-national regions	
(a) Standardized general rates of marriage and divorce	(a), (b): Based on census micro data, vital statistics and pooled survey datasets
(b) Standardized general rates of cohabitation union formation and dissolution	
(c) Total Fertility Rates (TFR) by parity	(c), (d), (e), (f): Based on estimates released by the Census Bureau and the National Center for Health Statistics
(d) Male and female life expectancies at birth	
(e) Total numbers of male and female migrants	
(f) Mean ages at first marriage and births	

(continued)

Table 3.1 (continued)

Contents of the data	e.g. data source for U.S. application
(g) Proportion of those aged 45–49 who do not live with parents (measuring level of race-sex-age-specific net rates of leaving the parental home; see Appendix 3 for the estimation procedure)	(g), (h), (i), (j): Census 5 % micro data files
(h) Age-sex-specific proportion of persons who live in group quarters	
(i) Age-sex-specific proportion of elderly living with child(ren) (see Appendix 3 for the estimation procedure)	
(j) Household size-specific average number of other relatives (than spouse/partner, parents and children) and non-relatives living in the same household	

Notes: The data categories of race or rural/urban are optional based on the actual demographic situation and data availability of the country or region under study. For example, the race dimension is distinguished in the U.S. households and living arrangements projections (see Chaps. 8, 9, 10 and 11), but it can be omitted in the other countries' applications if the race differentials are not crucial, the race-specific data are not available, or the sub-population sizes of the minority race groups are small. The rural–urban dimension can be included if the rural–urban differentials are substantial and the rural–urban specific data are available, such as the applications for China (see Chaps. 12, 13, 14 and 15). If the categories of race or rural/urban are adopted in the application, all data listed in this table will need to be race-specific or rural/urban-specific

(b) Age-sex-specific occurrence/exposure (o/e) rates of marriage/union formation and dissolution (see Figs. 2.1 and 2.2 in Chap. 2), derived from survey datasets.
(c) Age-parity-specific o/e rates of marital and non-marital fertility, derived from survey datasets. Users of the ProFamy software have a choice of providing either one set of age-parity-specific o/e rates only for married women (assuming non-marital births are negligible) or multiple sets of age-parity-specific o/e rates for women with different marital/union statuses.
(d) Age-sex-specific net rates of leaving the parental home, estimated based on two adjacent census micro data files and the intra-cohort iterative method Coale(1984, 1985; Stupp 1988; Zeng et al. 1994), using of the modules of the ProFamy software package.
(e) Age-sex-specific rates of international in-migration and out-migration, or age-sex-specific rates of international net-migration
(f) Age-sex- (and marital status, if possible) specific o/e rates of out-migrants from the region under study to the rest of the country; age-sex- (and marital status if possible) specific frequency distribution of in-migrants from the rest of the country to the region under study. If data on age-sex-specific out-migration and in-migration are not available, one can use the age- sex-specific net rates of domestic migration. The domestic age-sex-specific migration rates may be estimated for the nation and sub-national regions based on the census micro data files.

Normally, the model standard schedules of fertility, mortality, and marriage/union formation and dissolution and international migration (see (2) in Table 3.1) need to be estimated at the national level only, and then can be employed for projections at the sub-national level. The age-sex-specific rates of domestic in-migration and out-migration at the sub-national level can be estimated based on census or large survey micro data files.

3. Projected (or assumed) demographic summary parameters – Total Fertility Rate (TFR), life expectancy at birth (*e0*), standardized general rates of marriage, divorce, cohabitation and union dissolution (see Appendix 2 for definitions of these standardized general rates), total number of migrants, mean ages at first marriage and at births in future years, etc. – are needed for projections at both national and sub-national levels (see (3) in Table 3.1).

It is important to note that the data described above and listed in Table 3.1 (1) through (3) as required by the ProFamy extended cohort-component model for household and living arrangement projections are all available from conventional demographic data sources such as ordinary surveys, vital statistics, and censuses Zeng et al. (2006). As an illustrative example, data sources for the U.S. application are listed in the last column of Table 3.1.

In sum, using estimated or existing national model standard schedules and the ProFamy extended cohort-component method, household and living arrangement projections at the national or sub-national level require a census micro data file and the projected (or assumed) demographic summary parameters for the future years. If the option of rural–urban classification is chosen by the user, such as in application to China, the data specified in (1) through (3) in Table 3.1 are rural/urban- specific. In this case, age-sex- (and marital status, if available) specific frequency distribution of rural–urban net migration within the country or region under study will be needed; one will also need to specify the proportion of urban population in the future projected years. If race classifications are included for the projections, as in the U.S. applications, the data described in (1) through (3) of Table 3.1 are all race-specific.

It is ideal to have the age-specific demographic schedules described above in (1) and (2) observed in the recent past from the country or region under study. When some demographic age-sex-specific standard schedule(s) are not available, especially for provinces or states, one may use the age-specific standard schedule (s) based on data at national level as the model standard schedules. One may even use model standard schedules based on data from another country or region where the general age pattern of demographic processes is similar to that in the country or region under study. For example, if Canadian age-sex-specific schedules of cohabitation union formation and dissolution are not available, one may use the U.S. rates as a model standard schedule and the Canadian future years' anticipated general rates of cohabitation union formation and dissolution as summary measures. This will enable one to reasonably project the age-sex-specific rates of union formation and dissolution in the future years in Canada, because the general age pattern of union formation and dissolution in Canada is similar to that in the U.S. Such an approach is similar to the practice of jointly employing the regional model life tables as model standard schedules and projected life expectancy at birth as anticipated mortality levels (summary measure) to project age-specific death rates in the future years in population projections (see Sect. 2.4 of Chap. 2 for justification and discussion).

3.2 Estimation and Projections of Age-Sex-Specific Demographic Rates

As Keyfitz (1972) pointed out, projections with a trend extrapolation of each age-specific rate can result in an excessive concession to flexibility, and can readily produce erratic results. Thus, we focus on the estimation and projection of demographic summary measures. We also use age-sex-specific standard schedules of demographic rates to define the age patterns of demographic processes. The standard schedules can be assumed either to be stable or to include systematic changes in timing and shape during the projection years Zeng et al. (2000). When fertility is postponed to later ages or advanced to earlier ages, for example, one may shift the age-specific standard schedule of fertility to the right or the left based on the amount of increase or decrease in the mean age at childbearing; the shape of the fertility schedule, however, remains unchanged. One may also model assumptions that fertility will be delayed or advanced while the curve becomes more spread or more concentrated through parametric modeling Zeng et al. (2000).

Zeng et al. (2013a) estimated the U.S. race-age-sex-specific o/e rates of marital status transitions and the race-age-parity-marital-status-specific o/e rates of fertility in the 1970s, 1980s, and 1990s. This work is based on pooled data from 10 waves of four major national surveys conducted from 1980 to 1996 with a total sample size of 394,791 women and men. The estimates show empirically that the basic shapes of the demographic schedules remained reasonably stable from the 1970s to the 1990s, while the timing changed remarkably. We thus may reasonably assume that in normal circumstances the basic shape of the age-sex-specific standard schedules remain stable, while the changes in timing are modeled through the changing mean age at marriage and fertility in family household projections.

If the observed standard schedules of age-parity-specific o/e rates of fertility are available only for married women but non-marital births are not negligible, one may assume either that the age-pattern of fertility of non-married women is the same as that of married women or that there is some systematic difference in the age-pattern of fertility between married and non-married women, while the non-marital fertility level differs from the marital fertility level. You may proportionally modify the standard schedules of married women to match your estimated fertility level of non-married women. For example, one may multiply the age-parity-specific o/e rates of fertility for married women by the ratio of the general fertility rates of women with various non-married statuses to the general fertility rate of married women. The marital-status-specific general fertility rate is defined as the total number of births to women of a certain marital status divided by the total number of women of reproductive ages (15–49) and that marital status. If one believes that never-married and not-cohabiting women tend to give birth earlier than married women, one may shift the standard schedules to the left correspondingly to approximately match the estimated timing difference between these two groups of women. All of the calculations related to the above concerns could be done in an Excel worksheet.

In the ProFamy extended cohort-component approach, we adopted the simple approach to calculate the required time-varying and age-specific fertility, mortality, and migration rates in future years; namely, we proportionally inflate or deflate the age-sex-status-specific standard schedules of fertility, mortality and migration to get time-varying age-sex-status-specific rates that are consistent with the projected parity-specific TFR, life expectancy at birth, and total number of migrants in future years. We also use the mean age at births and mean age at first marriage to monitor the changes in timing of fertility and first marriage in the projection period. Calculation of age-sex-status-specific rates of marriage/union formation and dissolution in future years for family household projections are, however, not as simple; an two-step procedure to calculate the sex-age-specific rates while ensuring the consistency of the two-sex constraints and the projected standardized general rates of marriage/union formation and dissolution is presented in Appendix 4.

3.3 Pooling Data from Multiple Surveys to Estimate the Age-Sex-Status-Specific Standard Schedules: Rationale and Justification

Previous empirical research has shown that pooling/combining data from multiple surveys can provide enhanced estimates by increasing the sample size; pooled data consist of independently sampled observations, which largely rule out correlation in the error terms across different observations (Wooldridge 2003). According to Schenker and Raghunathan (2007), several projects have been conducted within and outside the National Center for Health Statistics to enhance estimation by combining different surveys to extend coverage, improve analysis on self-reported data, and increase the accuracy of measurements derived for smaller population groups and smaller areas. Researchers from various universities in Australia have conducted a project focused on "successful ageing" that uses pooled data observations from nine national and local longitudinal surveys with different sampling strategies (including random, stratified, or clustered sample designs) and age ranges (Anstey et al. 2010). The research team claims that, with appropriate weights, the pooled dataset is nationally representative and their analytical strategy overcomes the limitations of a single survey which include small numbers of persons with specific medical conditions and therefore lack statistical power for effective comparisons among groups with specific characteristics such as very old age, low-prevalence disorders, or co-morbidities (Anstey et al. 2010). The "Comparison of Longitudinal European Studies on Aging (CLESA)" has undertaken a similar pooling and harmonization approach to analyze data from six longitudinal studies (Minicuci et al. 2003; Anstey et al. 2010: 49). The United Nations Inter-agency Group for Child Mortality Estimation has pooled a number of datasets from vital registration systems, national population censuses, and household surveys to produce the best estimates of child and infant mortality rates for various countries (UNICEF, WHO, The World Bank and UN Population Division 2007). Pooling different survey datasets has also been a recognized practice in other fields, such as econometrics (e.g., Wooldridge 2003), public opinion research (e.g., Brace et al. 2002), and in biology studying species other than humans (e.g., Fancy 1997).

However, pooled datasets also have limitations. As not all contributing datasets may be equally and nationally representative, the samples needs to be re-weighted to produce population estimates. As with all retrospective and longitudinal surveys, data may be biased by memory errors, sample attrition, mortality, and other non-responses (Anstey et al. 2010: 49). Therefore, household projection studies should fully utilize census and population register vital statistics data whenever they are relevant, consistently measured, of good quality, and available.

3.4 Estimation of Demographic Summary Measures

Estimates of the required demographic summary parameters – *TFR*, life expectancy at birth, total number of migrants, mean age at first marriage and birth – are straightforward. In the previous version of the ProFamy model and program (Zeng et al. 1997, 1998), we used period multistate life table propensities of marital status transitions as summary measures of marriage formation and dissolution. This approach restricted the practical applicability of the model because the data needed to construct period multistate life tables and estimate the propensities are likely not available at the provincial/state level, and are often not available at the national level for some countries. We now use much more practically applicable summary measures of general rates of marriage/union formation and dissolution, which are usually available at national and provincial/state levels. The general rates of marriage/union formation and dissolution in year t are defined by dividing the total number of events of marriage/union formation and dissolution occurring in the year t by the total number of persons who are at risk of experiencing these events. A few important points must be clarified in defining such general rates in our family household projection model.

First, we use the most recent census-counted sex-age-marital/union status distributions (i.e., the base population of the household projection) as the "standard" to calculate standardized general rates in future projection years. Following the language used in Preston et al. (2001: 24), the standardized general rate in the future projection year *t* is the estimated general rate in year *t* if it retained its sex-age-specific o/e rates in year *t* but had the age distribution of the risk population in the most recent census year (i.e., the base year of the projection). By employing standardized general rates for future projection years, we eliminate distortions in levels of marriage/union formation and dissolution due to changes in population age structure. For example, the unstandardized general marriage (or divorce) rate would decrease/increase solely due to the structural growth/decline of the numbers of elderly persons even if the age-specific marriage (or divorce) rates did not change. This is because the risks of marriage (or divorce) of the elderly are substantially lower than those of younger people.

Second, we cannot employ sex-specific general rates of marriage/union formation and dissolution as projected (or assumed) summary measures in future years because it would be impossible to ensure that the projected sex-specific general rates are consistent with the two-sex constraints. This is because the two-sex constraints also depend on the unknown (to-be-projected) sex-age-marital/union status distributions in future years. We therefore define the general rates of marriage/union formation and dissolution for males and females combined. Consequently, gender differentials in the age-specific marriage/union formation and dissolution rates are determined by the sex-age-specific standard schedules of o/e rates of marital/union status transitions and projected future years' population structure by age, sex, and marital/union status, while meeting the two-sex constraints.

Third, we estimate overall (rather than marital/union-specific) summary measures of marriage/union formation. Never-married, widowed, and divorced men and women may marry each other; a cohabiting couple whose legal marital statuses are different may marry, or a cohabiting person may leave his or her partner to marry another person. Similarly, never-married, widowed, and divorced persons may form a cohabitation union with each other. Thus, employing separate summary measures of marriage/cohabitation for never-married, widowed, and divorced persons would make it impossible to ensure the two-sex consistency because of the cross-marriage/union-formation among people with different marital/union statuses. Thus, we define the overall summary measures of marriage/union formation ($GM(t)$, $GC(t)$) to include relevant events with different marital/union statuses before the onset of marriage or cohabitation. This implies that changes in the overall intensities of various marriages and cohabitations are proportional to changes in the overall summary measures. This assumption is reasonable because different kinds of marriages and cohabitations are all related to general social attitudes toward marriage and cohabitation. If one is not satisfied with such an assumption, one may simply inflate or deflate the standard schedules (estimated from survey data) of sex-age-specific rates of marriage/cohabitation for never-married, widowed, and divorced persons differently according to one's assumptions. This adjustment will reflect projected differentials in future years, while the overall summary measures reveal the general level of marriage/union formation. On the other hand, the sex-age-status-specific rates of marriage/union formation are calculated for persons with different marital/union statuses before the onset of marriage and cohabitation, respectively. Combining the detailed sex-age-status-specific rates with the overall summary measures of marriage/union formation is a reasonable approach to model differentials in marriage/cohabitation among different types of not-married and not-cohabiting persons while meeting the two-sex constraints. Furthermore, one can easily estimate the more detailed sex-specific summary measures of first marriage, remarriage, or cohabitation of never-married and ever-married persons in year *t* once the sex-age-status-specific rates of marriage/union formation and dissolution in year t $(m_{ij}(x,s,t))$ have been projected.

The procedures for estimation of the general rates of marriages, divorces, and cohabitation union formations and dissolutions in the starting year of the projection (using the U.S. application at the state level as an illustrative example) are presented in Appendix 5.

Appendix 1: Procedures to Ensure the Accuracy of the Base Population for the Projections

Procedure to Ensure Accurate Total Population Size and Age/Sex Distributions in the Starting Year of the Projection

Define. $W(k,m,p,c,x,s,T1)$ – age(x) and sex(s) specific number of persons with statuses k (co-residence with parents), m (marital/union status), p (parity), c (number of co-residing children) in the starting year ($T1$) of the projection, derived from the sample dataset.

$N(m,x,s,T1)$ – age-sex-marital/union-status-specific number of persons in the starting year of the projection, based on the 100 % census tabulations.

To ensure accurate total population size and age/sex distributions, $W(k, m, p, c, x, s, T1)$ must be adjusted:

$$W'(k,m,p,c,x,s,T1) = W(k,m,p,c,x,s,T1)\left[N(m,x,s,T1)/\sum_k\sum_p\sum_c \mathrm{W(k,m,p,c,x,s,T1)}\right] \quad (3.1)$$

If no age-sex-marital/union-status-specific number of persons based on the 100 % census tabulations are available, but age-sex-specific number of persons ($N(x,s,T1)$) in the starting year based on the 100 % census tabulations are available, Eq. 3.1 is modified as:

$$W'(k,m,p,c,x,s,T1) = W(k,m,p,c,x,s,T1)\left[N(x,s,T1)/\sum_k\sum_p\sum_c\sum_m \mathrm{W(k,m,p,c,x,s,T1)}\right] \quad (3.2)$$

Procedure to Ensure an Accurate Total Number of Households in the Starting Year of the Projection

As described in Sect. 3.1.1 above, we have obtained correct total (100 %) population classified by age, sex and *k*, *m*, *p*, *c* statuses in the staring year of the projection ($W'(k, m, p, c, x, s, T1)$). Using our ProFamy model accounting system, we first get a total number of households in the starting year of the projection, which may not be equal to the 100 % census count of the total number of households. For example, the difference between the ProFamy model count and the census count of the total number of households is 1.5–2.0 % using the U.S. 1980 and 1990 census micro data files and the 100 % census tabulations of population age and sex distributions from 1980 and 1990. The reason why there is such a discrepancy is that the sampling fractions of individual persons and household units are not exactly the same. Although the discrepancy is generally small, we need to do some adjustment to ensure an accurate total number of households in the starting year of the projection. We have done this using a simple procedure (Zeng et al. 2006) described below. Note that the following procedure assumes that we do not have census 100 % tabulations of number of households by age of reference persons, which is the usual case.

Define: $H1(j)$ – number of households with size j in the starting year, derived by the ProFamy model count, using both census sample data set and the census 100 % tabulations of the population age-sex (and marital status, if available) distributions.

$H2(j)$ – total number of households with size j in the starting year, based on the 100 % census tabulation.

$TH2$ – total number of all households in the starting year, based on the 100 % census tabulation.

$T(x,s)$ – age-sex-specific total number of persons including reference and non-reference persons in the starting year, according to the 100 % census tabulation.

$W1(x,s,j)$ – age-sex-specific total number of reference persons of the households with size j in the starting year, according to the ProFamy model count.

$NW1(x,s)$ – age-sex-specific total number of non-reference persons in the starting year, according to ProFamy model count, where $T(x,s) = \sum_j W1(x,s,j)$ $+NW1(x,s)$.

$W2(x,s,i)$ and $NW2(x,s)$ are the adjusted number of reference persons (with household size j) and non-reference persons, respectively.

$T(x,s)$, $W1(x,s,j)$, $NW1(x,s)$, $W2(x,s,j)$ and $NW2(x,s)$ are all 5-year age specific.

1. First adjustment to ensure that the household size distribution is consistent with the census 100 % tabulation:

$$W2'(x,s,j) = W1(x,s,j)(H2(j)/H1(j))$$

2. Second adjustment to ensure that the total number of all households is consistent with the census 100 % tabulation, while the relative distribution of household size remains unchanged as in step (1):

$$W2(x,s,j) = W2'(x,s,j)\left\{TH2/\left[\sum_x\sum_s\sum_j \mathrm{W2}'(\mathrm{x},\mathrm{s},\mathrm{j})\right]\right\}.$$

3. Adjust non-reference persons:

$$NW2(x,s) = NW1(x,s)\left\{\left[T(x,s) - \sum_j W2(x,s,j)\right]/[T(x,s) - W1(x,s)]\right\}$$

Proof.

$$\sum_x\sum_s\sum_j W2(x,s,j) = \sum_x\sum_s\sum_j W2'(x,s,j)\left\{TH2/\left[\sum_x\sum_s\sum_j W2'(\mathrm{x},\mathrm{s},\mathrm{j})\right]\right\}$$

$$= TH2$$

$$NW2(x,s) = NW1(x,s)\left\{\left[T(x,s) - \sum_j W2(x,s,j)\right]/[T(x,s) - W1(x,s)]\right\}$$

$$= NW1(x,s)\left\{\left[T(x,s) - \sum_j W2(x,s,j)\right]/[W1(x,s) + NW1(x,s) - W1(x,s)]\right\}$$

$$= T(x,s) - \sum_j W2(x,s,j)$$

so, $\sum_j W2(x,s,j) + NW2(x,s) = \sum_j W2(x,s,j) + T(x,s) - \sum_j W2(x,s,j) = T(x,s)$

Appendix 2: Standardized General Rates of Marriage/Union Formation and Dissolution

The standardized general rate of marriage/union formation and dissolution in the projection year t is defined as the total number of events that would occur if the age-sex-specific rates of occurrence of the events in year t were applied to the most recent census-counted sex-age-marital/union status distribution derived from the census data (Zeng et al. 2006).

Let $N_i(x,s,r,T1)$ denote the number of persons of age x, marital/union status i, race or rural/urban category r, and sex s counted in the most recent census year $T1$ (i.e., the starting population of our household projection);

$m_{ij}(x,s,r,t)$, sex-age-status-specific rates of transition from marital/union status i to j in year t $(i \neq j)$.

The $m_{ij}(x,s,r,t)$ are to be calculated by the ProFamy program, while ensuring the consistency of the two-sex constraints and the projected standardized general rates of marriage/union formation and dissolution, which are defined below, in the year t (see Appendix 4 for details on how to calculate $m_{ij}(x,s,r,t)$).

Let $GM(r,t)$ denote the projected race or rural/urban-specific standardized general rate of marriages including first marriage and remarriage for males and females combined.

$$GM(r,t) = \frac{\sum_{x=\alpha}^{\beta} \sum_{s=1,2} \sum_{i} N_i(x,s,r,T1) m_{i2}(x,s,r,t)}{\sum_{x=\alpha}^{\beta} \sum_{s=1,2} \sum_{i} N_i(x,s,r,T1)}, i = 1,3,4,5,6,7 \quad (3.3)$$

where α is the lowest age at marriage; β is the higher boundary of the age range in which the general rate of marriage/union formation and dissolution is defined.

Let $GD(r,t)$ denote the projected race- or rural/urban-specific standardized general divorce rate for males and females combined.

$$GD(r,t) = \frac{\sum_{x=\alpha}^{\beta} \sum_{s=1,2} N_2(x,s,r,T1) m_{24}(x,s,r,t)}{\sum_{x=\alpha}^{\beta} \sum_{s=1,2} N_i(x,s,r,T1)}. \quad (3.4)$$

Let $GC(r,t)$ denote the projected race- or rural/urban-specific standardized general rate of cohabiting of never-married and ever-married males and females combined.

$$GC(r,t) = \frac{\sum_{x=\alpha}^{\beta} \sum_{s=1,2} [N_1(x,s,r,T1) m_{15}(x,s,r,t) + N_3(x,s,r,T) m_{36}(x,s,r,t) N_4(x,s,r,T1) m_{47}(x,s,r,t)]}{\sum_{x=\alpha}^{\beta} \sum_{s=1,2} [N_1(x,s,r,T1) + N_3(x,s,r,T1) + N_4(x,s,r,T1)]}$$

(3.5)

Let $GCD(r,t)$ denote the projected race- or rural/urban-specific standardized general union dissolution rate for males and females combined.

$$GCD(r,t) = \frac{\sum_{x=\alpha}^{\beta} \sum_{s=1,2} [N_5(x,s,r,T1) m_{51}(x,s,r,t) + N_6(x,s,r,T1) m_{63}(x,s,r,t) + N_7(x,s,r,T1) m_{74}(x,s,r,t)]}{\sum_{x=\alpha}^{\beta} \sum_{s=1,2} [N_5(x,s,r,T1) + N_6(x,s,r,T1) + N_7(x,s,r,T1)]}$$

(3.6)

Appendix 3: Procedure to Estimate Proportions of Those Aged 40–44 in Year *t* Who Do Not Live with Parents and Proportions of Elders Aged *x* in Year *t* Living with Adult Child(ren), While Taking into Account the Effects of Large Changes in Fertility

The procedures presented in this Appendix are designed for those populations in which the fertility level has been largely reduced in the past a few decades, implying that the availability of children for old parent(s) to co-reside with (if desired) has been substantially reduced (e.g. the case of China). Although the procedures are applicable to all populations, they may not be necessary for populations such as the U.S. and European countries which did not experience such large reduction in fertility level in recent decades. In that case, one may simply project or assume that the future years' proportions of those aged 45–49 who do not live with parents and proportions of the elderly living with adult child(ren) will remain constant or by trend extrapolation or expert opinions.

Let's define the following variables:

$L(42,t)$ – Proportion of those aged 40–44 (on average aged 42) in year t who do not live with parents;

$S(42,t)$ – Proportion of those aged 40–44 (on average aged 42) in year t who live with old parents ($S(42,t) = 1.0–L(42,t)$);

$N(x,t)$ – Proportion of elderly aged x in year t living with adult child(ren) (and the child's spouse if the child is married);

$n_0(t - x + 25)$ – Probability of dying of the elderly cohort members aged x in year t before their children reach average age at childbearing; As detailed cohort mortality data are usually not available, we may reasonably assume that $n_0(t - x + 25)$ is approximately equal to cumulative mortality rate up to average age at childbearing (e.g., age 25) in year $t–x + 25$;

$n_1(x,t)$ – Proportion of life-time infecundity of the elderly aged x in year t;

$n_2(x,t)$ – Proportion of old parents aged x in year t who do not live with adult child among those who have at least one adult child, due to preference of independent living or children's mobility or other socioeconomic reasons;

$n_3(x,t)$ – Proportion of old parents aged x in year t who are not able to live with adult child(ren) even if they wish to do so among those who have at least one adult child, due to shortage of children (i.e., child generation size is smaller than parental generation size);

$M(t–x + 40)$ – Proportion of eventually ever-married among adult children of the elderly aged x in year t (assuming the highest age at first marriage is 40; one may adopt a different assumption);

$P(t - x + 25)$ – Male and female combined probability of surviving up to average age at childbearing for the adult children of the elderly aged x in year t; $P(t–x + 25)$ is equal to cumulative survival probability up to average age at childbearing (assuming the average age at childbearing is 25; one may adopt a different assumption) in year $t-x + 25$.

G(x,t) – Index of offspring resource with respect to potential of co-residence between old parents and adult children; *G(x,t)* is defined as the sum of half of the average number of married children (as married children may also possibly live with their spouse's parents if they wish) and the average number of adult children who were never married for whole life, among elderly aged x in year t;

$$G(x,t) = 0.5M(t-x+40)\cdot TFR(t-x+25)\cdot P(t-x+25) + (1-M(t-x+40))\cdot TFR(t-x+25)\cdot P(t-x+25)$$

Estimation of Proportions of Those Aged 40–44 Who Do Not Live with Parents (L(42,t))

$$L(42,t) = 1.0 - \frac{1-n_0(t-x+25)-n_1(x,t)-n_3(x,t)-[1-n_0(t-x+25)-n_1(x,t)-n_3(x,t)]\cdot n_2(x,t)}{G(x,t)}, \tag{3.7}$$

$$S(42,t) = 1 - L(42,t); \tag{3.8}$$

$$n_2(x,t) = 1.0 - \frac{S(42,t)G(x,t)}{1-n_0(t-x+25)-n_1(x,t)-n_3(x,t)} \tag{3.9}$$

If $G(x, T1) \geq (1.0 - n_0(T1 - x + 25) - n_1(x, T1))$, $n_3(x, T1) = 0$;
If $G(x,t) \leq (1.0 - n_0(t - x + 25) - n_1(x,t))$,

$$n_3(x,t) = (1.0 - n_0(t-x+25) - n_1(x,t)) - G(x,t); \tag{3.10}$$

Therefore, we only need $P(t-x + 25)$, $M(t-x + 40)$, $n_0(t-x + 25)$, $n_1(x,t)$, $n_2(x,t)$ and $TFR(t-x + 25)$ to estimate $L(42,t)$ and $S(42,t)$, which are needed for family household projection for the countries in which fertility declined substantially in recent decades. $P(t-x + 25)$, $M(t-x + 40)$, $n_0(t-x + 25)$, $n_1(x,t)$ and $TFR(t-x + 25)$ can be easily estimated from demographic data sources, which is straightforward, but estimation of $n_2(x,t)$ needs some more discussion. We can estimate the $n_2(x, T1)$ of the elderly aged x in the census year $T1$ (i.e., starting year of the projections), based on the observed proportion of those aged 40–44 (on average aged 42) in census year $T1$ ($S(42,T1)$) who live with old parents, using the formula (3.9) in either the case (1) or (2) as follows:

1. If $G(x, T1) \geq (1.0 - n_0(T1 - x + 25) - n_1(x, T1))$, $n_3(x, T1) = 0$, and

$$n_2(x, T1) = 1.0 - \frac{S(42, T1)G(x, T1)}{1 - n_0(T1 - x + 25) - n_1(x, T1)};$$

2. If $G(x, T1) \leq (1.0 - n_0(T1 - x + 25) - n_1(x, T1))$, $n_3(x, T1) = 1.0 - n_0(T1 - x + 25) - n_1(x, T1) - G(x, T1)$, and
   $S(x, T1) = 1.0 - n_2(x, T1)$ [derived based on replacing $n_3(x, T1)$ in Eq. 3.9 by $(1.0 - n_0(T1 - x + 25) - n_1(x, T1) - G(x, T1))$.

$$n_2(x, T1) = 1.0 - S(x, T1),$$

Once we estimated the $n_2(x, T1)$ in the census year $T1$ (i.e., starting year of the projections), we can estimate or project (or assume) the $n_2(x, T1)$ in the future years based on trend extrapolation or expert opinions, and then estimate the $L(42,t)$ and $S(42,t)$ in the corresponding future years.

Estimating Proportions of Elderly Living with Adult Child(ren) (N(x,t))

$$N(x, t) = 1 - n_1(x, t) - n_3(x, t) - [1 - n_1(x, t) - n_3(x, t)] \cdot n_2(x, t), \qquad (3.11)$$

We can estimate the $N(x,t)$, using the formula (3.11) in either the case (1) or (2) as follows:

1. If $G(x,t) \geq (1.0 - n_0(t - x + 25) - n_1(x,t))$, $n_3(x,t) = 0$, and

$$N(x, t) = 1 - n_1(x, t) - [1 - n_1(x, t)] \cdot n_2(x, t)$$

2. If $G(x,t) \leq (1.0 - n_0(t - x + 25) - n_1(x,t))$, $n_3(x,t) = 1.0 - n_0(t - x + 25) - n_1(x,t) - G(x,t)$, and

$$N(x, t) = G(x, t) - G(x, t)n_2(x, t)$$

$N(x,t)$ is an average proportion of the old parents aged x who live with an adult child (and the child's spouse if the child is married), and $N(x,t)$ represents the overall level of co-residence between old parents aged x in year t and their adult children. In the same time, we estimate the sex-age-marital/union status-specific proportions of the elderly living with children as a standard schedule based on the census (or survey) data. Using these standard schedules and the estimated $N(x,t)$ in the future years, we can estimate the sex-age-marital/union status-specific proportions of elderly living with children in the future years.

Appendix 4: Procedure to Calculate Sex-Age-Specific Rates While Ensuring the Consistency of the Two-Sex Constraints and the Projected Standardized General Rates of Marriage/Union Formation and Dissolution

Input:

GM(r,t), GD(r,t), GC(r,t), GCD(r,t): projected (or assumed) race- or rural/urban-specific standardized general rates of marriage/union formation and dissolution in the projection year *t*; *r* stands for race or rural/urban dimension;

$m^s_{ij}(x,s,r)$: the sex-age-specific standard schedules of o/e rates of transition from marital/union status i to marital/union status j between age x and x + 1.

Output:

$m_{ij}(x,s,r,t)$: the sex-age-specific o/e rates of transition from marital/union status *i* to marital/union status $j(i \neq j)$ between age *x* and $x + 1$ in the projection year *t*, that is consistent with the two-sex constraints and the projected standardized general rates of marriage/union formation and dissolution.

One important conceptual note must be clarified – we adjust the initial standard schedules of age-sex-status-specific o/e rates rather than probabilities of marital/union formation and dissolution to achieve consistency with the projected summary measures and the two-sex constraints. The age-sex-status-specific o/e rate is defined as the number of events that occurred in the age interval divided by the number of person-years lived at risk of experiencing the event. The age-specific rates can be analytically translated to the age-sex-status-specific probabilities using the matrix formula in the context of multiple increment-decrement models (see, e.g., Preston et al. 2001; Schoen 1988; Willekens et al. 1982). This approach could adequately handle the issues of competing risks. Furthermore, adjusting probabilities directly may result in an inadmissible value that is greater than one; adjusting age-specific o/e rates would not yield such an inadmissible probability value, however.

The procedure consists of two steps (refer to: Zeng et al. 2004).

Step 1. Adjustment to comply with the two-sex constraints, following the harmonic mean approach

We use the harmonic mean approach to ensure two-sex consistency in household projections in monogamous societies. The harmonic mean satisfies most of the theoretical requirements and practical considerations for handling consistency problems in a two-sex model Keilman (1985; Pollard 1977; Schoen 1981).

In order to calculate the number of events that occurred in year *t*, we need to calculate the mid-year population $\left(\overline{N'_i}(x,s,r,t)\right)$, classified by age, sex, marital/union status and race or rural/urban status, if it is distinguished. The $\overline{N'_i}(x,s,r,t)$ are the averages of the populations at the beginning and the end of the year *t* and can be considered as an approximation of the person-years lived in status *i* (i.e., at risk of experiencing the event of transition from status *i* to *j*).

Let $N_i(x,s,r,t)$ denote the number of persons of age x, marital/union status i, r status, and sex s at the beginning of year t, which are known through the preceding year's projection. When t refers to the starting year of the projection, the $N_i(x,s,r,t)$ are derived from the census data. The sex-age-specific rates $m_{ij}(x,s,r,t)$ and sex-age-specific probabilities $P_{ij}(x,s,r,t)$ were defined earlier and their relationship can be expressed in the matrix formula (ref. to Willekens et al. 1982). We seek to estimate $m_{ij}(x,s,r,t)$ through adjusting $m_{ij}(x, s, r, t - 1)$, which are known through the preceding year's estimation. When t refers to the starting year of the projection, $m_{ij}(x, s, r, t - 1)$ are equal to the standard schedules. The estimated $m_{ij}(x,s,r,t)$ must be consistent with the two-sex constraints of all race groups combined or rural/urban combined and the projected race or rural/urban-specific standardized general rates of the marriage/union formation and dissolution in year t.

$$N'_i(x+1,s,r,t+1) = \sum_j P'_{ij}(x,s,r,t)N_j(x,s,r,t). \tag{3.12}$$

$$\overline{N'_i}(x,s,r,t) = 0.5\left[N_i(x,s,r,t) + N'_i(x+1,s,r,t+1)\right]. \tag{3.13}$$

Keep in mind for later consideration that $\overline{N'_i}(x,s,r,t)$ (the average of $N_i(x,s,r,t)$ and $N'_i(x + 1, s, r, t + 1)$) is only a first approximation, since $N'_i(x + 1, s, r, t + 1)$ is based on the $P'_{ij}(x,s,r,t)$, which are not the final estimates for year t.

The total number of new marriages of persons of sex s ($s = 1, 2$, referring to females and males, respectively) who were not cohabiting before marriage for all race groups combined or for rural/urban combined in year t ($TM(s,t)$) is estimated as follows:

$$TM(s,t) = \sum_r \left[\sum_i \sum_{x=\alpha}^{\omega} \overline{N'_i}(x,s,r,t)m_{i2}(x,s,r,t-1)\right], \quad i = 1,3,4$$

where ω is the highest age considered in the family household projection; α is the lowest age at marriage. To meet the two-sex constraint, the sex-age-specific rates of marriage among persons who were not cohabiting before marriage need to be adjusted:

$$m'_{i2}(x,s,r,t) = m_{i2}(x,s,r,t-1)\left[\frac{2TM(1,t)TM(2,t))}{TM(1,t)+TM(2,t)}/TM(s,t)\right], \quad i = 1,3,4 \tag{3.14}$$

The estimated total number of new divorces of persons of sex s for all race groups combined or for rural/urban combined in year t ($TD(s,t)$) is

$$TD(s,t) = \sum_r \left[\sum_{x=\alpha}^{\omega} \overline{N'_2}(x,s,r,t)m_{24}(x,s,r,t-1).\right.$$

To meet the two-sex constraint, the sex-age-specific rates of divorce need to be adjusted:

$$m'_{24}(x,s,r,t) = m_{24}(x,s,r,t-1)\left[\frac{2TD(1,t)TD(2,t))}{TD(1,t)+TD(2,t)}/TD(s,t)\right]. \tag{3.15}$$

The rates of widowhood depend on spouses' death rates, which are calculated before the two-sex constraints adjustments, based on the standard mortality schedules and the projected life expectancy at birth in year t. The already projected spouses' death rates should not be adjusted again; they must be used as a "standard". Thus, instead of employing the harmonic mean approach, we simply adjust the rates of widowhood to be consistent with the total number of spouses who die in year t. The total number of persons (i.e., spouses) of sex s who died for all race groups combined or for rural/urban combined in year t with an intact marriage before death ($TDM(s,t)$) based on already projected sex-age-specific death rates is

$$TDM(s,t) = \sum_r \left[\sum_i \sum_{x=\alpha}^{\omega} \overline{N'_2}(x,s,r,t) d_2(x,s,r,t)\right]$$

where $d_2(x,s,r,t)$ is the already projected death rate of married persons of age x and sex s in year t.

The estimated total number of newly widowed persons of sex s for all race groups combined or for rural/urban combined in year t ($TW(s,t)$) is

$$TW(s,t) = \sum_r \left[\sum_i \sum_{x=\alpha}^{\omega} \overline{N'_2}(x,s,r,t) m_{23}(x,s,r,t-1)\right]$$

To meet the two-sex constraint, the sex-age-specific rates of widowhood need to be adjusted using $TDM(s,t)$ as a "standard":

$$m'_{23}(x,s,r,t) = m_{23}(x,s,r,t-1)\left[\frac{TDM(s^{-1},t)}{TW(s,t)}\right]. \tag{3.16}$$

where "s^{-1}" indicates the opposite sex of "s".

The estimated total number of newly cohabiting persons of sex s for all race groups combined or for rural/urban combined in year t ($TC(s,t)$) is

$$TC(s,t) = \sum_r \left[\sum_{x=\alpha}^{\omega} \overline{N'_1}(x,s,r,t) m_{15}(x,s,r,t-1) + \sum_{x=\alpha}^{\omega} \overline{N'_3}(x,s,r,t) m_{36}(x,s,r,t-1)\right.$$
$$\left. + \sum_{x=\alpha}^{\omega} \overline{N'_4}(x,s,r,t) m_{47}(x,s,r,t-1)\right].$$

To meet the two-sex constraint, the sex-age-specific rates of cohabiting need to be adjusted:

$$m'_{15}(x,s,r,t) = m_{15}(x,s,r,t-1)\left[\frac{2TC(1,t)TC(2,t))}{TC(1,t)+TC(2,t)}/TC(s,t)\right]. \quad (3.17)$$

$$m'_{36}(x,s,r,t) = m_{36}(x,s,r,t-1)\left[\frac{2TC(1,t)TC(2,t))}{TC(1,t)+TC(2,t)}/TC(s,t)\right]. \quad (3.18)$$

$$m'_{47}(x,s,r,t) = m_{47}(x,s,r,t-1)\left[\frac{2TC(1,t)TC(2,t))}{TC(1,t)+TC(2,t)}/TC(s,t)\right]. \quad (3.19)$$

The estimated total number of new marriages of persons of sex s who were cohabiting before marriage for all race groups combined or for rural/urban combined in year t *(TCM(s,t))* is

$$TCM(s,t) = \sum_r \left[\sum_i \sum_{x=\alpha}^{\omega} \overline{N'_i}(x,s,r,t) m_{i2}(x,s,r,t-1)\right], \quad i = 5,6,7.$$

To meet the two-sex constraint, the sex-age-specific rates of marriage of persons who were cohabiting before marriage need to be adjusted:

$$m'_{i2}(x,s,r,t) = m_{i2}(x,s,r,t-1)\left[\frac{2(TCM(1,t)TCM(2,t))}{TCM(1,t)+TCM(2,t)}/TCM(s,t)\right],$$

$$i = 5,6,7. \quad (3.20)$$

The estimated total number of events of cohabitation union dissolution of persons of sex s for all race groups combined or for rural/urban combined in year t *(TCD(s,t))* is

$$TCD(s,t) = \sum_r \left[\sum_{x=\alpha}^{\omega} \overline{N'_5}(x,s,r,t) m_{51}(x,s,r,t-1) + \sum_{x=\alpha}^{\omega} \overline{N'_6}(x,s,r,t) m_{63}(x,s,r,t-1) + \sum_{x=\alpha}^{\omega} \overline{N'_7}(x,s,r,t) m_{74}(x,s,r,t-1)\right].$$

To meet the two-sex constraint, the sex-age-specific o/e rates of cohabitation union dissolution need to be adjusted:

$$m'_{51}(x,s,r,t) = m_{51}(x,s,r,t-1)\left[\frac{2TCD(1,t)TCD(2,t))}{TCD(1,t)+TCD(2,t)}/TCD(s,t)\right]. \quad (3.21)$$

$$m'_{63}(x,s,r,t) = m_{63}(x,s,r,t-1)\left[\frac{2(TCD(1,t)TCD(2,t))}{TCD(1,t)+TCD(2,t)}/TCD(s,t)\right]. \quad (3.22)$$

$$m'_{74}(x,s,r,t) = m_{74}(x,s,r,t-1)\left[\frac{2(TCD(1,t)TCD(2,t))}{TCD(1,t)+TCD(2,t)}/TCD(s,t)\right]. \quad (3.23)$$

The sex-age-specific rates, $m'_{ij}(x,s,t)$, are adjusted for consistency with the two-sex constraint as described above, but they need to be further adjusted to be consistent with the projected race or rural/urban-specific standardized general rates of marriage/union formation and dissolution in year t, and will be described in Step 2 as follows.

Step 2. Adjustment for consistency with the projected race- or rural/urban-specific standardized general rates of marriage/union formation and dissolution in year t

To calculate the $m_{ij}(x,s,r,t)$ while ensuring consistency with the race or rural/urban- specific standardized general rates $GM(r,t)$, $GD(r,t)$, $GC(r,t)$ and $GCD(r,t)$, we first estimate the $GM'(r,t)$, $GD'(r,t)$, $GC'(r,t)$ and $GCD'(r,t)$, based on $N_i(x, s, r, T1)$ derived from the most recent census and the $m'_{ij}(x,s,r,t)$, which were consistent with the two-sex constraints and estimated in Step 1 described above, and using the formulas presented in Appendix 2. We then use the same adjustment factor to adjust male and female rates as follows.

$$m''_{i2}(x,s,r,t) = \frac{GM(r,t)}{GM'(r,t)} m'_{i2}(x,s,r,t), i = 1,3,4,5,6,7. \quad (3.24)$$

$$m''_{24}(x,s,r,t) = \frac{GD(r,t)}{GD'(r,t)} m'_{24}(x,s,r,t). \quad (3.25)$$

$$m''_{15}(x,s,r,t) = \frac{GC(r,t)}{GC'(r,t)} m'_{15}(x,s,r,t). \quad (3.26)$$

$$m''_{36}(x,s,r,t) = \frac{GC(r,t)}{GC'(r,t)} m'_{36}(x,s,r,t). \quad (3.27)$$

$$m''_{47}(x,s,r,t) = \frac{GC(r,t)}{GC'(r,t)} m'_{47}(x,s,r,t). \quad (3.28)$$

$$m''_{51}(x,s,r,t) = \frac{GCD(r,t)}{GCD'(r,t)} m'_{51}(x,s,r,t). \quad (3.29)$$

$$m''_{63}(x,s,r,t) = \frac{GCD(r,t)}{GCD'(r,t)} m'_{63}(x,s,r,t). \tag{3.30}$$

$$m''_{74}(x,s,r,t) = \frac{GCD(r,t)}{GCD'(t)} m'_{74}(x,s,r,t). \tag{3.31}$$

Note that the adjustments described in Step 1 use the mid-year populations $\left(\overline{N'_i}(x,s,r,t)\right)$, which are the preliminarily estimated average of the populations at the beginning and the end of year t and approximations only, since they are not based on the final estimates of the sex-age-specific rates of marriage/union formation and dissolution. Although we use the same adjustment factors for males and females, the rates adjusted in Step 2 may not be exactly consistent with the two-sex constraints mainly because $\overline{N'_i}(x,s,r,t)$ are not the final estimates. We, therefore, need to repeat the adjustment procedure described in Step 1 by using the $m''_{ij}(x,s,t)$ estimated in Step 2. More specifically, we calculate $N''_i(x+1, s, r, t+1)$ and $\overline{N''_i}(x,s,r,t)$, using the $m''_{ij}(x,s,t)$ and employing the formulas 3.12 and 3.13. We then use $\overline{N''_i}(x,s,r,t)$ and $m''_{ij}(x,s,r,t)$ to replace $\overline{N'_i}(x,s,r,t)$ and $m_{ij}(x, s, r, t-1)$ in the formulas in Step 1 to get the new estimates $m'''_{ij}(x,s,r,t)$, which satisfy the two-sex constraints. We then use the new estimates of $m'''_{ij}(x,s,r,t)$ to calculate the new estimates of race- or rural/urban-specific standardized general rates of marriage/union formation and dissolution: $GM''(r,t)$, $GD''(r,t)$, $GC''(r,t)$, $GCD''(r,t)$. If the absolute values of the relative difference between the new estimates of the standardized general rates and the corresponding projected general rates are all less than a selected criterion (e.g., 0.01 or 0.001), we have achieved our goals for computing the sex-age-specific (and race- or rural/urban-specific if so distinguished) rates of marriage/union formation and dissolution in year t. Otherwise, we need to repeat the adjustment procedures described in Step 2 and Step 1 until the selected criterion is met.

To provide numerical examples, we used the procedure described above in Step 1 and Step 2 with the standardized general rates as summary measures to calculate the time-varying sex-age-specific rates of marriage/union formation and dissolution in the projection years. The standard schedules are based on the estimates of the U.S. sex-age-specific rates of marriage/union formation and dissolution in 1990–1996 (Zeng et al. 2012b). The sex-age-marital/union status distributions of the starting year of the projection were derived from the U.S. 2000 census micro sample data file. We estimated models with seven marital/cohabiting statuses including cohabitation for the four race groups respectively and combined.

The required number of repetitions of Step 1 and Step 2 using standardized general rates as summary measures is between 2 and 4, as indicated in Table 3.4.1.

The results of the illustrative numerical applications listed in Table 3.4.1 demonstrate that the iterative procedures expressed in Steps 1 and 2 are valid for

Table 3.4.1 Number of repetitions of Step 1 and Step 2 in an illustrative example (*GM(r,t)* decrease by 4 %, *GD(r,t)* increase by 5 %; *GC(r,t)* increase by 8 %; *GCD(r,t)* increase by 6 %)

Criterion (relative difference): 0.01		Criterion (relative difference): 0.001	
All races combined	Four race groups respectively	All races combined	Four race groups respectively
2	3	3	4

Source: Zeng et al. (2004)

practical applications. Based on the final estimates of $m_{ij}(x,s,r,t)$ in year *t*, one can also construct multi-state marital/union status life tables for males and females separately and calculate the detailed sex-specific period life table propensities of transitions from marital/union status *i* to *j* ($PP_{ij}(s,r,t)$) in the year *t*. $PP_{ij}(s,r,t)$ are informative to reflect the gender differentials of the intensities of transitions among various marital/union statuses, which are consistent with the two-sex constraints and the projected standardized general rates of marriage/union formation and dissolution in the projection years.

Appendix 5: Procedure to Estimate General Rates of Marriage/Union Formation and Dissolution at the Starting Year of the Projections

As an illustration of the application, we present procedures to estimate the U.S. race-specific general rates of marriages, divorce, cohabitation, and union dissolution at the state level (Zeng et al. 2013a). The procedures presented here are applicable to other countries and regions; the race dimension (denoted as "*r*") may be replaced by a rural–urban dimension, or eliminated if no race and no rural/urban dimension is distinguished in your applications.

Estimating the U.S. State-Race-Specific General Rates of Marriage and Divorce at the Starting Year of the Projections

Given that we have the published total numbers of marriages and divorces for each of the 50 states and DC for all races combined but not for specific races, we employ the following procedure to estimate the U.S. state-race-specific general rate of marriage ($GM(r,T1)$) and divorce ($GD(r,T1)$) in the census year *T1* which is the starting year of the projection. The marital/union status codes i and j are defined earlier in Sect. 1 of Chap. 2. To simplify the presentation, we omit the state dimension in all variables and formulas in this Appendix.

Let: $N_i(x,s,r,T1)$ denote the number of persons of age x, race r, marital/union status i and sex s counted in the census year $T1$ in the state;

$M_{ij}(x,s,r)$, the national model standard schedules of the race-sex-age-specific o/e rates of transition from marital/cohabiting status i to $j (i \neq j)$, where i and j represent the seven marital/union statuses.

$m_{ij}\ (x,s,r,T1)$, the estimated race-sex-age-specific o/e rates of transition from marital/union status i to j in the census year $T1 (i \neq j)$ in the state; and $TM(T1)$, the published all-races-combined total number of marriages (newly married couples) including first marriages and remarriages that occurred in the state in the census year $T1$.

We assume that the race-sex-age-specific o/e rates of first marriage and remarriage in the state in the census year are proportional to the corresponding national model standard schedule rates,

$m_{i2}(x, s, r, T1) = \gamma(T1) M_{i2}(x,s,r)$; $i \neq 2$ (The subscript 2 represents currently married status); where the $\gamma\ (T1)$ is estimated as:

$$\gamma(T1) = \frac{2TM(T1)}{\sum_{x=\alpha}^{\beta}\sum_{s=1,2}\sum_{r}\sum_{i} N_i(x,s,r,T1) m_{i2}(x,s,r)}, \quad i \neq 2 \tag{3.32}$$

where α (usually taken as 15) and β are the low and the upper boundary of the age range in which the events of marriage/union formation and dissolution occur.

We then use the estimated $m_{i2}(x,s,r,T1)$ and $N_i(x,s,r,T1)$ to calculate the estimated race-specific $GM\ (r,T1)$ in year $T1$ for the state, employing the formula 3.3 in Appendix 2.

Let $TD(T1)$ denote the published all-races-combined total number of couples who divorced in the state in the census year $T1$. We assume that the race-sex-age-specific o/e rates of divorce in the state in the census year are proportional to the corresponding model standard schedule rates of divorce, namely, $m_{24}(x,s,r,T1) = \delta\ (T1)\ M_{24}(x,s,r)$; (The subscript 2 and 4 represent currently married and divorced) where the $\delta\ (T1)$ is estimated as:

$$\delta(T1) = \frac{2TD(T1)}{\sum_{x=\alpha}^{\beta}\sum_{s=1,2}\sum_{r}\sum_{i} N_2(x,s,r,T1) M_{24}(x,s,r)}, \tag{3.33}$$

We then use the estimated $m_{24}(x,s,r,T1)$ and $N_2(x,s,r,T1)$ to calculate the estimated race-specific $GD\ (r,T1)$ in year $T1$ for the state, employing the formula 3.4 in Appendix 2.

Estimating the U.S. Race-Specific General Rates of Cohabitation and Union Dissolution at the State Level

Because we do not have published data on the total numbers of cohabitation formation and dissolution events at the state level, we cannot estimate the U.S. race-specific general rate of cohabitating (*GC(r,T1)*) and union dissolution (*GCD(r,T1)*) at the state level directly; thus, we need to employ an indirect estimation approach by iterative proportional fitting. First, we use the previous census data as the base population and the race-specific model standard schedules and the other estimated demographic parameters as input to project the household distributions (race-specific) from the previous census year to the most recent census year (*T1*), which is the starting year of the projection. Through such projections, we obtain the projected all-races-combined proportions of households with a cohabiting couple among all households in the most recent census year, denoted as *PC*. Second, we then compare *PC* with the observed all-races-combined proportion of the households with a cohabiting couple among all households observed in the most recent census, denoted as *CC*. If the *PC* is higher (or lower) than *CC* by a margin of a pre-determined criterion (e.g., 1 %), we proportionally adjust the race-sex-age-specific rates of cohabitation union formation and dissolution by the following formulas:

$$m_{15}(x,s,r,T1) = M_{15}(x,s,r)\ (2 - PC/CC); \quad (3.34)$$

$$m_{36}(x,s,r,T1) = M_{36}(x,s,r)\ (2 - PC/CC); \quad (3.35)$$

$$m_{47}(x,s,r,T1) = M_{47}(x,s,r)\ (2 - PC/CC); \quad (3.36)$$

$$m_{51}(x,s,r,T1) = M_{51}(x,s,r)\ (PC/CC); \quad (3.37)$$

$$m_{63}(x,s,r,T1) = M_{63}(x,s,r)\ (PC/CC); \quad (3.38)$$

$$m_{74}(x,s,r,T1) = M_{74}(x,s,r)\ (PC/CC); \quad (3.39)$$

We then use the above-adjusted rates and the other data to redo the projection from the previous census year to the starting year *T1*, and calculate the new projected *PC* and compare it with *CC*. If the new *PC* is still higher (or lower) than *CC* by a margin of a pre-determined criterion, we repeat this iterative proportional fitting procedure until the projected *PC* and the observed *CC* are reasonably close to each other (e.g., with a relative difference between −1 % and +1 %), and we then estimate the *GC(r,T1)* and *GCD(r,T1)*, employing the formulas 3.5 and 3.6 in Appendix 2

Given the cohabitation data constraints at the state level, this procedure produces reasonably good estimates of the *GC(r,T1)* and *GCD(r,T1)*, as shown by the results of the validation testing projections from 1990 to 2000 for each of the 50 states and DC presented in Chap. 4.

Chapter 4
Empirical Assessments and a Comparison with the Headship Rate Method

4.1 Empirical Assessments: Comparisons of Projections and Census Enumerations at the National and Sub-National Levels

One useful way to validate a projection model and computer program is to project between two past dates for which the observations are known, and then compare the observed data with the projected data. We assessed the accuracy of the ProFamy method and program by projecting: (1) U.S. households by race from 1990 to 2000 (Zeng et al. 2006), (2) Chinese households by rural and urban areas from 1990 to 2000 (Zeng et al. 2008), and (3) Chinese households by rural and urban areas and Eastern, Middle, and Western regions from 2000 to 2010 (Zeng et al. 2013b).

We used 1990 U.S. census data to calculate the U.S. starting population for the projections to 2000. We then conducted two kinds of tests. The first was to apply the ProFamy method and program and the race-sex-age specific standard schedules observed in the 1980s together with projected demographic summary measures in the 1990s via extrapolations based on time series data from 1970 to 1990 (Gu et al. 2005). This test assumes that we have no data after 1990 and bases the forecast solely on data before 1991 and the ProFamy model. This exercise tests the accuracy of forecasts using the ProFamy model in the real world, assuming the accuracy of the 2000 census observations. The second test used the ProFamy method and program and the race-sex-age specific standard schedules and summary measures observed in the 1990s as input to project U.S. households from 1990 to 2000. This test validates the simulation properties of the ProFamy model based on the assumptions that the input data (observed in the 1990s) and the 2000 census observations (outcome in this exercise) are correct.

Comparisons between the census-observed and model-projected main measures of U.S. household distributions in 2000, derived from the above described testing exercises, show that the differences are within reasonable ranges (see Tables 4.1 and 4.2). More specifically, in the first and second tests, respectively, the relative differences between the observed and projected total number of households are

Y. Zeng et al., *Household and Living Arrangement Projections*,
The Springer Series on Demographic Methods and Population Analysis 36,
DOI 10.1007/978-90-481-8906-9_4,

Table 4.1 Comparison between census-observed and ProFamy-projected U.S. household and population indices in 2000 based on data before 1991

	Households				Population		
	Census	ProFamy	Diff. %		Census	ProFamy	Diff. %
Total number of household	105,480,101	106,474,544	0.9	Total population	281,421,906	277,170,688	−1.5
Average household size	2.59	2.52	−2.7	Percent of			
				Children age < 18	25.69	25.53	−0.6
Percent of				Elderly aged 60+	16.27	16.87	3.7
1 person household	25.82	26.05	0.9	Elderly aged 65+	12.43	13	4.6
2 person household	32.63	33.01	1.2	Oldest-old aged 80+	3.26	3.6	10.3
3 person household	16.53	18.06	9.3	Group quarters	2.76	2.72	−1.5
4 person household	14.20	13.57	−4.4				
5+ person household	10.83	9.30	−14.1	Dependent ratio of			
				Children	0.42	0.42	1.2
				Old	0.20	0.21	4.5
Married couple family %	51.66	53.76	4.1	Children and old	0.62	0.63	2.3

Table 4.2 Comparison between census-observed and ProFamy-projected U.S. household and population indices in 2000 based on data before and during 1990s

	Households				Population		
	Census	ProFamy	Diff. %		Census	ProFamy	Diff. %
Total number of household	105,480,101	105,901,696	0.4	Total population	281,421,906	276,417,600	−1.8
Average household size	2.59	2.53	−2.4	Percent of			
				Children age < 18	25.69	25.33	−1.4
Percent of				Elderly aged 60+	16.27	16.92	4.0
1 person household	25.82	25.19	−2.4	Elderly aged 65+	12.43	13.04	4.9
2 person household	32.63	33.81	3.6	Oldest-old aged 80+	3.26	3.61	10.6
3 person household	16.53	18.09	9.4	Group quarters	2.76	2.73	−1.3
4 person household	14.20	13.80	−2.8				
5+ person household	10.83	9.11	−15.9	Dependent ratio of			
				Children	0.42	0.41	−1.2
				Old	0.20	0.21	4.5
Married couple family %	51.66	53.69	3.9	Children and old	0.62	0.62	0

0.9 % and 0.4 %, and the projected average household sizes are 2.7 % and 2.4 % smaller than the observed ones. In the first and the second tests, the absolute values of the relative difference between the projected and observed proportions of households with 1, 2, 3, or 4 persons, which constitute a large majority of

Table 4.3 Comparing ProFamy projections from 2000 to 2010 and census-observed households and population in 2010 in China at national level

	Census	ProFamy	Diff %
Population size	13.33	13.29	−0.31
Ages 0–9	1.46	1.56	6.27
Ages 10–14	0.75	0.76	1.20
Ages 15–19	1.00	1.00	0.61
Ages 15–59	9.34	9.28	−0.59
Ages 60+	1.78	1.69	−4.93
Ages 65+	1.19	1.11	−6.45
Coll. hh. pers.	0.93	0.92	−1.16
Total households #	4.02	4.03	0.18
% Ages 65+	8.92	8.37	−6.16
Average hh. size	3.08	3.07	−0.43

Notes: (1) Coll. hh. pers. – Number of persons living in collective households; hh. – household; (2) The units of Population size, Age groups 0–9, 10–14, 15–19, 15–59, 60+ and 65+, Coll. hh. pers., and Total households # are 100 million

U.S. households, are 0.9–9.3 % and 2.4–9.4 %; the differences between projected and observed percentages of married couple households are 4.1 % and 3.9 %; and the relative differences between the observed and projected percentages of persons who live in group quarters are −1.5 % and −1.3 %. The absolute values of the relative discrepancies between the observed and projected total population sizes, percentages of children younger than 18 years old, percentages of the elderly aged 65+, and the dependency ratios are 0.6–4.6 % and 0.6–4.9 %, respectively. The discrepancy rates of two measures concerning the smaller groups of oldest-old and big households are relatively large: the difference percentages of the oldest-old aged 80 or older are 10.3 % and 10.6 %; the percentages of big households with five or more persons are −14.1 % and −15.9 %.

We also performed a similar validation test of family household projections and the population from 2000 to 2010, comparing the results of the projections with the 2010 census observations in China at the national level (see Table 4.3). Again, the differences are within a reasonable range (Zeng et al. 2013a).

To further test whether the ProFamy extended cohort-component method and software work reasonably well at the sub-national level, we conducted a set of empirical validation tests of household and living arrangement projections for each of the 50 U.S. states and DC, all using national race-sex-age-specific model standard schedules estimated based on pooled national survey data,[1] except that the race-age-sex-specific domestic migration rates are estimated based on the census 5 % micro data files for each of the U.S. states and DC (Zeng et al. 2013a). The tests were based on projections from 1990 to 2000 using the 1990 census data as

[1] (a) National Survey of Family Households (NSFH) conducted in 1987–1988, 1992–1994, and 2002; (b) National Survey of Family Growth (NSFG) conducted in 1983, 1988, 1995, and 2002; (c) Current Population Surveys (CPS) conducted in 1980, 1985, 1990, 1995; (d) Survey of Income and Program Participation (SIPP) conducted in 1996 (see Zeng et al. (2012b) for discussions on justifications of pooling data from the four surveys).

the base population and summary parameters based on data before 1991, with comparisons of the projected estimates and the census observations in 2000. These tests assume that we have no data after 1990 when projecting 1990 to 2000 and assess the accuracy of the sub-national projections using the ProFamy model in the real world, assuming the 2000 census data are accurate.

We use the percent error (PE), Mean Absolute Percent Error (MAPE), Mean Algebraic Percent Error (MALPE) and Median Absolute Percent Error (MEDAPE), which are the most commonly used measures of forecast errors Smith et al. (2001: 302–304), to assess the validity of the household and living arrangement projections at sub-national levels using the ProFamy approach. More specifically, the PE is defined as the difference between the ProFamy projection in 2000 and the census observation in 2000 divided by the census observation in 2000 and multiplied by 100 for each of the 50 U.S. states and DC. The MAPE and MEDAPE are the average and median of the absolute values of PEs across all of the 50 U.S. states and DC, and MALPE is the algebraic mean of PEs (in which positive and negative values offset each other) across all of the 50 U.S. states and DC.

Forecast errors based on comparisons of ProFamy projections from 1990 to 2000 and census observations in 2000 for total number of households, average household size, % of households of 1, 2–3, and 4+ persons, % couple-households, total population size, % children, % elderly aged 65+, % oldest-old aged 80+, and dependency ratios for the 50 U.S. states and DC are summarized in Fig. 4.1a, b and Table 4.4. Among the sets of tests between projections and observations for 306 main indices[2] of household and living arrangements in the 50 U.S. states and DC, 29.1 %, 33.9 %, 17.4 %, 12.9 % and 6.7 % of the forecast errors are <1.0 %, 1.0–2.99 %, 3.0–4.99 %, 5.0–9.99 % and ≥10 %, respectively (Fig. 4.1a). The percentage distributions of forecast errors of the 306 main indices of population in the set of comparisons among projections and 2000 census observations in the 50 U.S. states and DC are also low: 29.7 %, 43.4 %, 16.5 %, 9.5 % and 0.8 % are <1 %, 1.0–2.99 %, 3.0–4.99 %, 5.0–9.99 % and ≥10.0 %, respectively (see Fig. 4.1b).

The Mean Absolute Percent Error (MAPE) and Median Absolute Percent Error (MEDAPE) of the main household indices in comparisons between the projections and census-observations in 2000 for the 50 U.S. states and DC are all within reasonable small ranges of 1.6–4.7 % and 1.1–3.5 % (see the 2nd and 3rd columns of panel (A) in Table 4.4). The Mean Algebraic Percent Error (MALPE) of average household size and percent of 2–3 persons household are negative, −0.56 % and −1.06, respectively, and all of the other MALPEs for household projections are positive, within a range of 0.04–2.91 % (see the 4th column of panel (A) in Table 4.4). Similar to those error rates of comparing the main indices of household projections between projections and census observations, the ranges of all of the forecast errors of the main population indices for the 50 U.S. states and DC are all reasonably small

[2] We compare six main indices of household projections and six main indices of population projections for each of the 50 U.S. states and DC and thus both the total number of household indices and the total number of population indices under comparison is 306, respectively.

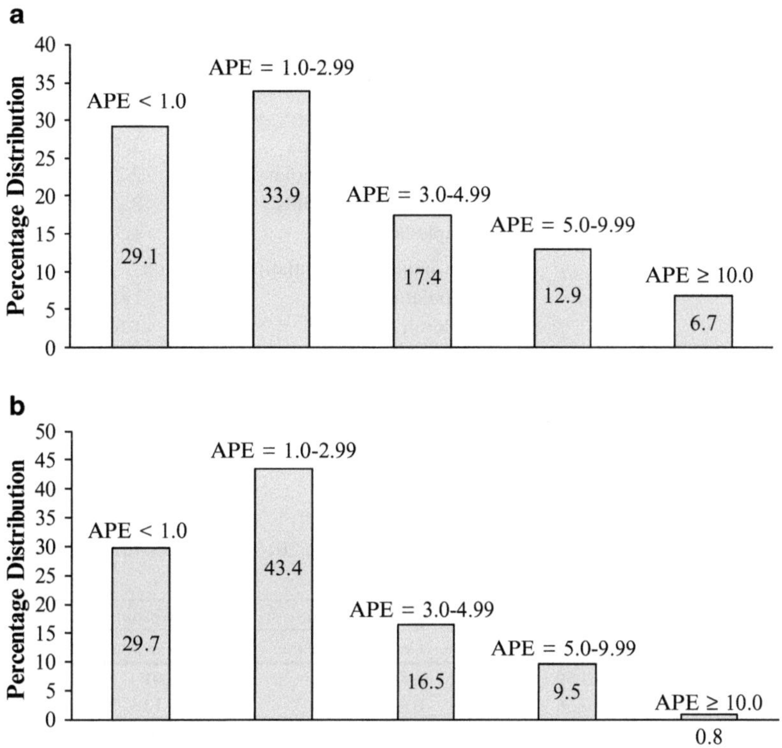

Fig. 4.1 (**a**) Percentage distributions of the absolute percent errors (*APE*) of comparisons between the ProFamy projections from 1990 to 2000 and the Census observations in 2000, *six main indices of households* for each of the 50 U.S. states and DC. (**b**) Percentage distributions of the absolute percent errors (*APE*) of comparisons between the ProFamy projections from 1990 to 2000 and the Census observations in 2000, *six main indices of population* for each of the 50 U.S. states and DC

(see the 2nd, 3rd and 4th columns of panel (B) in Table 4.4).[3] No significant associations between the forecasts errors and population sizes of the states were found. This is similar to what was found in some other projections ESRI (2007).

[3] We performed another set of tests of projections from 2000 onwards using the ProFamy approach and data *prior to 2001* by comparing the projections and ACS observations in 2006 for the 50 states and DC. It turns out that 34.2, 35.0, 21.9, and 9.0 percent of the percent errors of the 306 indices of the household projections are <1.0 %, 1.0–2.99 %, 3.0–4.99 %, and 5.0–9.99 %, respectively, and none is over 10 %. A similar scale and pattern of forecast errors were also found in tests of projections from 2000 onwards using ProFamy approach and data prior to 2001 and comparing projections and ACS observations in 2006 and 2009 for the six counties of South California and the Minneapolis-Saint Paul Area Wang (2009a, b, 2011a, b). Apart from space limitations, we did not present detailed results from these additional tests here, mainly because the 2006 and 2009 ACS data may not be accurate enough to serve as a benchmark standard for the validation tests (Alexander et al. 2010; Swanson 2010).

Table 4.4 Mean forecasting errors based on comparisons between the ProFamy projections from 1990 to 2000 (based on data before 1991) and the census observations in 2000 across the 50 U.S. states and DC

	MAPE	MEDAPE	MALPE
(A) Main indices of household projection			
Total number of households	1.63	1.07	0.04
Average household size	1.75	1.16	−0.56
% of 1-person households	4.73	3.45	2.91
% of 2–3-person households	2.63	2.2	−1.06
% of 4+ person households	4.08	2.76	0.06
% of couple households	2.07	1.35	0.56
(B) Main indices of population projection			
Total population size	1.35	1.09	−0.58
% of children aged ≤18	1.96	1.82	1.39
% of elderly aged 65+	2.52	2.08	−1.45
% of oldest-old aged 85+	3.44	2.81	−2.02
Dependency ratio of children	2.55	2.21	1.56
Dependency ratio of elderly	2.8	2.19	−1.05

Table 4.5 Comparing ProFamy projections from 2000 to 2010 and census-observed households and population in 2010 in China by regions

	Eastern region			Middle region			Western region		
	Census	ProFamy	Diff %	Census	ProFamy	Diff %	Census	ProFamy	Diff %
Population size	5.53	5.50	−0.52	6.88	6.90	0.26	0.91	0.8	−3.39
Ages 0–9	0.51	0.56	8.55	0.83	0.86	4.04	0.12	0.14	11.78
Ages 10–14	0.28	0.28	0.84	0.40	0.41	0.99	0.07	0.07	4.00
Ages 15–19	0.39	0.38	−2.85	0.53	0.55	2.86	0.08	0.08	2.51
Ages 15–59	3.96	3.92	−1.08	4.75	4.78	0.50	0.63	0.59	−5.74
Ages 60+	0.78	0.75	−4.14	0.90	0.86	−4.79	0.10	0.08	−12.64
Ages 65+	0.52	0.50	−4.95	0.60	0.56	−6.62	0.07	0.05	−16.98
Coll. hh. pers.	0.47	0.47	0.05	0.41	0.40	−2.41	0.05	0.05	−2.29
Total households #	1.75	1.75	0.16	2.02	2.03	0.49	0.25	0.25	−2.20
% Ages 65+	9.47	9.05	−4.46	8.72	8.12	−6.86	7.15	6.14	−14.06
Average hh. size	2.90	2.88	−0.79	3.21	3.20	−0.09	3.46	3.40	−1.72

Notes: (1) Coll. hh. pers. – Number of persons living in collective households; hh. – household; (2) The units of Population size, Age groups 0–9, 10–14, 15–19, 15–59, 60+ and 65+, Coll. hh. pers., and Total households # are 100 million

We also conducted a similar validation test for projections of family households and population from 2000 to 2010 by comparing the results of the projections with the 2010 census observations in the Eastern, Middle, and Western regions in China (see Table 4.5); the differences are again within a reasonable range (Zeng et al. 2013b).

Note that there are no fixed guidelines for the evaluation of the accuracy of population forecasts, but we may compare ours with others. Our household and population forecast errors from 1990 to 2000 at the U.S. and Chinese national and sub-national levels are close to, or even smaller than, the population forecast errors of the U.S. Census Bureau (Campbell 2002) and some other institutions

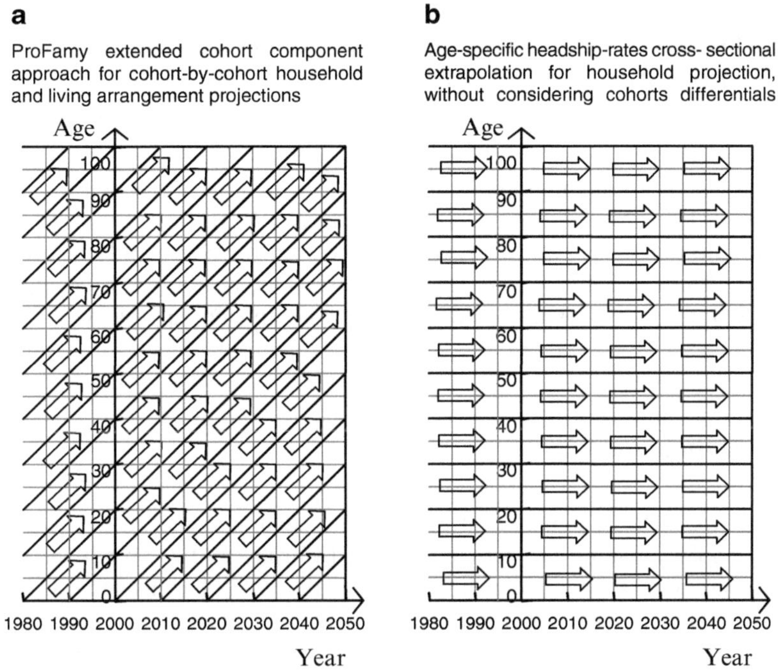

Fig. 4.2 A Comparison of the basic framework of the ProFamy extended cohort-component approach and the headship-rates method

(e.g., ESRI 2007). According to previous studies, it is fairly common for some countries in the United Nations population projections to have 2–5 % forecast errors for the total population and 5–10 % forecast errors for age-specific sub-populations for a 10-year projection period (e.g., Khan and Lutz 2008). These prior forecasts and forecast evaluations provide the frame to interpret the validation test results summarized in Figs. 4.1 and 4.2 and Tables 4.1, 4.2, 4.3, 4.4 and 4.5: that the forecast errors of household and population projections at the national and sub-national levels using the ProFamy extended cohort-component method are within a reasonably and relatively small range, and even mostly smaller than the forecasting errors reported in other assessments studies. We are not sure whether the discrepancies listed in Figs. 4.1 and 4.2 and Tables 4.1, 4.2, 4.3, 4.4 and 4.5 are due mainly to the model specification, or to inaccuracies of the census and survey data, or to a combination of these. It is clear, however, that the ProFamy extended cohort-component approach for simultaneously projecting households, living arrangements, and population age/sex distributions works reasonably well not only at the national level, but also at the sub-national level.

4.2 A Comparison Between the Classic Headship Rate Method and the ProFamy Extended Cohort-Component Approach

The classic headship rate method suffers serious shortcomings and has been criticized widely by demographers for more than two decades (Bell and Cooper 1990; Burch 1999; Mason and Racelis 1992; Murphy 1991; Spicer et al. 1992). The headship rate method is, however, still widely used for household projections by statistical offices and market analysis agencies. Thus, it deserves a detailed comparison with the new ProFamy extended cohort-component approach.

4.2.1 *Conceptual Issues*

Headship rate: The designation of a household head is vague, ill-defined, and an arbitrary choice, making projections difficult (Murphy 1991). Trends in headship rates are thus not easy to model (Mason and Racelis 1992). For instance, an increase in female headship rates may occur because the census or survey was carried out in the daytime, when more women were available to complete the questionnaire than were men. Some of these women also may have wished to show their power by classifying themselves as the household head. Or the increase may be due to an actual increase in women's socioeconomic status. But in either case, it is not due to real changes in demographic conditions.

ProFamy has no such conceptual problems.

4.2.2 *Linkage with Demographic Rates*

Headship rate: There is no way to link headship rates to demographic rates; it is impossible to incorporate the projected or assumed propensity/timing of demographic processes into headship rates (Mason and Racelis 1992; Spicer et al. 1992).

ProFamy uses demographic rates as input for household projections, and thus facilitates analysis of the effects of changes in demographic rates on family household structure (see Table 3.1 and associated discussions in Chaps. 2 and 3). As Morgan (2004), for example, indicates, ProFamy approach provides a framework and tool to assess which of the demographic changes in marriage, divorce, fertility, mortality, migration and so on, may affect family households; thus, the ProFamy approach allows one to rank the demographic components most responsible for recent changes and most likely to impact future family households.

4.2.3 Cross-Sectional Extrapolations Versus Cohort-by-Cohort Projections

Headship rate: Given the nature of cross-sectional extrapolations of the headship rate method (see panel B in Fig. 4.2), it cannot model and project differentials of older and younger cohorts, which likely experience very different demographic events and have substantially different household structure and size. Consequently, employing the headship rate method may result in incorrect forecasts. For example, an extrapolative projection of headship rates of the elderly 20–30 years into the future based on headship rates of elderly cohorts today may be misleading because cohorts who will be elderly in the future but are younger today experienced, and may continue to experience, higher rates of marriage/union disruption and lower fertility than the elders today who have already completed most of their family life course.

In contrast, *ProFamy* extended cohort-component approach projects all individuals grouped by cohorts and specified attributes (e.g., a group of persons of the same age, race, sex, marital/union status, and co-residence status with parents and children). The calculations of the ProFamy model proceed iteratively, cohort-by-cohort, group-by-group, and time-period-by-time-period, using demographic rates as input (see panel A in Fig. 4.2). Clearly, the ProFamy extended cohort-component approach is theoretically and realistically more robust than the headship rate method which relies on cross-sectional extrapolation without taking cohort differentials into consideration.

4.2.4 Household Members Other than Heads

Headship rate: One of the most problematic features of the headship rate method is that it lumps all household members other than heads into one category "non-head" with no projected information (Burch 1999). This makes it impossible to study family households, marital status and living arrangements of the elderly, adults, and children who are "non-head" but constitute the large majority of the population. This is a disadvantage for business, academic, and policy analysis and planning.

ProFamy projects the marital status and living arrangement of all members of the entire population. For example, this includes the number and percentage of the elderly living alone, with spouse only, with children and/or others, in a private household or institution, and children living with two, one or no parents, and so on. This is useful for business and governmental planning and analysis of elderly and children care needs, poverty, welfare, social security, insurance, banking, credit card services, and the like.

4.2.5 Information Produced and Adequacy for Planning

Headship rate: The information on households produced by headship rate projections is very limited and inadequate for purposes of more detailed planning

Table 4.6 Household types projected by the classic headship rate method U.S. Census Bureau (1996)

Type code	Household types	Household sizes
1	Married couple household	Not available
2	Female-headed and no spouse household	Not available
3	Male-headed and no spouse household	Not available
4	Female non-family household	Not available
5	Male non-family household	Not available

and analysis (Bell and Cooper 1990). For example, the U.S. Census Bureau (1996) national household projection using the headship rate method and regression trends extrapolation projected only five household types by age group of household heads or householders, with no projected household sizes available (see Table 4.6). This is, again, disadvantageous since households with various sizes differ substantially in their needs for products and services. A recent extension of headship rate method can project household size, but with limited household types (Ediev 2007; Ediev et al. 2012). Census data could provide a more detailed stratification (also including household size, for example) of the headship rates in household projections than what was done by the U.S. Census Bureau (1996), if one assumes that the sex-age specific headship rates are constant over the projection time horizon, given the fact that it may not be realistic for regression trend extrapolation with detailed stratification of the headship rates. This would produce more detailed household projections, but the static approach of assuming more detailed constant sex-age specific headship rates departs from the real world and may not be accepted by researchers and policymakers.

ProFamy provides much more detailed projected household types and sizes by age of reference person than the headship rate method (see Table 4.7) and uses time-varying demographic rates as input. For example, the "female-headed and no spouse" households projected by the headship rate method (U.S. Census Bureau 1996) mix households of one woman only and households of a not-married mother with children into one group without marital status and household size information. In contrast, ProFamy classifies female-headed and no spouse into different types that incorporate sizes of households (e.g., one woman only, a divorced or widowed and not-cohabiting mother with child(ren)), all by a woman's marital/union status (Zeng et al. 2006).

Previous research found that, as compared to the ProFamy approach, the headship rate method yields seriously misleading results regarding increases in automobile use in Austria (Prskawetz et al. 2004) and housing demands in the U.S.A (Zeng et al. 2013a). This is because future Austrian and American households will comprise many more one- and two-person households (which mostly need only one car and 1–2 bedroom housing units) compared to today's households, but the headship rate method usually projects household numbers without information on household sizes (U.S. Census Bureau 1996). Such disadvantages of the headship rate method and advantages of the ProFamy method also apply to other business analyses, for example, consumption of home-based energy and other products and services in many countries for which future trends indicate many more one- and

Table 4.7 Household types and sizes projected by the ProFamy extended cohort component method

Type code	Household types	Household sizes
One generation households		
1–3	One man only by marital status	1
4–6	One woman only by marital status	1
7–9	One man and other/non-relative by marital status of the man	2,3,4,5,or 6+
10–12	One woman and other/non-relative by marital status of the woman	2,3,4,5,or 6+
13	One married couple only	2
14	One cohabiting couple only	2
15	One married couple and other/non-relative	3,4,5,6,or 7+
16	One cohabiting couple and other/non-relative	3,4,5,6,or 7+
Two-generation households		
17	Married couple and children/other	3,4,5,6,7,8,or 9+
18	Cohabiting couple and children/other	3,4,5,6,7,8,or 9+
19–21	Single-mother and children/other by marital status of the single mother	2,3,4,5,6,7,8,or 9+
22–24	Single-father and children/other by marital status of the single father	2,3,4,5,6,7,8,or 9+
Three-generation households		
25	Married (or cohabiting) couple with children and 1 or 2 grandparents	4,5,6,7,8,or 9+
26	Single-parent and children and 1 or 2 grandparents	3,4,5,6,7,8,or 9+

two-person households. Two articles published in *Nature* show that, for example, a rapid increase in households of smaller size, which results in higher per capita and total energy consumption, implies a threat of a larger demand for resources (Keilman 2003) and poses serious challenges to biodiversity conservation (Liu et al. 2003). This further supports the usefulness of household forecasts including size using the ProFamy model, in contrast to the headship rate method which likely excludes or with limited household size.

4.2.6 Methodology

Headship rate: The projection conducted by the U.S. Census Bureau (1996), which is a typical well-done household projection using the headship rate method, performed 100 sets of time series regression models to project age-sex specific headship rates in future years (the 100 sets = 10 age groups × 2 marital statuses × 5 household types). The 10 age groups were 15–17, 18–19, 20–24, 25–29, 30–34, 35–44, 45–54, 55–64, 65–74, 75+, the two marital statuses were never-married and ever-married (U.S. Census Bureau 1996: 18), and there were five household types (see Table 4.6). The dependent variables in the 100 sets of regression models are logistic transformations of the headship rates, and the independent variable is time. The future trends of headship rates are based purely on the regression of calendar time (with no connection to demographic rates), which were in some cases unreasonably extrapolated into future years. Therefore, the household projection was arbitrarily adjusted using the slopes of the regression line that were less extreme than those obtained from the 100 regression models. For example, slopes indicating changes in the percentage of never-married under age 35 were reduced by two-thirds; slopes indicating changes in the percentage of married-couple households for all ages were reduced by one-third; slopes indicating changes in the remaining household types were simply left at their 1990 levels. The adjustments made the projection look more reasonable, but the mechanisms behind these adjustments appear arbitrary (U.S. Census Bureau 1996).

The ProFamy model does not include any arbitrary adjustment of the slopes of the regression line. One needs either to prepare age-sex specific standard schedules of demographic rates or to employ the existing age-sex specific model standard schedules of demographic rates (see (2) of Table 3.1). One then projects or assumes the demographic summary measures (see (3) of Table 3.1) based on time series analysis or expert opinions. The standard schedules formulate the age pattern of demographic processes. The ProFamy model can take into account anticipated changes in the age pattern, such as delaying or advancing marriage and fertility, by adjusting the schedules to match the projected mean ages of the demographic events in future years. Based on standard schedules and demographic summary measures, ProFamy generates estimates of the age-sex-specific demographic rates needed to project households and the population in future years. Projecting future demographic summary measures can be done using the statistical software of time series analysis or expert opinion approaches. Users may even want to include time series data of other related socioeconomic covariates (e.g., GDP per capita, average income, labour force participation, education, urbanization) in the projection of demographic summary measures. Projections based on time series analysis or assumptions based on expert opinion are made about the components of change in demographic factors that produce household distributions in future years. This is analogous to, and a substantive extension of, the classic cohort-component population projection model.

4.2.7 Data Requirements, Time and Resource Costs

The *headship rate* method requires less data than does the ProFamy model. It is very simple and requires little time and resources if one assumes all sex-age specific headship rates remain constant over time. But such a static approach may not provide projections of acceptable accuracy, especially in societies where demographic and socioeconomic changes are underway. As shown by the empirical evidence of "family household momentum" in Sect. 3.3 in Zeng et al. (2006) and in Chap. 8 of this book, even if U.S. demographic rates were assumed to remain constant in future years, the age-specific headship rates would not be unchanged because the older cohorts, who had more traditional family patterns, would be replaced by younger cohorts with modern family patterns in the next couple of decades. In order to accommodate such non-constant rates, the headship rate method requires a large amount of data, time and resources if one follows the usual approach of regression trends extrapolation, for example, estimating the 100 regression equations, as the U.S. Census Bureau (1996) did.

By comparison, the ProFamy model takes a substantial amount of time and resources to prepare age-sex specific standard schedules. Once the age-sex specific standard schedules for a country have been estimated (and updated every 5–10 years or so, when new data become available) by a researcher, however, others could simply employ these standard schedules as "model standard schedules" for household forecasting at the national and sub-national levels. This is similar to the widely practiced application of model life tables (e.g., Coale et al. 1983; United Nations 1982), the Brass logit relational life table model (e.g. Murray et al. 2003), the Brass relational Gompertz fertility model (Brass 1974), and other parameterized models (e.g. Coale and Trussell 1974; Rogers 1986) in population projections and estimations. As discussed and justified in Sect. 2.2.4 of Chap. 2, numerous studies have demonstrated that the relational parameterized models consisting of a model standard schedule and a few summary parameters offer an efficient and realistic way to project or estimate demographic age-specific rates (Booth 1984; Brass 1978; Paget and Timaeus 1994; Zeng et al. 2000).

Using existing model standard schedules and projected (or assumed) demographic summary measures such as the TFR, life expectancy at birth, general rates of marriage, divorce, cohabitation, and union dissolution, as well as the ProFamy software, one can conveniently perform household forecasting at national and sub-national levels.

While we agree with many demographers' criticisms of the headship rate method (e.g., Bell and Cooper 1990; Burch 1999; Mason and Racelis 1992; Murphy 1991; Spicer et al. 1992), we believe that the choice between methods with different degrees of comprehensiveness depends on the user's needs. For a simple static, constant rate projection (without regression trends extrapolation) of the number of households with limited type/size information using easily accessible cross-sectional data at a low cost of time and resources, the headship rate approach may be satisfactory. For more detailed and realistic projections and analyses of household types, sizes, elderly living arrangements, and home-based products and service needs using various demographic rates as input, the ProFamy extended cohort-component approach is preferable. As an illustration, we next describe a comparison of housing demand forecast errors using the two methods.

4.3 A Comparison of Housing Demand Forecast Errors Between the Headship Rate Method and the ProFamy Extended Cohort-Component Approach

As discussed above, compared with the still-widely-used classic headship rate method, the ProFamy approach is theoretically advantaged and projects much more detailed household types, sizes, and living arrangements. However, the ProFamy approach needs substantially more data than does the classic headship rate method. This raises the question: Is it worthwhile to employ the ProFamy approach rather than the classic headship rate method if users simply need projections of home-based consumption demands, such as numbers of housing units by number of bedrooms, but do not care about the details of the household characteristics including detailed household types/sizes, marital/union status, co-residence status with parents, and children of the reference persons? The following assessments are designed to answer this question.

We projected from 1990 to 2000 the number of housing units by: (1) the number of bedrooms for each of the 50 U.S. states and DC, employing constant headship rates (see Sect. 1.1 in Appendix 1) and (2) the ProFamy extended cohort-component approach using data before 1991(see Sect. 1.2 in Appendix 1). By comparing the projected and census-observed number of housing units occupied by private households in 2000, we employed the two approaches to estimate the error rates of forecasts of the housing units by number of bedrooms. As shown in Table 4.8, the Mean Algebraic Percent Error of forecasts for the 0–1 bedroom,[4] 2-bedroom, 3-bedroom, and 4-bedroom housing units employing the constant headship rates are −18.67 %, 5.01 %, 4.30 % and −3.23 %, as compared to −6.25 %, 2.51 %, 1.38 % and 1.15 % for the ProFamy approach. The Mean Absolute Percent Error and Median Absolute Percent Error of forecasts of the 0–1 bedroom, 2-bedroom, 3-bedroom, and 4-bedroom housing units employing constant headship rates are about 114.4 %~128.4 %, 20.4 %~37.0 %, 13.2 %~27.9 % and 24.4 %~53.4 % higher than that employing the ProFamy approach, respectively.

Even if one uses the changing headship rates based on regression or other trend extrapolation method to more accurately project the numbers of households, it is

Table 4.8 Mean forecasting errors of housing demand projections from 1990 to 2000 (compared to the 2000 census observations) for the 50 U.S. states and DC, produced by the ProFamy cohort-component approach and constant headship rates

	0–1 bedroom units			2-bedroom units			3-bedroom units			4-bedroom units		
	ProFamy	Headship	% Diff.	ProFamy	Headship	% Diff.	ProFamy	Headship	% Diff.	ProFamy	Headship	% Diff.
MALPE	−6.25	−18.67	198.7	2.51	5.01	99.6	1.38	4.3	211.6	1.15	−3.23	−380.9
MAPE	8.71	18.67	114.4	5.87	7.07	20.4	3.94	5.04	27.9	6.52	8.11	24.4
MEDAPE	8.11	18.52	128.4	4.86	6.66	37.0	3.10	3.51	13.2	5.19	7.96	53.4

[4] The 0-bedroom housing unit term means that the bedroom is mixed with the living room.

Table 4.9 Mean forecasting errors of housing demand projections from 1990 to 2000 (compared to the 2000 census observations) for the 50 U.S. states and DC, produced by the ProFamy cohort-component approach and adjusted changing headship rates

	0–1 bedroom units			2-bedroom units			3-bedroom units			4-bedroom units		
	ProFamy	Headship	% Diff.	ProFamy	Headship	% Diff.	ProFamy	Headship	% Diff.	ProFamy	Headship	% Diff.
MALPE	−6.37	−15.35	141.0	2.46	5.96	142.3	1.41	3.71	163.1	1.22	−4.23	−446.7
MAPE	8.24	15.45	87.5	5.48	7.53	37.4	4.23	4.37	3.3	6.97	8.27	18.7
MEDAPE	7.73	16.50	113.5	5.06	7.88	55.7	2.80	3.03	8.2	4.57	7.64	67.2

still possible that the headship rates may result in biased projections of household consumption demands, which largely depend on household size (Myers et al. 2002), because the classic headship rate method may exclude household size. To test this hypothesis, we conducted another assessment in which the changing headship rates are assumed to produce the same numbers of households as those observed in the 2000 census in each of the 50 U.S. states and DC (see Sects. 1.3 and 1.4 in Appendix 1). We estimated average forecast errors by comparisons between the 2000 census observations and the adjusted projections of the housing units in 2000 employing the changing headship rate method and the ProFamy approach, respectively, across the 50 U.S. states and DC; see Table 4.9.

The empirical assessments show that, after the adjustments described above, the average negative-forecast error of 0–1 bedroom housing units by the headship rate method is reduced by 3.3 percentage points, but its forecast error is still substantially larger than that of the ProFamy approach. More specifically, the forecast errors of 0–1 bedroom housing units measured by MALPE, MAPE and MEDAPE are −15.35 %, 15.45 %, and 16.50 % for the headship rate method, in contrast to −6.37 %, 8.24 % and 7.73 % for the ProFamy approach (see Table 4.9). The positive forecast error rates of MALPE for 2-bedroom and 3-bedroom housing units by the headship rate method are 5.96 % and 3.71 %, in contrast to 2.46 % and 1.41 % by the ProFamy approach. On the algebraic average criterion, the headship rate method downwardly projected 4-bedroom housing units by −4.23 %, while the error rate for 4-bedroom housing units by the ProFamy approach is 1.22 %. The forecast errors listed in Table 4.9 show that the headship rate method produced substantially more serious negative forecast errors for the 0–1 bedroom and 4-bedroom units and positive forecast errors for the 2-bedroom and 3-bedroom housing units, as compared to the ProFamy approach.

Decennial census data facilitate understanding and interpretation of these results. As compared to 1990, the 1-person, 2-person, 3-person, 4~5 person, and 6+ person households in 2000 increased by 20.6 %, 16.9 %, 9.2 %, 9.3 % and 15.1 %, respectively. Clearly, American households with 1 person (which are more likely to need 0–1 bedrooms), 2 persons (which are more likely to need 0–1 bedrooms or 2 bedrooms) and 6+ persons (which are more likely to need 4-bedrooms)[5] increased

[5] Our research indicates that the increase in the proportion of American households with 6+ persons in 2000 as compared to 1990 is due to the changing racial composition of the population, given the fact that Hispanic, Asian and other non-White and non-Black minority groups have higher proportions of large households with 6+ persons and are growing substantially faster, especially the Hispanic group.

substantially faster during this decades than 3-person and 4~5 person households (which are more likely to need 2~3 bedrooms). Consequently, the headship rate method, which does not include household size, resulted in substantially more serious negative forecast errors for the 0–1 bedroom and 4-bedroom units and upward forecast errors for the 2-bedroom and 3-bedroom housing units, as compared to the ProFamy approach which projects detailed household size. This is consistent with what Prskawetz et al. (2004) found for vehicle projections for Austria, using the ProFamy approach with comparisons to the headship rate method.

Appendix 1: Procedures to Project Housing Demands Based on Household Projections Employing the Headship Rate Method or the ProFamy Approach

The projections are for each of the 50 U.S. states and DC, but we omit the state dimension in all variables and formulas in this Appendix to simplify the presentation.

Housing Demand Projections by the Constant Headship Rate Method

The projection conducted by the U.S. Census Bureau (1996), which is a typical well-done household projection using the headship rate method, projected age-sex-household type-specific headship rates in the future. The five household types listed in Table 4.6 and ten age groups were distinguished in the U.S. Census Bureau (1996) projection.

Let *Hr(x,s,h)* denote the age-sex-household type-specific headship rates estimated based on the 1990 census data, where x and s refer to age and sex of the household head; h refers to the household type.

P2000(x,s), number of persons aged x with sex s in year 2000, projected by the conventional cohort-component method for population projection by age and sex which is also part of the ProFamy model output.

HH(x,s,h), the number of households by age/sex of the household head and household type in 2000 projected by the headship rate method.

b(x,s,h,i), the age-sex-household type-specific proportions of households with i bedroom(s) (i =1, 2, 3, and 4, referring to 0–1, 2, 3, 4+ bedrooms), estimated using the 1990 census data; $\sum_i b(x,s,h,i) = 1.0$

HU(x,s,h,i), the number of housing units by number of bedrooms and by age/sex of the household head and household type in 2000 projected using the headship rate method.

The estimators for projecting *HH(x,s,h)* and *HU(x,s,h,i)* are:

$$HH(x, s, h) = Hr(x, s, h)\ P2000(x, s)$$

and

$$HU(x, s, h, i) = HH(x, s, h)\ b(x, s, h, i)$$

Housing Demand Projections by the ProFamy Extended Cohort-Component Approach

Let *EH(x,s,m,z)* denote the age-sex-household type-size-specific number of households in 2000, projected employing the ProFamy approach based on data before 1991; where *x*: 5-year age group; *s*: sex; *m*: household type (marital/union status of the reference person); *z*: household size – 1, 2, 3, 4, 5, 6+ persons.

eb(x,s,m,z,i), the age-sex-household type-size specific proportions of households with *i* bedroom(*s*) (*i* =1, 2, 3, and 4, referring to 0–1, 2, 3, 4+ bedrooms, respectively), estimated using the 1990 census data; $\sum_i eb(x, s, m, z, i) = 1.0$,

EHU(x,s,m,z,i), the age-sex-household type-size specific number of housing units by number of bedrooms in 2000 projected employing the ProFamy approach.

The estimator for projecting *EHU(x,s,,m,z,i)* is:

$$EHU(x, s, m, z, , i) = EH(x, s, m, z)\ eb(x, s, m, z, i)$$

Adjusted Housing Demand Projections by Changing Headship Rates

Instead of assuming constant headship rates, time-varying headship rates in future years may be projected by regression or other trend extrapolation methods based on time series data. For example, the U.S. Census Bureau (1996) performed time series regression models to project time-varying age-sex-household type-specific headship rates in future years. However, even if numbers of households are assumed to be correctly projected using the changing headship rates based on regression or other trend extrapolation method, it is possible that the headship rates still may result in biased projections of household consumption demands, which largely depend on household size (Myers et al. 2002), because the classic headship rate method excludes household size (U.S. Census Bureau 1996). To test this hypothesis, we conducted another assessment in which the changing headship rates are assumed to produce the same number of households as those observed in the 2000 census in each of the 50 U.S. states and DC. More specifically, we proportionally adjust the age-sex-household type-specific headship rates observed in 1990 by multiplying the ratio of the 2000-census observed total number of households to the total number of households in 2000 projected by the constant headship rates; we use these adjusted and changing age-sex-household type-specific headship rates to project the number of households (whose sum is equal to the census-observed total number of households), and then to project the housing units by number of bedrooms in 2000.

Let *TH2000* denote total number of households counted in the 2000 census of the state.

The estimator to project age-sex-household type-specific number of housing units by number of bedrooms in 2000, based on adjusted age-sex-household type-specific headship rates, which result in the projected total number of households being equal to the total number of households observed in the 2000 census, is:

$$HU'(x,s,h,i) = HH(x,s,h) \frac{TH2000}{\sum_x \sum_s \sum_h HH(x,s,h)} b(x,s,h,i)$$

Adjusted Housing Demand Projections by the ProFamy Extended Cohort-Component Approach

To ensure comparability, we proportionally adjusted the age-sex-household type-size-specific number of households projected by the ProFamy approach to result in the same projected total number of households as the 2000-census observations; we then produced the adjusted housing unit projections. The estimator to project age-sex-household type-size-specific number of housing units by number of bedroom in 2000, based on the adjusted age-sex-household type-size-specific projected number of households, whose sum is equal to the total number of households observed in the 2000 census, is:

$$EHU'(x,s,m,z,i) = EH(x,s,m,z) \frac{TH2000}{\sum_x \sum_s \sum_m \sum_z EH(x,s,m,z)} eb(x,s,m,z,i)$$

Chapter 5
Extension of ProFamy Model to Project Elderly Disability Status and Home-Based Care Costs, with an Illustrative Application

5.1 Introduction

With the well-documented, continuing increase in life expectancy and decline in fertility rates, population and household aging has become a serious challenge in most countries around the world. Rapid population aging may bring about a heavy burden to families and societies, and may erode the foundation of home-based care which has been the major resource supporting older adults in many countries for a few thousand years.

Rapid population aging will result in a declining labor supply and an increasingly higher proportion of GDP transferred into the long-term care for the elderly, both of which would negatively affect development at the macroeconomic level. At the microeconomic level, population aging may lead to increasingly heavy burden on home-based caregivers, resulting in increased precautionary saving and reduced consumption (Li and Chen 2006).

In many developing countries (e.g., China, Korea, Indonesia, India, Brazil, Mexico, etc.), older adults aged 65+ today have about 5–6 surviving children on average; the baby boomers, who were born during the 1950s and 1960s and will become elders after 2015, have fewer or slightly more than two children on average due to family planning programs and substantial changes in fertility attitudes along with quick socioeconomic development (Zeng 2007). This greatly decreased availability of children, plus the increased economic mobility of working age people, will produce a rapid increase in empty-nest elderly households (without co-resident children). Many elders will face the problems of lack of daily caregiving, which may deteriorate the foundation of home-based care for the elderly population and aggravate the economic burden and opportunity costs of adult children.

A number of prior quantitative studies have projected home-based care needs and costs for the disabled elderly in developed countries (e.g., CMMS 2004), but most prior studies in developing countries (e.g., China) have been qualitative (e.g., Hu et al. 2003; Li 1998; Jiang 2008). Generally speaking, three types of care needs

Y. Zeng et al., *Household and Living Arrangement Projections*,
The Springer Series on Demographic Methods and Population Analysis 36,
DOI 10.1007/978-90-481-8906-9_5, © Springer Science+Business Media Dordrecht 2014

and costs for older adults are described in quantitative studies: home-based care, measured in cash expenditures; non-cash home-based care delivered by family members, measured in care time (opportunity cost); and inpatient and outpatient care, measured in cash expenditures. It is widely accepted that care needs and costs are closely related to the health status of older adults. Clinical data based on diagnosis of illnesses in hospital are extremely difficult to access in developing countries such as China and reflect only one aspect of health status. Health survey data based on activities for daily living (ADL), including bathing, dressing, eating, indoor transferring, toileting, and continence, are readily available and have proven in many studies to be a significant predictor of health, care needs and costs, and mortality among older adults. As a result, many scholars use ADL status to measure health care needs for older adults (e.g., Liang 1999).

Projections of ADL status among older adults in developed countries have been relatively popular in recent decades and generally follow two methodologies. The simple proportional distribution projection method multiplies the age-sex-specific proportions of ADL statuses for older adults at baseline by projected age-sex-specific numbers of older adults in future years (Suthers et al. 2003). The multi-state transition projection method estimates the age-sex-specific transition probability matrices of ADL statuses for older adults based on survey data and combines them with matrices of status-specific population forecasts (e.g., Lakdawalla et al. 2003). In general, most demographers calculate projections of elderly home-based care needs and costs by multiplying the ADL status-specific costs per person at baseline by the projected ADL status-specific population distribution in future years (Mayhew 2000). A few scholars have projected care expenditures separately for the surviving elders and for the deceased elderly whose care costs rise sharply in the last few months of life (Serup-Hansen et al. 2002).

A series of studies have indicated that home-based care needs and costs are closely correlated with age, gender, marital status, and the family structure of older adults. For example, disabled elders who are unmarried or living alone have much higher needs for paid services in the home compared to those who live with children and/or a spouse (Grundy 2001). Older adults who co-reside with children tend to receive more informal, non-cash home-based care than those with an empty nest (Zhang 2004). However, almost all previous projections of home-based care needs and costs did not consider simultaneously the different ADL statuses dynamic transitions, family structure and living arrangements of older adults. The present study overcomes this limitation by introducing multistate projections of ADL status transition dynamics for older adults into the ProFamy extended cohort-component method for family household and living arrangement projection (Zeng et al. 1998, 2006, 2013a; Chaps. 2, 3, and 4 of this book). The theoretical framework, method, and computer programs of the ProFamy multistate model have been employed and extended in this chapter to simultaneously project family structure, ADL status transitions, and home-based care needs and costs for the elderly population, with an illustrative application to China (Zeng et al. 2014a).

5.2 The Further Extended ProFamy Method Including Projections of Elderly Disability and Home-Based Care Costs

Figure 5.1 presents the theoretical and demographic framework for our projections of households, living arrangements, elderly disability, and home-based care costs, highlighting interactions among home-based care needs, resources, and costs, as well as their socioeconomic implications. Based on this framework, we further extended the ProFamy extended cohort-component method into a model for projecting home-based care needs and costs for older adults.

As noted above, almost all of the previously published projection models for the elderly disabled population and home-based care needs and costs ignored the family household structure and living arrangements of older adults; this was due to lack of reliable methods using available demographic data to project households and living arrangements. Currently, as discussed in detail in Chap. 4, the classic headship rate method is the most commonly used approach for household projections although it has been criticized widely by demographers for about two decades (Mason and Racelis 1992; Murphy 1991; Spicer et al. 1992). In addition to other limitations, the headship rate method lumps all household members other than heads into one category "non-head" and does not project information for them. This last limitation makes it impossible to study the household status and living arrangements of the elderly, adults, and children who are not the head of household (Burch 1999). Consequently, the headship rate method is not appropriate for projections of disability and home-based care costs, which are directly linked to older adults in the population and their family household structure and living arrangements.

In contrast to the classic headship rate method, the ProFamy extended cohort-component model uses demographic rates as input to project household types, sizes, and living arrangements for all individuals in households (including older adults) grouped by cohort and specified attributes (e.g., sex, age, rural/urban, marital/union

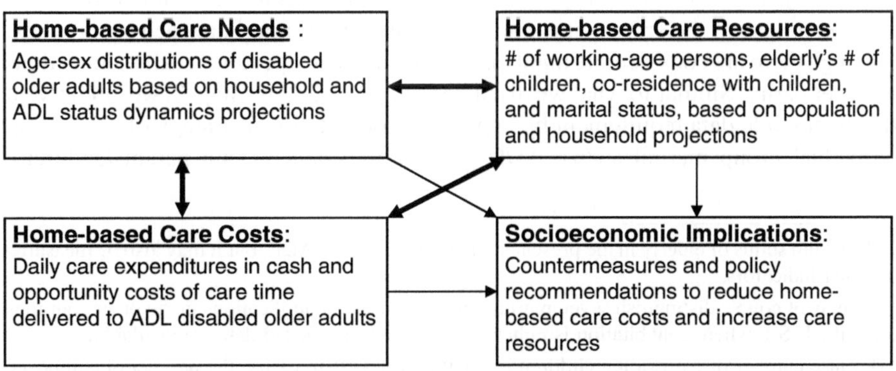

Fig. 5.1 Basic theoretical and demographic framework for projecting elderly disability status and home-based care costs

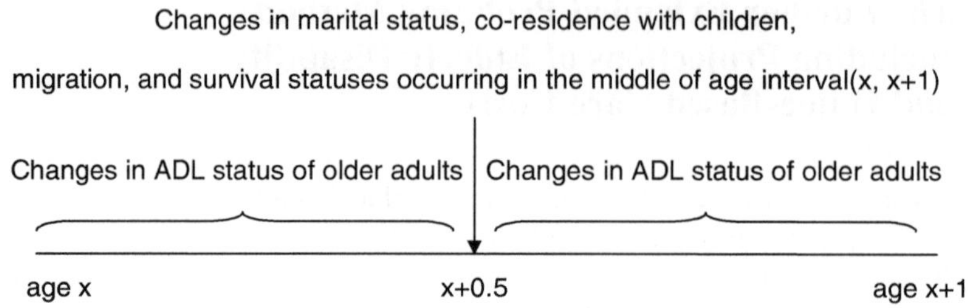

Fig. 5.2 Computational strategy to calculate changes in demographic, familial, and ADL statuses of older adults by single year of age

status, co-residence status with children and parents). Previous chapters have dealt solely with projections of household and living arrangements; the analyses presented in this chapter project dynamic changes in disability status of older adults (measured by ADL[1]) and their home-based care costs, classified by age, gender, rural/urban residence, marital status, and number of co-residing children and co-residing parents. The key point of this analysis is to introduce and estimate changes in older adults' disability status as well as related home-based care costs.

Figure 5.2 depicts the basic ProFamy computational strategy for calculating changes in family structures and ADL statuses of the older population, which is a substantial extension of the basic ProFamy model depicted in Fig. 2.3 of Chap. 2. In our illustrative application to China, we distinguished four marital statuses,[2] two ADL statuses (active and disabled), six parities (0, 1, 2, 3, 4, 5+), six co-residence with children statuses, and three co-residence with parents statuses. If the conventional multistate computation strategy were adopted, we would have to estimate a cross-status transition probabilities matrix with 254,016 (=504 $\times$ 504; where $504 = 4 \times 2 \times 3 \times \sum_{p=0}^{5} (p+1)$) elements[3] for each sex at each age for the elderly population. This is certainly not practical, as it would be impossible to have a sufficiently large dataset with appropriate sub-sample sizes to reasonably estimate so many elements of the cross-status transition probabilities matrix at each age and sex, although there are considerable numbers of structural zero elements such as transitions to lower parity. Similar to what we discussed in Sect. 2.2.2 of Chap. 2, we circumvent this impossible data requirement by applying a computational strategy of calculating individual group marital status, ADL status, co-residence (with parents/children),

[1] The health status of elderly in the present study is measured by ADL, but it may also be measured by other indicators.

[2] We may of course distinguish seven marital/union statuses for applications in the other countries (e.g., the U.S.), where cohabitation is rather common and the needed data are available.

[3] Because number of co-residing children is equal to or less than parity, the number of composite statuses of parity and co-residing children is $\sum_{p=0}^{5} (p+1)$ rather than (6×6).

migration, and survival status changes for older adults that assumes: (a) ADL status transitions occur throughout the first and second half of the single-year age interval, and (b) marital/union status changes, children's leaving parental home, migration, and death occur in the middle of the age interval (see Fig. 5.2). This strategy was originally proposed by Bongaarts (1987) and further justified mathematically and numerically by Zeng (1991a: 61–63 and 80–84).

In addition, our extended model combines the projection of family structure, living arrangement, and ADL status for elders aged 65+ (depicted in Fig. 5.2) with the projection of family structure and living arrangement for the younger population aged 0–64, including fertility, mortality, changes in marital status and co-residence with children and parents, and migration (see Fig. 2.3 of Chap. 2), in order to form a dynamic projection model including all of the individuals in the population. The extended model in this chapter projects not only ADL statuses and home-based care needs and costs for older adults, but also age-sex-specific numbers and family household structures of the working-age population, i.e., the caregivers for the disabled elderly.

As we discussed in Chaps. 2 and 3, the basic projection module for institutional living arrangements for older adults is also included in the present extended projection model. But the projection of institutional-based care costs for the elderly was not included in the present illustrative application to China because the sub-sample size for the institutionalized elderly in the Chinese survey data was too small to estimate meaningful ADL status transition rates and care costs. Furthermore, one single illustrative study cannot deal with too many issues given that projections for institutionalized elders and their costs are a complicated sub-field related to many other factors such as economic income, social facilities, and Chinese cultural attitude, which are out of the scope of this illustrative study. Thus, we focus on home-based care need and cost projections and analysis. Specifically, the present study focuses on how changes in households and living arrangements may affect home-based care expenditures in cash and care workdays devoted to caring for disabled elders; inpatient or outpatient medical care expenditures are not included, as the needed data are not available for China and they are out of the scope of this chapter.

5.3 An Illustrative Application to China

5.3.1 Data Sources and Estimates

The population of China classified by rural/urban residence, single year of age, sex, marital status, number of co-residing children and parents, and whether living in private household or institutional residence at the baseline of projection, was extracted from the micro-data file of the 2000 census[4] and updated with published

[4] The micro-data of the sixth census of China conducted in 2010 are still not available for scholars and the public to use so far.

rural/urban-single-year-age-sex-specific data from the 2010 census. Based on the micro-data file of the 2000 census and the 2005 mini-census for 1 % of the population, we estimated the distributions of rural/urban-single age-sex-specific occurrence/exposure (o/e) rates of first marriage and fertility by parity, and age-sex-specific net migration rates between rural and urban areas. The model standard schedules of age-sex-specific o/e rates of divorce and remarriage were estimated from two nationwide surveys: the Chinese In-depth Fertility Survey in 1987 and the Chinese Longitudinal Healthy Longevity Survey conducted in 2002, 2005 and 2008–2009. We also estimated the rural/urban-specific general marriage rate and divorce rate based on the age-sex-specific standard model schedules of marriage and divorce rates, 2000 census micro data, and published total numbers of marriage and divorce in 2000–2010.

Estimates of transition probabilities for ADL statuses among older adults and average home-based care costs for ADL-disabled elders, which are the input of our projections, are based on the Chinese Longitudinal Healthy Longevity Survey (CLHLS). The CLHLS was conducted in a randomly selected half of the counties and cities of 22 provinces (out of a total of 31 provinces) in China since 1998, covering 85 % of the total population. We used the older adult samples from the 2002 and 2005 CLHLS waves, which contain 15,983 and 16,566 elderly respondents, respectively, as well as their follow-up ADL status and survival/death data collected in the 2005 and 2008–2009 waves, respectively. The data collected through the in-home face-to-face interviews contain basic individual and household demographic characteristics, ADL status, socioeconomic background, health, psychological characteristics, cognitive function, lifestyle, eating habits, economic resources, daily received, etc. The CLHLS data have been widely used by demographers inside and outside of China and its reasonably good quality is widely recognized (e.g., Gu 2008; Goodkind 2009).

Activities of daily living (ADL) are measured by six separate questions on bathing, dressing, eating, indoor transferring, toileting, and continence; we classify a respondent as ADL disabled if he/she needs help in any one of the above six ADL items, following international standards. Based on survival status and change in ADL status at the follow-up survey, we applied multiple regression and curve fitting methods to estimate the transition probabilities of ADL status and probabilities of mortality for older adults classified by age, sex, ADL status, rural/urban residence, marital status, and living arrangement with children.

If a CLHLS respondent reported needing help with an ADL, two further questions are asked about "How much is the total direct cost in cash paid for caregiving last week?"[5] and "How many hours (not-paid) in total did your spouse, children, grandchildren and their spouses, or others help you last week?". For respondents who were interviewed in the previous survey and died before the

[5] Following the widely adopted international standards in the elderly population heath surveys, the reference period of one week for care expenditures and time is intended to reduce recall errors of relatively long periods of time.

follow-up wave, one of his/her family members was interviewed to estimate home-based care expenditures in the last month before death.[6] Home-based care and associated costs included nursing, door-to-door services, and items or goods for care provided in the home, but not inpatient and outpatient medical care and expenditures. We used the 2008–2009 CLHLS data and multiple regression to estimate the average home-based care expenditures in cash (yuan) for ADL disabled survivors (per year) and decedents (in the last month before death), as well as the non-cash home-based care workdays (opportunity costs) delivered by family members for ADL disabled survivors per year.

To illustrate the overall relative burdens of home-based care in future years, we also projected the proportion of home-based elderly care cash costs among the national GDP, as forecasted by the World Bank and the Development Research Center of State Council of China (2013).

5.3.2 Scenario Design and Parameters

Uncertainties in mortality rates, ADL status changes, and home-based care costs per disabled person should be considered in the design of projection scenarios. The medium mortality scenario, which is widely used in Chinese population forecasts by scholars and agencies, assumes a life expectancy of about 78.1 years for both sexes combined in 2050. This is rather conservative, given that the average life expectancy in Shanghai in 2011 was already 82.5 years old (Health Bureau of Shanghai 2012). Some recent research indicates that there may be a significant improvement in mortality in the first half of this century because of biomedical breakthroughs and better personal health practices, such as healthy diets, smoking cessation, and exercise (see, e.g., Shekell et al. 2005). Therefore, we modelled another more optimistic *low mortality scenario*, namely, life expectancy for both sexes combined was assumed to approach 84.8 in 2050, which is only 2.3 years higher than that in Shanghai in 2011. This optimistic mortality scenario is subject to uncertainty, but we believe that it is not impossible.[7] Despite the uncertainty, the medium and low mortality scenarios (see Table 5.1) bracket an informative range of possible average life expectancy at birth in China in the next decades.

Demographers and health researchers have posited three different hypotheses about the future trends of disability in relation to increasing lifespan. The compression of morbidity theory assumes that morbidity will decline and become compressed into a shorter duration of disability before death, associated with improvements in

[6] According to the widely adopted international practice, the appropriate period for collecting information on care costs before death is one month on average.

[7] The Chinese life expectancy at birth was 71.4 in 2000 and 73 years in 2005, based on the census and mini-census data. Assuming the same future annual rate of increase as that in 2000–2005, the life expectancy at birth in China would be 87.4 years in 2050. Therefore, the low mortality scenario assumption of a life expectancy at birth of 84.8 in 2050 and 88 years old in 2080 in China may not be too optimistic.

Table 5.1 Main demographic parameters used in the projections

	Rural					Urban				
Year	2010	2015	2030	2035	2050	2010	2015	2030	2035	2050
TFR-medium fertility with 2-child policy	2.01	2.09	2.09	2.27	2.27	1.24	1.67	1.67	1.80	1.80
Life expectancy at birth										
Male: medium mortality	69.38	70.03	72.20	72.90	75.00	73.24	73.91	75.90	76.60	78.60
Male: low mortality	69.38	70.69	74.84	75.98	79.40	73.24	75.16	80.91	87.84	81.9
Female: medium mortality	73.35	74.06	76.20	76.80	78.90	77.18	77.86	79.90	80.60	82.50
Female: low mortality	73.35	75.05	80.16	81.52	85.6	77.18	78.85	83.86	85.17	89.10
General marriage rate	0.0674	0.0674	0.0674	0.0674	0.0674	0.0601	0.0601	0.0601	0.0601	0.0601
General divorce rate	0.0022	0.0022	0.0022	0.0022	0.0022	0.0056	0.0056	0.0056	0.0056	0.0056
Average age at birth	25.20	25.20	26.50	26.50	26.50	26.00	26.00	27.30	27.30	27.30
Urban population %						37	52.3	60	64	75

healthy life styles and increases in life expectancy (Fries 1980). Conversely, the second theory argues that morbidity and disability will generally expand with the decline of mortality because it enhances the survival probabilities of unhealthy elderly groups (Olshansky et al. 1991). Manton (1982) proposed the "dynamic equilibrium model", which is in the middle between the optimistic and pessimistic hypotheses. It is, however, not clear which theory is more reasonable to predict changes in morbidity and disability among older adults in China. Gu and Zeng (2006) estimated that the prevalence of ADL disability among older adults in China had been declining 0.98 % every year. On the other hand, Du and Wu (2006) declared that the proportion of disabled elderly population increased by 1 % every year. Huang (2006) argued that it is most likely that the age-specific disability of the elderly will remain unchanged across future years in China.

Based on this body of prior research, we designed the following four projection cost scenarios (low, medium, high(a), and high(b)) for combinations of mortality decline and ADL status change scenarios among the Chinese elderly.

The *low cost scenario* assumes that there will be slow increases in life expectancy (medium morality) with a generally greater improvement in the prevalence of ADL disability among the elderly population (compression of morbidity). Specifically, this cost scenario assumes that the age-sex-rural/urban-marital status-coresidence with children-specific probabilities of ADL status transition from "active" to "disabled" will decline by 1 % annually in the projection period after 2010, while transitions from "disabled" into "active" will increase by 1 % annually.

The *medium cost scenario* assumes that there will be slow increases in life expectancy (medium morality) and the general health of the elderly population will remain stable (dynamic equilibrium). This medium cost scenario assumes that the age-sex-rural/urban-marital status-coresidence with children-specific ADL status transition probabilities will remain unchanged.

The *high(a) cost scenario* assumes that there will be more rapid increases in life expectancy (low morality) and the general health of the elderly population will remain stable (dynamic equilibrium). This scenario specifies that the age-sex-rural/urban-marital status-coresidence with children-specific ADL status transition probabilities will remain unchanged while life expectancy increases rapidly.

The *high(b) cost scenario* assumes that there will be more rapid increases in life expectancy (low morality) and the general health of the elderly population will deteriorate in the future (expansion of morbidity). More specifically, the age-sex-rural/urban-marital status-coresidence with children-specific ADL status transition probabilities from "active" into "disabled" will increase by 1 % annually, while the transition probabilities from "disabled" into "active" will decrease by 1 % annually.

We projected future cash costs of home-based care for disabled older adults in two different ways: (1) projecting the annual growth rate of cash costs of home-based care based on time series data analysis and trend extrapolation[8]; and (2) assuming that the annual growth rate of home-based care costs for the elderly would be the same as that for GDP during 2010–2050. We assume that in all of the scenarios, the average number of non-cash home-based care workdays provided by family members per disabled elder per year remains the same as that in the baseline 2010. Such a constant assumption concerning the care cost per disabled elder may not be true, but it enables us to more appropriately investigate the impacts of changes in elderly health level (measured by ADL) and mortality on the home-based care costs rather than mixing them with the effects of changes in per person costs.

In this Chapter, we adopted the medium fertility assumption based on an universal two-child with encouraging adequate spacing policy scenario, which assumes a smooth transition period to around 2015 when an average couple in urban and rural areas would have 1.8 and 2.27 children in their lifetime, respectively. We assume that the average age at first and second or higher order births will increase by 0.75 and 1.5 years in 2030 as compared to 2015, which constitutes an annual growth rate of about 0.05 and 0.1 years, respectively, during the years 2015–2030, due to delays of marriages and births under rapid socioeconomic development and the encouragement of governmental policies. As a result, the period TFRs of the first and second or higher order births would be 5 % and 10 % lower than the parity-specific lifetime cohort TFRs, respectively (Bongaarts and Feeney 1998). This would lead to the period TFRs in rural and urban areas being 2.09 and 1.67 during the period 2015–2030. We assume that after 2030 there will be no fertility quantity and timing limitation policy in China any more and the

[8] We use a quadratic curve to smooth the fluctuation of the annual growth rates during 1990–2008 and linear curve fitting during 2009–2030; we assume the care costs and GDP grow at the same annual rate after 2030.

decline in mean age at birth will cease, and thus rural and urban period TFR will be the same as the assumed cohort TFR (2.27 and 1.8) (see Table 1).

Analysis of the impacts of different fertility policy scenarios on future care provider resources and home-based care costs for disabled elderly per working-age person is presented in Sect. 14.4.4 of Chap. 14.

5.3.3 Results

5.3.3.1 Trajectories of ADL Status Transitions and Home-Based Care Costs

Figure 5.3 shows comparisons of age-specific transition probabilities between ADL statuses for older adults in different groups classified by rural/urban residence, gender, marital status, and co-residence with children, which are inputs for our projections. The age-specific transition probabilities from "active" to "disabled" are relatively low for those elders aged 65–74, but grow quickly after age 75. Differences in transition probabilities from "active" to "disabled" between all groups of rural/urban residence, gender, and co-residence status with children are small before age 80, but become larger after age 80. The transition probabilities from "active" to "disabled" after age 80 are higher for urban residents, females, and elders coresiding with children as compared to those who are rural residents, males, and those not living together with children. There are no significant differences in transition probabilities from "active" to "disabled" between the groups who are currently married and not currently married.

The curves of ADL status transition probabilities from "disabled" to "active" have an inverted J-type shape which reaches a peak at age 70–74 and then declines sharply with increase in ages. Combining the information in Fig. 5.3a–d about older adults' transition probabilities of ADL statuses, we learn that elders living in rural areas have advantages over elders in urban areas,[9] male elders have advantages over females, and elders not coresiding with children have advantages over those coresiding with children.[10] In addition, married elders have some advantages over unmarried elders but the difference is small.

Figure 5.4 gives the average home-based care cash expenditures per disabled elder per year by rural/urban residence, gender, marital status, and coresidence with

[9] Poorer facilities may force rural older persons to perform daily activities by themselves; this frequent exercise may enable them to better maintain or recover their capacities for daily living than their urban counterparts. Furthermore, the harder life and higher mortality at younger ages in rural areas may have resulted in a population of older persons who are more selected than their counterparts in cities and towns are.

[10] Perhaps those who live alone may more likely have active ADL capacity; such selection may result in elders having disadvantages in ADL status being more likely to coreside with children compared to those living alone.

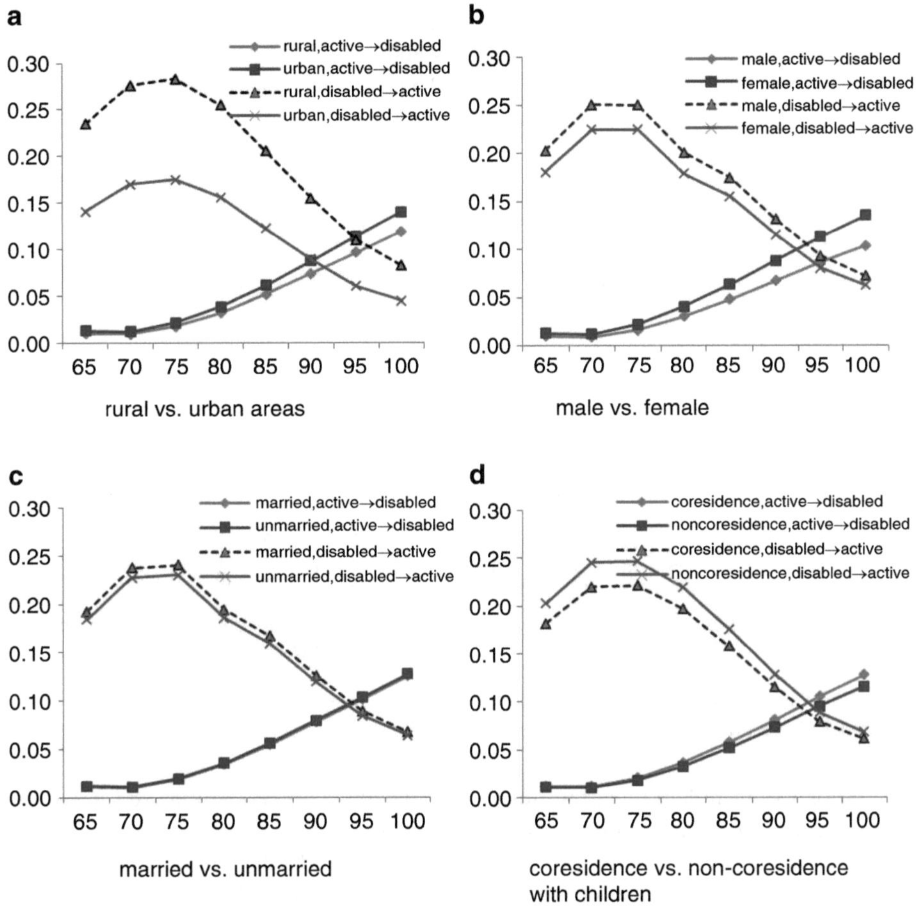

Fig. 5.3 Transition probabilities between ADL statuses for older adults, by rural/urban residence, gender, marital status, and coresidence with children

children, which are estimated based on the CLHLS 2008–2009 survey data and used as input for projection in our study. The results show that home-based care cash expenditures per disabled elder per year are substantially higher among those who live in urban areas, males, the unmarried, and those not living together with children compared to those who are in rural areas, females, currently married, and living together with children, respectively.

5.3.3.2 Trends Under the Medium Cost Scenario

Tables 5.2, 5.3 and 5.4 show the outcomes of the projections under the medium cost scenario, which is in general reasonable and helpful to understand the basic characteristics of future trends (Smith et al. 2001), while the high and low cost

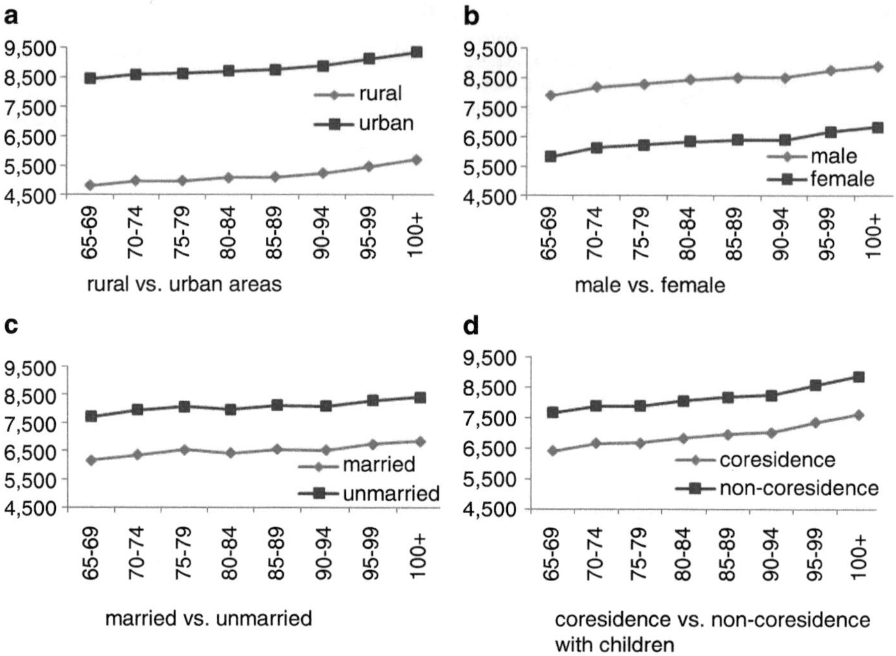

Fig. 5.4 Annual home-based care costs in cash (yuan) per ADL disabled elder, by rural/urban residence, gender, marital status, and coresidence with children

scenarios would reflect uncertainties. A few insights can be summarized based on the massive information in our projection results.

First, the annual growth rate of the number of disabled elders (3.5 %) is remarkably higher than that of the total elderly population (2.8 %), and the number of the disabled oldest-old will increase much faster as compared to the disabled young-old (see Table 5.2). The total number of old-age population in 2050 will be 3.0 times as large as that in 2010 and the amount of disabled elders will be 4.5 times as many as that in 2010. The annual growth rate of the disabled oldest-old population aged 80+ in 2010–2050 is 4.8 %, which is nearly twice that of the disabled young-old aged 65–79 (2.7 %). The number of the disabled oldest-old in 2050 will be 6.5 times as many as that in 2010 (see Table 5.2), in contrast to 2.9 times for the disabled young-old. The comparatively fast increase in the number of the disabled elderly, especially the disabled oldest-old, is due to the vast size of the baby boomer cohorts, who were born during the 1950s and 1960s and will become elderly soon and enter oldest-old ages around 2030–2040. Another reason is the accelerated decline in mortality of the elderly, especially at the oldest-old ages, along with the expansion of the human lifespan, which may cause the much faster growth of the oldest-old in the next few decades. At the same time, the oldest-old have substantially higher likelihood of being ADL disabled as compared to the young-old.

Second, the annual growth rate of disabled elders who do not coreside with children is remarkably higher than that of disabled elders who are living together with children. The annual growth rates for the disabled young-old and the oldest-old

Table 5.2 Projected number of ADL disabled elders (in thousands) and growth over time, by age, marital status, and living arrangement with children, under the medium cost scenario, 2010–2050

	Disabled young elders aged 65–79					Disabled oldest-old aged 80+					Total disabled elders				
	Married		Unmarried			Married		Unmarried			Married		Unmarried		
Year	Living with child	Not with child	Living with child	Living alone	Total	Living with child	Not with child	Living with child	Living alone	Total	Living with child	Not with child	Living with child	Living alone	Total
2010	1,688	1,178	1,367	476	4,709	461	347	2,266	595	3,670	2,149	1,526	3,634	1,071	8,380
2020	2,714	1,882	1,812	644	7,052	752	531	3,327	1,012	5,622	3,466	2,413	5,138	1,657	12,674
2030	3,983	2,802	2,692	977	10,454	1,296	891	4,821	1,519	8,526	5,278	3,693	7,512	2,496	18,980
2040	5,423	3,823	3,634	1,489	14,369	2,177	1,525	7,837	2,539	14,078	7,601	5,347	11,471	4,028	28,447
2050	4,895	3,539	3,154	1,845	13,432	4,096	2,899	12,578	4,291	23,864	8,991	6,438	15,732	6,135	37,296
r	2.7	2.8	2.1	3.4	2.7	5.6	5.5	4.4	5.1	4.8	3.6	3.7	3.7	4.5	3.8
Ratio 2030/2010	2.4	2.4	2.0	2.1	2.2	2.8	2.6	2.1	2.6	2.3	2.5	2.4	2.1	2.3	2.3
Ratio 2050/2010	2.9	3.0	2.3	3.9	2.9	8.9	8.4	5.6	7.2	6.5	4.2	4.2	4.3	5.7	4.5

Note: "r" means annual growth rate (%) in the period 2010–2050

Table 5.3 Projected home-based care costs (in cash) for ADL disabled elders and home-based care workdays delivered by family members for disabled elders under the medium cost scenario, 2010–2050

	% of total home-based care costs (in cash) for disabled elders among national GDP		Home-based care workdays for disabled elders delivered by family members (in millions)		
Year	Home-based care costs grow at the same rate of GDP growth	Home-based care costs growth rates estimated by time series analysis	Aged 65–79	Aged 80+	Subtotal
2010	0.24	0.24	1,210	1,128	2,338
2020	0.37	0.45	1,795	1,740	3,535
2030	0.57	0.75	2,694	2,646	5,340
2040	0.88	1.16	3,747	4,413	8,160
2050	1.18	1.58	3,505	7,494	11,000
Annual growth rate %	4.04	4.79	2.69	4.85	3.95
Ratio of # 2030 vs. 2010	2.4	3.1	2.2	2.3	2.3
Ratio of # 2050 vs. 2010	4.9	6.5	2.9	6.6	4.7

who are un-married and living alone are 3.4 % and 5.1 % on average, respectively, which are much higher than those who are un-married and living together with children (see Table 5.2). The number of the disabled young-old and the oldest-old who will be un-married and live alone in 2050 will be 3.8 times and 7.2 times as large as that in 2010, while corresponding figures for those who are unmarried but co-reside with children will be 2.8 times and 5.0 times. Note that the medium cost scenario assumes that the prevalence of co-residence with children, which is affected by social attitudes, job opportunities in other areas, and migration, remains

Table 5.4 Projected rural/urban distributions of young-old and oldest-old ADL disabled elders (in thousands), under the medium cost scenario, 2010–2050

	Disabled young-old aged 65–79						Disabled oldest-old aged 80+					
	Urban		Rural		Total		Urban		Rural		Total	
Year	#	%	#	%	#	%	#	%	#	%	#	%
2010	2,356	50.0	2,354	50.0	4,709	100	1,677	45.7	1,993	54.3	3,670	100
2020	3,785	53.7	3,266	46.3	7,052	100	2,870	51.0	2,752	49.0	5,622	100
2030	6,358	60.8	4,096	39.2	10,454	100	4,682	54.9	3,844	45.1	8,526	100
2040	9,598	66.8	4,771	33.2	14,369	100	8,518	60.5	5,560	39.5	14,078	100
2050	10,320	76.8	3,112	23.2	13,432	100	15,841	66.4	8,023	33.6	23,864	100

unchanged at the current level. Thus, the trend reflected in Table 5.2 that the annual growth rate of disabled elders who are un-married and live alone is much higher than those who are un-married and live with children is totally due to the great reduction of fertility rates in the past four decades that will likely lead to a shortage of children to care for the future elderly population.

Third, the annual growth rate in percentage of home-based care costs in cash for disabled elders as a fraction of total GDP in the first half of this century will be about 4.0–4.8 %, which is about 1.5–2.0 times as high as the annual growth rate of the elderly population (see Table 5.3). Such a trend may be mainly due to increases in the ratios of disabled older adults living alone in urban areas to the total disabled elderly population, and the higher care costs per disabled elder living alone in urban areas.

Fourth, data in Table 5.4 show that the urban and rural distribution of young-old disabled elders is consistent with the proportion of urban residents in the total population; however, the percentage of the disabled oldest-old in urban areas in 2050 is 66.4 %, which is 8.6 percentage points lower than that of the projected proportion of urban residents among the total population. The main reason is that most rural peasants, who were middle-aged in the 1980s and 1990s and will become oldest-old after 2030, will stay in their villages while the large-scale population migration from rural to urban areas will continue to occur among younger people.

5.3.3.3 Possible Ranges of Trends

The projection results under the low, medium, high(a), and high(b) cost scenarios presented in Fig. 5.5 show that the total number of disabled elders in China will rapidly climb from 8.4 million in 2010 to 12.0–14.6 in 2020, 16.6–24.9 in 2030, and 29.9–61.8 million in 2050.[11] Figure 5.6 shows that the non-cash workdays of care provided by family members will increase rapidly from 2.3 billion workdays in 2010 to 3.3–4.1 billion in 2020, 4.7–7.1 billion in 2030, and 8.8–18.6 billion workdays in 2050. Figure 5.7a shows that the percentage of home-based care cash expenditures among total GDP in China will grow up from 0.24 % in 2010

[11] In our study, the gap between the high (b) and low scenarios will increasingly enlarge as the projection period is prolonged, which is similar to results in other demographers' high, medium, and low projections Lee and Tuljapurkar (2001: 22).

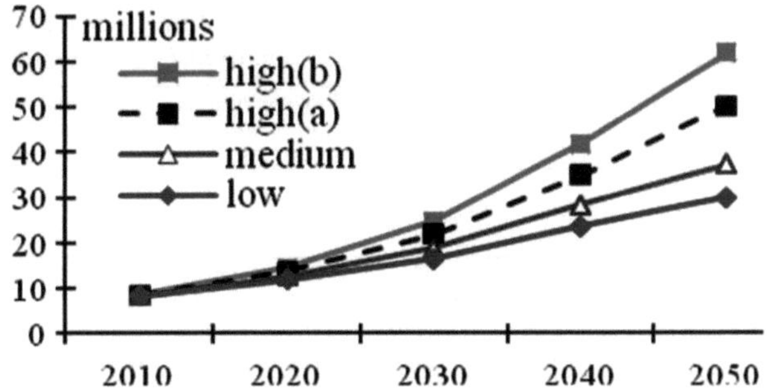

Fig. 5.5 Projected number of ADL disabled elders under different scenarios (in millions)

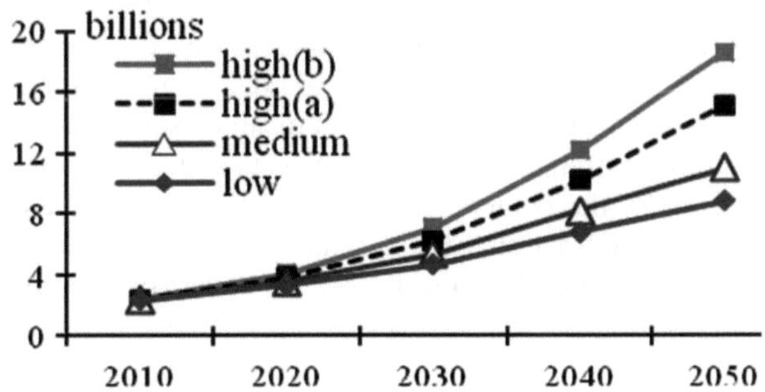

Fig. 5.6 Projected number of home-based work days for disabled elders delivered by members under different scenarios (in billions)

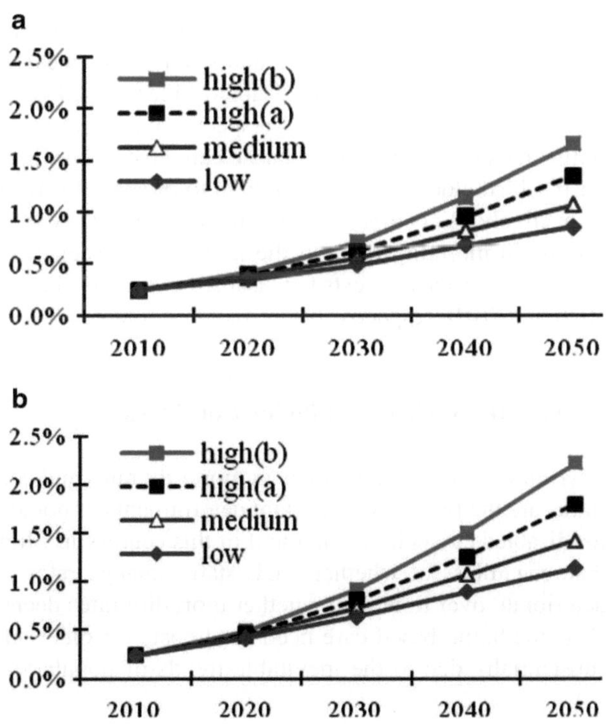

Fig. 5.7 (**a**) Projected percentage of total home-based care costs in cash for disabled elders among national GDP, assuming service wages grow at same rate of GDP. (**b**)Projected percentage of total home-based care costs in cash for disabled elders among national GDP, assuming service wages follow trend extrapolation

Table 5.5 Sensitivity analysis of the impacts of future changes in mortality rate and the prevalence of ADL disability on home-based care needs/costs for elderly

	High(a) vs. medium scenarios: *impact of mortality rates declining more rapidly*			High(b) vs. high(a) scenarios: *impact of ADL disability deteriorating*			Low vs. medium scenarios: *impact of ADL disability improving*		
Year	# of disabled elders	Costs in cash	Opportunity costs	# of disabled elders	Costs in cash	Opportunity costs	# of disabled elders	Costs in cash	Opportunity costs
2020	+8.36 %	+6.17 %	+9.24 %	+6.09 %	+5.23 %	+6.08 %	−5.75 %	−4.84 %	−5.75 %
2030	+14.97 %	+11.88 %	+16.56 %	+14.23 %	+12.57 %	+14.20 %	−12.54 %	−10.85 %	−12.60 %
2040	+22.94 %	+18.87 %	+25.19 %	+19.52 %	+17.68 %	+19.40 %	−16.74 %	−14.81 %	−16.78 %
2050	+34.56 %	+29.61 %	+37.69 %	+23.18 %	+21.40 %	+22.92 %	−19.74 %	−17.83 %	−19.75 %

to 0.35–0.41 % in 2020, 0.49–0.71 % in 2030, and 0.85–1.66 % in 2050, assuming that the annual growth rates of home-based care costs for disabled elderly are the same as the annual growth rates of GDP. When estimating annual growth rates for home-based care expenditures per disabled senior based on time series data analysis and trend extrapolation, with care costs and GDP growing at the same rate of speed after 2030, the percentage of home-based care expenditures in cash for disabled elders as a fraction of the total GDP will increase to 0.41–0.49 % in 2020, 0.63–0.92 % in 2030, and 1.14–2.21 % in 2050 (see Fig. 5.7b).

5.3.3.4 Sensitivity Analysis

The sensitivity analysis presented in Table 5.5 shows that, as compared to the medium cost scenario which assumes constant ADL status transition rates, the low cost scenario (improved ADL status transition rates) would cause payment costs and non-paid workdays for disabled elders in 2030 and 2050 to be reduced by 10.9–12.6 % and 17.8–19.8 %, respectively; the high cost scenario (worsened ADL status transition rates) would cause care costs of payments and non-paid workdays for disabled elders in 2030 and 2050 to increase by 12.6–14.2 % and 21.4–22.9 %, respectively. As compared to the medium mortality scenario, the more remarkable decline in mortality rates in the high(b) cost scenario would increase the home-based care needs and costs for disabled elderly by 11.9–16.6 % and 29.6–37.7 % in 2030 and 2050, respectively.

5.3.3.5 Discussions and Policy Considerations

This study shows that declines in mortality rates and changes in older adults' ADL status are the two most important determinants of home-based care needs and costs for disabled elders in the first half of this century in China. Our analysis also shows that, regardless of whether ADL status change rates are assumed to improve or deteriorate over time, and whether mortality rates decrease moderately or remarkably, the home-based care needs and costs for older adults in China will increase substantially due to the inevitable trends of population aging and extremely fast increase of the oldest-old. Consequently, relevant reforms in policies and socioeconomic planning should respond to the serious challenges of large and rapid increases in home-based care needs and costs.

Our projections found that if ADL status change rates substantially improve (under the low cost scenario) or substantially deteriorate (under the high(b) cost scenario), the care needs and costs for disabled elders would remarkably decrease or increase accordingly. Note that the most effective way to improve elderly ADL statuses is not only to cure disease, but also to effectively prevent illnesses. The latter may be more important since elders who are cured of one disease are likely to suffer another illness and long-term ADL disability if there is no improvement in their ability to prevent disease. Therefore, studies of why some elders remain happy and healthy up to oldest-old ages, and how to reach such pathways of healthy aging, should be further strengthened.

Our data analysis shows that female elders have significant disadvantages in ADL status transitions compared with male counterparts (see Fig. 5.3b). However, the home-based care expenditures for female disabled elders are much lower than that for male disabled elders (see Fig. 5.4b); this is an important issue that should receive much more attention from the Chinese government and the public. The government and society should try their best to guarantee that both male and female older adults enjoy equal benefits. The finding that home-based care expenditures for disabled elders without a spouse are much higher than expenditures for married disabled elders (see Fig. 5.4c) indicates that, in order to reduce home-based care costs for the elderly, remarriage for widows and widowers should be strongly encouraged and a series of measures should be taken to remove obstacles to remarriage for the not-married elders, especially in rural areas, in terms of laws, regulations, and social attitudes.

5.4 Concluding Remarks

This chapter further develops and empirically applies the ProFamy extended cohort-component model to project future home-based care needs and costs for disabled older adults, with integration of multistate projections of elderly disability status transitions, household structures, and living arrangements, based on commonly available demographic data. This integrated multistate model projects the dynamics of ADL statuses and yearly workdays and payments for disabled elders, providing relatively more realistic and detailed information on future trends compared to previous studies that excluded the crucially important determinant of elderly living arrangements. This is an innovative step forward in the field as it integrates multistate dynamic projections of elderly activity of daily living status transitions, household structure, living arrangements, home-based care needs and costs for disabled elders with projections for working age care providers simultaneously in one model. The detailed projections we have presented could prove useful for governmental policy analysis, strategic plans for future public services, and private sector market potentials research.

We emphasize that projections for time horizons of less than 20 years may be used as forecasting for business and governmental planning, but any results beyond that should be considered to be simulations only, due to large uncertainties after more than 20 years. Thus, the projection results of the middle years of this century presented in this chapter should be mainly regarded as simulations. Such simulations are useful for academic and policy analysis to answer the "what, if" questions about effects of changes in demographics and disability status transition rates on future general trends and patterns of elderly disability and home-based care needs and costs for disabled elders in China, but they cannot be considered to be accurate forecasts.

Chapter 6
Household and Living Arrangement Projections at the Small Area Level

6.1 Basic Concepts to Apply the ProFamy Approach in Combination with Ratio Methods for Small Areas

It is usually very difficult to obtain adequate data at the small area level[1] to estimate the demographic parameters that are necessary to apply the classic cohort-component approach for population projection and to apply the ProFamy extended cohort-component method for household and living arrangement projections. Indeed, even census micro datasets, although valuable, cannot provide full information at the small area level for all the characteristics that are needed for cohort-component projections. Therefore, most researchers use "indirect" methods that "borrow strength" based on a projection of the parental region[2] in which the small area is located; this indirect method increases the stability and accuracy of population projections for smaller counties and cities (Rao 2003; Smith and Morrison 2005; Smith et al. 2001). The ratio method (e.g., extrapolating a small area's share of the region population) is frequently used for small area projections because its data requirements are minimal, it is easy to apply, and its projections are often reasonably accurate (Smith 2003).

In household and living arrangements projections at the small area level using the ProFamy extended cohort-component model, we employ the ratio method with either a constant-share or a shift-share specification (Smith et al. 2001), based on household and living arrangement projections of the parental region.[3] After household and living arrangements are projected for the parental region, we then calculate the

[1] Small areas refer to the smaller counties, cities and towns, as well as places, possibly even tracts or block groups, which have a small population size.

[2] The parental region may be a state in the United States, a province in other countries, or another kind of sub-national administrative district (including a large county, city, municipality, etc.).

[3] If an area (county, city, or town) has a reasonably large population size and has the needed data to estimate the demographic parameters, one could apply the ProFamy extended cohort-component method directly, and no need to apply it indirectly through combination with the ratio method.

Y. Zeng et al., *Household and Living Arrangement Projections*,
The Springer Series on Demographic Methods and Population Analysis 36,
DOI 10.1007/978-90-481-8906-9_6,

Table 6.1 Illustrative example of 11 categories of households by type and size for projections at the small area level, using combined ProFamy and ratio methods

Category	Type (h)	Size (s)	Characteristics of the household category
1	1	1	One-single-man only, household size 1
2	1	2–3	One-single-man with child/other, household size 2–3
3	1	4+	One-single-man with child/other, household size 4+
4	2	1	One-single-woman only, household size 1
5	2	2–3	One-single-woman with child/other, household size 2–3
6	2	4+	One-single-woman with child/other, household size 4+
7	3	2	One-couple only, household size 2
8	3	3–4	One-couple household with child/other, household size 3–4
9	3	5+	One-couple household with child/other, household size 5+
10	4	1	Men living in group quarters
11	5	1	Women living in group quarters

Note: "Single" refers to not-married (including never-married, divorced, or widowed) and not-cohabiting persons. "One-couple" refers to a married or cohabiting couple

race-sex-age-specific proportions of households with various types and sizes in the county/city among the corresponding households in the parental region. We assume that the proportions are constant or changing based on past trends or projected new trends, and then multiply the existing household and living arrangement projections of the parental region by the proportions in the small area to derive the household and living arrangement projections for a small area within the parental region. The assumption imposed and the rationale of constant-share or shift-share approaches in household and living arrangement projections for small areas are the same as those generally used for small area population projections (Smith et al. 2001). The formulas for the constant-share and shift-share ratio methods, adopted from Smith et al. (2001: 177–179) for household and living arrangement projections of small areas with the ProFamy approach, are presented in Sect. 6.2.

Given the sample size limitations for small areas, we classify the household and living arrangement projection output into 11 categories by household type and size and by age, sex, and race (if sample size allows) of the reference person; this classification will be illustrated through an application in Sect. 6.3. The classification of these 11 categories, as shown in Table 6.1, is for illustration only, as one may group the available more detailed outcomes of household and living arrangement projections by type and size[4] into larger (more detailed) or smaller (more simplified) numbers of categories, depending on the population size of the small area under study and the purpose of the analysis.

Note that household and living arrangement projections using the ratio method approach can be done for a small area based on its parental state's household and living arrangement projections produced either by the new ProFamy approach or the classic headship rate method. However, as discussed in Chap. 4, a typical well-done national household projection based on the classic headship rate method projected only five

[4] Ref. to Table 4.6 of Chap. 4 for the available more detailed outcomes of household and living arrangement projections by type and size, employing the ProFamy approach.

household types by age groups of household head without household size (U.S. Census Bureau 1996). By comparison, the ProFamy approach projects much more detailed household types, sizes, and living arrangements for all members of the population (see a comparison in Tables 4.5 and 4.6 in Chap. 4). Consequently, the ProFamy approach, using the same ratio method and parental state's projections, would produce much more detailed household and living arrangement projections for small areas. Therefore, the practical usefulness of the ratio method is strengthened if it is applied in combination with the ProFamy approach compared to the headship rate method.

6.2 The Constant-Share and Shift-Share Ratio Methods

Let *T1* denote the starting year of the projection; *T1* may be the most recent census year or a year after the most recent census if one decides to use more recent large survey data such as the American Community Survey (ACS) as a baseline for the projection.[5] Denote t as future years of the projection. We use the subscript $_p$ and $_c$ to represent the variables for the parental region and the small area, respectively.

Input:

$H_p(h, s, r, x, T1)$: the observed number of households of type h and size s (see Table 6.1) with a reference person of race r and age x in the parental region in the baseline year *T1*; Age x can be 5-year, 10-year, 15-year, or 20-year age groups, depending on the user's choice based on the population sample size.

$H_p(h, s, r, x, t)$: the projected number of households of type h and size s with a reference person of race r and age x in the parental region in year t.

$g(h, s, r, x, T1)$: the observed proportions of the households of type h and size s with a reference person of race r and age x of the small area among the corresponding households of the parental region in the baseline year *T1*;

$P_p(T1)$: the observed total population size in the parental region in the baseline year *T1;*

$P_p(t)$: the projected total population size in the parental region in year t;

$P_c(T1)$, the observed total population size in the small area in the baseline year *T1*;

$H_p(h, s, r, x, T1)$, $g(h, s, r, x, T1)$, $P_p(T1)$ *and* $P_c(T1)$ are derived from census or ACS datasets;

$H_p(h, s, r, x, t)$ and $P_p(t)$ are available from the already-conducted household and living arrangement projections for the parental region.

Output:

$H_c(h, s, r, x, t)$: projected number of households of type h and size s with a reference person of race r and age x in the county/city in the year t;

[5] For the major demographic indicators, the annual ACS is representative at the state level, and 2-, 3-, 4-, or 5-year moving averages of the ACS data are representative for the sub-state areas, depending on their population size.

$P_c(t)$: projected total population size in the future year t in the small area.

Computation:
Based on the constant-share method (Smith et al. 2001):

$$H_c(h,s,r,x,t) = H_p(h,s,r,x,t)g(h,s,r,x,T1)$$

$$P_c(t) = P_p(t)\ [P_c(T1)/P_p(T1)]$$

Based on the shift-share method (Smith et al. 2001):

$$\begin{aligned}H_c(h,s,r,x,t) = H_p(h,s,r,x,t)\{g(h,s,r,x,T1) \\ + [((t-T1)/(T1-T0))(g(h,s,r,x,T1) - g(h,s,r,x,T0))]\},\end{aligned}$$

$$\begin{aligned}P_c(t) = P_p(t)\{\ P_c(T1)/P_p(T1) + [((t-T1)/(T1-T0)) \\ (P_c(T1)/P_p(T1) - P_c(T0)/P_p(T0))]\},\end{aligned}$$

where $T0$ refers to the previous census year preceding to the starting year of the projection $(T1)$.

Additional adjustments are necessary to maintain consistency between the total population size implied by the projected number of households by type, size, race, sex, and age and the projected total population size in the small area. Let $H'_c(h, s, r, x, t)$ denote the finally adjusted projected number of households of type h and size s with a reference person of race r and age x in the small area in year t.

$$H'_c(h,s,r,x,t) = H_c(h,s,r,x,t)\left\{(P_c(t)/\sum_s\sum_h\sum_r\sum_x [H_c(h,s,r,x,t)\times s]\right\}$$

6.3 Empirical Assessment and Illustrative Applications

To assess the accuracy of the combined ratio method and ProFamy approach and present illustrative applications, we calculated projections from 1990 to 2000 and compared projected estimates with census-observed counts in 2000 for sets of randomly selected 25 small counties and 25 small cities which were more or less evenly distributed across the United States. The comparisons show that, in general, most forecast errors are reasonably small – mostly less than or slightly more than 5 %. More specifically, as shown in Table 6.2, the Mean Algebraic Percent Error (MALPE) for the small counties and small cities are all within a very small range (means: 0.69 and –0.58), and there is no large difference between the forecast

Table 6.2 MAPE, MALPE and MEDAPE percent error estimates between the indices projected from 1990 to 2000 and census observations in 2000 for 25 randomly selected small counties and 25 randomly selected small cities, based on the ProFamy and ratio method

	Population size	Total number of households	Average household size	% 1-person household	% 2–3 person household	% 4+ person household	% married couple household	Average error rate
MALPE of the 25 counties	0.40	1.18	−1.55	3.05	−1.87	1.36	2.23	0.69
MAPE of the 25 counties	7.69	8.62	2.99	5.78	3.45	6.68	4.23	5.63
MEDAPE of the 25 counties	4.49	4.57	2.32	4.69	2.79	6.29	3.34	4.07
MALPE of the 25 cities	−4.39	−0.72	−1.24	4.60	−1.89	−0.59	0.16	−0.58
MAPE of the 25 cities	14.78	13.37	5.29	10.73	4.78	9.77	6.60	9.33
MEDAPE of the 25 cities	9.81	9.30	4.97	7.58	2.66	9.77	5.53	7.09

Note: MAPE (*Mean Absolute Percent Error*), MALPE (*Mean Algebraic Percent Error*) and MEDAPE (*Median Absolute Percent Error*) are the most commonly used measures of forecast errors (Smith et al. 2001: 302–304)

Table 6.3 Population size distributions of the 25 randomly selected small counties and the 25 randomly selected small cities

Randomly selected 25 cities			Randomly selected 25 counties		
Pop size	Number	%	Pop size	Number	%
<2,000	8	32.0	<10,000	8	32.0
2,000–4,999	8	32.0	10,000–49,999	9	36.0
5,000–29,999	6	24.0	50,000–99,999	4	16.0
≥ 30,000	3	12.0	≥ 100,000	4	16.0
Total	25	100	Total	25	100

errors of the small counties and small cities. However, the differences in the forecast errors measured by Mean Absolute Percent Error (MAPE) and Median Absolute Percent Error (MEDAPE) between small counties and small cities are substantial: the MAPE and MEDAPE are within the ranges of 2.99–8.62 (mean: 5.63) and 2.32–6.29 (mean: 4.07), respectively, for the 25 randomly selected small counties, and 4.78–14.78 (mean: 9.33) and 2.66–9.81 (mean: 7.09), respectively, for the 25 randomly selected small cities. It is clear that the mean absolute percent error rates in the validation tests for the small cities are substantially larger than those for the small counties. This might be because the constant-share assumption is more likely to be violated in the cases of small city household and living arrangement projections because the demographics in some cities changed substantially during the 1990s due to economic events such as opening, closing, or reallocating large enterprises or institutions.

As shown in Table 6.3, about one-third of the 25 randomly selected small cities have a very small population size of less than 2,000 residents; the population size in about one-third of the 25 randomly selected small counties is less than 10,000 persons. We analyzed the forecast errors of the test projections from 1990 to 2000 by population size among the 25 randomly selected small counties and 25 randomly selected small cities and did not find significant correlation between population size and forecast errors.

The validation tests and illustrative application results at the small area levels summarized in Tables 6.2 and 6.3 show that the forecast errors are within a relatively small range and the illustrative applications are satisfactory. It is uncertain what portions of the errors are due to the model specification or due to inaccuracies of the data. It is clear, however, that the ProFamy extended cohort-component approach for simultaneously projecting households, living arrangements, and population age/sex distributions work reasonably well, not only at the national level as shown in Zeng et al. (2006) and Zeng et al. (2013a), but also at the sub-national level. The ProFamy approach combined with the ratio method also works reasonably well at the small area level.

Chapter 7
A Simple Method for Projecting Pension Deficit Rates and an Illustrative Application

7.1 Introduction

Scholars, policymakers, and the public are very much concerned about possible annual pension deficits in future years induced by unavoidable population aging. The ever-growing number and proportion of the older population is the consequence of rapid fertility decline, baby boomers born in the late 1940s, 1950s and 1960s entering old age, and continued substantial decline in mortality rates.

In general, existing actuarial models such as the World Bank's Pension Reform Options Simulation Tool-kit (PROST) need the following data for the baseline year and each forecasted year: (1) age-sex-specific labor force participation rates, unemployment rates, earnings profiles, contributors and pensioners, years of employment up to retirement, pension payments, retirement rates, fertility, mortality, and migration,[1] and (2) some macroeconomic indicators, such as GDP (Gross Domestic Product) and its growth rate, pension fund contribution rate, and replacement rate etc. (Sin 2005; World Bank 2003; Becker and Paltsev 2001). Some economic/financial forecasting models may also need age-sex-specific disability, survivor, and evasion and exemption rates (Becker and Paltsev 2001). However, this large data requirement is unlikely to be met in developing and transitional countries or at the regional level in developed countries, as it is extremely difficult to get good age-sex-specific data for decentralized systems that are run locally with different components and requirements. For example, a well-supported World Bank study on China's pension system reform used the PROST approach for only six municipalities (Chongqing, Guangzhou, Huaibei, Tianjin, Wuhu, and Ziyang) and three provinces (Liaoning, Fujian and Zhejiang – all of which are in the more-developed coastal regions) because the required detailed data from other regions of China were not available (Sin 2005).

[1] A few actuarial models for pension deficit forecasts do not necessarily need the age-sex-specific data, but they are relatively more complicated and need much more statistical knowledge (e.g., Becker and Paltsev 2001; Bedard 1999; Cairns and Parker 1997; Haberman and Wong 1997; Hamayon and Legros 2001); thus, they are not applied as widely as the other models.

Y. Zeng et al., *Household and Living Arrangement Projections*,
The Springer Series on Demographic Methods and Population Analysis 36,
DOI 10.1007/978-90-481-8906-9_7,

This regional data limitation might result in projected pension deficits that do not accurately represent the whole country. Moreover, complicated actuarial models for pension projections, which include many age-sex-specific variables and complex relationships among the variables, increase the stochasticity and measurement errors to which the model outcomes are subject. Some scholars call this kind of bias "specification randomness", which can increase forecast errors (Van Imhoff and Post 1998: 111). For example, Smith and Sincich (1992: Exhibit 2) compared the forecast errors of the population size for the 50 states of the United States for forecast horizons of 5, 10, 15, and 20 years from the mid-1950s to the early 1980s using five simple extrapolation techniques, a cohort-component model, and two more complicated structural models; they found that a more complicated projection model may not forecast the population size as accurately as a simpler model does. Many other analyses reached a similar conclusion (e.g., Ahlburg 1992). Thus, the choice between a simple or a complex model mainly depends on the purpose of the study. If the purpose is to analyze detailed age-sex-specific pension contributions, benefits, expenditures and revenues, a complex actuarial model is appropriate (e.g. West 1999). However, it may not be necessary for demographers and policy analysts, who are not interested in such details, to deal with the huge amount of data required to run complex actuarial models.

The main goal of this chapter is to present a simple and practical method associated with ProFamy extended cohort-component projection approach to explore how demographic and retirement age policies may affect future pension deficits, using application to China as an illustration.

7.2 The Method

The concept of *annual pension deficit* used in this study refers to the difference between total premiums collected annually from all kinds of workers/employers/ assets in the program and total pension benefits paid annually to all retirees plus administrative expenditures by the government-managed pension program. The pension program may be a defined benefit (DB) plan,[2] a defined contribution (DC) plan,[3] or a mixture of the two. In fact, for the purpose of assessing the overall annual pension deficit (or surplus), the co-existing DB and DC programs can be integrated into one general pension system of premium contribution and retirement wage payments. If there is a pool of assets in a funded DC program, it will appreciate at the rate of interest or investment return and be included as part of

[2] A Defined Benefit (DB) plan promises the participant a specific monthly benefit at retirement.

[3] A Defined Contribution (DC) plan provides an individual account for each participant. The benefits are based on the amount contributed into the plan and are also affected by income, expenses, gains, and losses of the contributed pension funds. There are no promises of a fixed monthly benefit at retirement.

the premium contribution to the pension system, recognizing that the rate of return to capital is net of the administrative costs.

To simplify the presentation, we present only the basic formulas of the simple method with discussions about its input parameters in this section; the mathematical derivation of the formulas is included in Appendix 1.

Let $P(t)$ denote the contribution rate in year t, namely, the average proportion of total contributions to the pension fund by workers and employers among the total wages of all workers in year t;[4] $P(t)$ includes contributions by workers and employers in DB or DC schemes as well as the interest or investment return of the pool of assets in a funded DC system.

$B(t)$, the replacement rate in year t, namely, the ratio of the average retirement wage per retiree to the average wage per worker in year t;[5] $n(t)$, the annual pension deficit rate, which is defined as the ratio of the total amount of annual pension deficit (i.e., the difference between the annual total pension payments/costs and the annual total pension premium contributions/assets' income) to the total wages in year t. We pull all income and payments of the co-existing DB and DC programs into one overall "input–output" accounting system managed by the government. The $n(t)$ can be positive (pension fund deficit), zero (balanced), or negative (pension fund surplus).

$d_2(t)$, the dependency ratio of elderly, namely, the ratio of the total number of elderly persons over the average age of retirement to the total number of persons of labor force age (e.g., from age 18 to the average age at retirement), which can be derived by the ProFamy extended cohort-component projection method or the classic cohort-component population projection method.

$r(t)$, the retirement rate, i.e., the proportion of retirees who receive pensions among the total number of elderly persons over the average age at retirement; $e(t)$, the pension program participation rate, i.e., proportion of the persons who participate in the pension programs among the total number of persons of labor force ages;

As shown in the Appendix, we can mathematically derive the following analytic formula:

$$n(t) = B(t) \cdot d_2(t) \frac{r(t)}{e(t)} - P(t) \tag{7.1}$$

Equation 7.1 presents a simple method for projecting annual pension deficit rate as a percentage of the total wages in future years, based on one demographic indicator, the dependency ratio of the elderly ($d_2(t)$), and the pension program parameters of the replacement rate ($B(t)$), contribution rate ($P(t)$), retirement rate ($r(t)$), and pension program participation rate ($e(t)$).

When time series data with reasonably good quality for $r(t)$ *and* $e(t)$ are available, it is recommended to use Eq. 7.1 to predict the annual pension deficits rate

---

[4] We use the "average proportion" and "average wage" to define the average contribution rate and average replacement rate, for simplicity and to avoid the difficulties of data unavailability.

[5] The wages for retirees and workers include in kind benefits such as housing subsides and periodic distributions of free goods.

in future years. However, if good data for $r(t)$ *and* $e(t)$ are not available (as is the case in China and many other developing and transitional countries), one may use the available and relatively reliable statistics of $n(t)$, $d_2(t)$, $B(t)$, and $P(t)$ to estimate the ratio of the retirement rate to the pension program participation rate ($r(t)/e(t)$) based on the analytical relationship expressed in Eq. 7.1. In fact, $r(t)/e(t)$ (denoted as $c(t)$ hereafter) in Eq. 7.1 is itself a valid indicator of the difference in participation in the pension program between the older generations who have currently reached retirement ages and the younger working generations. If $c(t)$ is equal or close to one, the equilibrium of pension program participation between older and younger generations is reached; if $c(t)$ is significantly greater (smaller) than one, the pension participation rate in older generations is higher (lower) than that in younger generations. Consequently, when reliable data for estimating/projecting $r(t)$ *and* $e(t)$ separately are not available, one may simply use $c(t)(=r(t)/e(t))$ as one of the projected or assumed input parameters. Thus,

$$n(t) = B(t)d_2(t)c(t) - P(t) \tag{7.2}$$

Note that $d_2(t)$, the dependency ratio of the elderly, can be readily derived from standard population forecasting or household and population forecasting following the ProFamy approach based on demographic parameters of fertility, mortality, and migration; the other required parameters ($B(t)$, $P(t)$, $r(t)$ and $e(t)$ (or $c(t)$ instead of $r(t)$ and $e(t)$)) all have clear policy or program meanings, and they can be projected by application of trend extrapolation methods based on available time series data or by experts' opinions.

7.3 Illustrative Application to China

7.3.1 The Pension System in China

In China, the pension system was introduced in 1952 and supported only employees of state-owned enterprises and institutions, with a large majority in urban areas; it was a typical DB plan under the socialist planning economy. The coverage of the Chinese pension system by the end of the twentieth century included about 140 million persons among a total population of about 1270 million (Poston and Duan 2000: 721). According to a 1992 nationwide representative survey of the elderly, only 5.9 % of the rural elderly aged 60 and older were pension recipients, in contrast to 73.7 % in urban areas (CRCA 1994). The compulsory age at retirement in China has been 60 for men and 52.2 for women.[6] However, there is some variation in the actual age at retirement. Exceptionally skilful professionals and high-ranking

[6] The compulsory retirement age is 50 for female workers and 55 for female cadres (including teachers, medical personnel, other professionals, and administrators). The weighted average of compulsory retirement age for women is 52.2.

officers are allowed (or requested) to retire later; there is also an early retirement scheme due to disability certified by a medical doctor.

The old pension system has been undergoing reform since the early 1990s. The pension system reform was characterized by a grandfather clause, whereby previous employees followed the old approach and the new employees follow the new approach. More specifically, state-owned enterprise (SOE) workers who were employed before the reform still follow the DB plan; new SOE workers who were employed after the pension reform, employees of private and collective enterprises, and the self-employed follow the new program of a combination of "individual account" and "social plan". In the new program, the individual contribution/account is substantially subsidized and backed-up by the government-managed social plan He (1998, 2001). Thus, the new Chinese pension program is a mixture of the standard DC and DB programs.

The Chinese rural old age insurance program, in which individuals' premium contributions were subsidized by local collective funds[7] and government, was first launched as an experimental project in Shandong province in the early 1990s and quickly spread to all over the country. By the end of 1995, 61.2 million rural peasants aged 20–60 participated in the program, and the participation rate among the population aged 20–60 was 14.2 %; there were about 80 million participants and about 890,000 rural peasants aged 60+ starting to receive monthly payments from the old age insurance program.[8] However, this program stagnated and shrank from 1999 to 2008. By the end of 2004, there were about 53.9 million participants, a decline of 32.6 % from 1999; about 10 % of counties completely discontinued the rural old age insurance program (Zeng 2005b). The official saying for such a situation was "*Zhen Dun Gai Ge* (consolidate and reform)", but the major cause was actually attributed to important policymakers and scholars' arguments that the pension deficit problems in urban China would be very serious due to rapid population aging and there would be no resources left to devote to the pension program in rural areas.

The most recent development in rural old age insurance is that in September 2009, the Chinese State Council announced that 10 % of all counties in China will launch the "New Rural Old Age Insurance Program (NROAIP)" by the end of the year 2009 (Ye 2009). As compared to the previous rural old age insurance program, the governmental subsidy and back-up for the NROAIP substantially increased; it is explicitly stated that the premium will be jointly paid by the individuals, local and central governments, and the state will ensure the basic and minimum income level for all elderly who participate in the program. It was reported by the governmental agency's new release that the basic NROAI had almost universally covered all rural residents by the end of 2012. The new and promising policy actions are being taken

---

[7] The local collective funds in China include the accumulated income and assets collectively owned by the local community.

[8] Data obtained from Ministry of Civil Affairs, see Zeng (2002).

and the new rural old age insurance program, a typical combination of DC and DB, is expected to be quickly developed, although it is still premature.

Obviously, the current Chinese pension system is very diverse – between urban and rural sectors, between state-owned and fast-growing private enterprises, and between previously and newly employed workers. The system is decentralized and funds are being managed at the province level or lower.[9] This is similar to other developing and transitional countries. It is thus extremely hard to apply a sophisticated and complicated actuarial model to predict the Chinese pension deficit for future years, because detailed age-sex-sector-specific actuarial input data that are compatible across various sectors (regions and groups) are not available. However, as illustrated below, we are able to investigate the general impacts of possible changes in fertility and retirement age policies on the annual pension deficit in future years using the simple method proposed in Zeng (2011) and summarized in Sect. 7.2, an approach which requires data on only a few measurable and predictable overall summary parameters. These summary parameters are *averages,* including various sectors of state-owned and private/collective enterprises, previously and newly employed workers, and urban and rural areas. This simplified approach is not only motivated by the lack of detailed actuarial data, but also because it serves well the purpose of macro policy analysis.

7.3.2 *Assumptions of Parameters for Different Scenarios*

7.3.2.1 Medium and Low Fertility

The 2000 Chinese census reported an extremely low period Total Fertility Rate (TFR) of 1.22. Adjusted for under-reporting, the period TFR estimated by demographers and statistical offices around the year 2000 ranged from approximately 1.6–1.8. Based on various studies (e.g., Guo 2004; Zhang and Zhao 2006), it is likely that China's TFR is close to the low bound of this range. Using the census-observed rural–urban TFR differential, we estimate that the period TFRs in 2000 were 1.9 for rural areas, 1.15 for urban areas, and 1.63 for all of China.

We designed two fertility policy scenarios in consideration of the opposing effects of relaxing fertility policy and rapid economic development, which tend to reduce fertility, and the tremendous differences in socioeconomic development and fertility attitudes/behavior between rural and urban areas.

The *low fertility scenario* assumes that the current fertility policy in China will remain unchanged. The current fertility policy in China, which is often simplified as the "one-child policy," allows 63.0 % of Chinese couples to have one child only, 35.6 % to have two children, and 1.3 % to have three children; this implies that the

[9] Refer to Johnson (2000), West (2000) and Yin et al. (2000) for more detailed discussions on the previous/current status and reform of Chinese pension programs.

Table 7.1 Assumptions for cohort and period Total Fertility Rates (*TFR*) associated with medium and low fertility policy scenarios

	Medium fertility with two-child policy						Low fertility with current policy unchanged					
	Cohort TFR			Period TFR			Cohort TFR			Period TFR		
Year	Rural	Urban	Total	Rural	Urban	Total	Rural	Urban	Total	Rural	Urban	Total
2000	1.98	1.20	1.70	1.90	1.15	1.63	1.98	1.20	1.70	1.90	1.15	1.63
2015	2.27	1.80	2.05	2.09	1.67	1.89	1.98	1.20	1.61	1.98	1.20	1.61
2030	2.27	1.80	1.98	2.09	1.67	1.83	1.98	1.20	1.50	1.98	1.20	1.50
2035	2.27	1.80	1.96	2.27	1.80	1.96	1.98	1.20	1.47	1.98	1.20	1.47
2050	2.27	1.80	1.92	2.27	1.80	1.92	1.98	1.20	1.40	1.98	1.20	1.40
2080	2.27	1.80	1.85	2.27	1.80	1.85	1.98	1.20	1.28	1.98	1.20	1.28

current overall Chinese fertility policy is on average about 1.47 children per couple Guo et al. (2003). However, actual fertility significantly differs from fertility intended by policy. Under the current policy, couples are allowed to have two children if both parties are an only child, and in some provinces the second birth is allowed if at least one party is an only child. The number of such only child couples will significantly increase in due course, which may result in more second births even if the current policy is unchanged. Also, considering the opposite effect of rapid socioeconomic development on young people's fertility attitudes, which may lead couples to have fewer or even no children, it is assumed in the "low fertility" scenario that the period rural and urban TFRs in 2015 would be 1.98 and 1.2 (an increase by about 4.3 %, as compared to 2000) and remain constant thereafter (See Table 7.1).

The *medium fertility scenario* assumes a smooth transition period to around 2015 when all rural and urban couples in China would be universally allowed to choose to have a second child with appropriate spacing of 4–5 years. Such a spacing program should be implemented on a totally voluntary basis, with socioeconomic incentives but no coercion. Considering the much lower level of socioeconomic development in rural areas than in urban areas and the fact that fertility outcomes may differ significantly from policy, the cohort TFRs in and after 2015 are assumed to be 2.27 in rural areas and 1.8 in urban areas. Given that appropriate birth spacing in conjunction with rapid socioeconomic development would result in a delay of marriages and births, we assume that the age at first- and second- or higher-order births will increase by 0.75 and 1.5 years in 2030 as compared to 2015,[10] which constitutes an annual growth rate of 0.05 and 0.1 years, respectively, during the years 2015–2030. According to the method proposed by Bongaarts and Feeney (1998), the projected period TFR of the first- and second- or higher-order births in

[10] This assumption concerning the increase in the Chinese mean ages at births during the soft-landing period 2015–2030 is reasonable (or may be conservative) based on the fact that Chinese mean ages at first marriage, first and second births increased by 1.6, 1.6 and 3.1 years old in 2000 as compared to 1990.

Table 7.2 Life expectancies at birth under the medium and low mortality assumptions

	Medium mortality						Low mortality					
	Rural		Urban		Total		Rural		Urban		Total	
Year	Men	Women	Men	Women	Men	Women	Men	Women	Men	Women	Men	Women
2000	68.0	72.0	72.0	76.0	69.4	73.4	68.0	72.0	72.0	76.0	69.4	73.4
2050	73.5	77.5	77.0	81.0	76.1	80.1	79.4	85.6	81.9	89.1	81.3	88.2
2080	76.0	80.1	79.6	83.7	79.2	83.4	82.0	88.3	84.6	91.9	84.4	91.5

the years between 2015 and 2030 will be 5 % and 10 % lower than the cohort parity-specific TFR, due to fertility tempo effects. The period TFR of all parities combined in rural and urban areas in 2015–2030 will be 2.09 and 1.67 (in contrast to the cohort TFR 2.27 and 1.8), respectively. Around 2030–2035, we expect that China will "soft-land" to allow its citizens to freely choose family size and fertility timing without any policy restriction on reproduction; the period TFR will be the same as the cohort TFR in and after 2035 (see Table 7.1).

7.3.2.2 Medium and Low Mortality

The medium mortality scenario adopted in this study assumes that there will be rather slow progress in reducing mortality in China during this century – from a life expectancy of 71.4 years for both sexes combined in 2000, to 78.1 in 2050, and 81.3 in 2080, which is similar to the medium mortality assumed in population projections for China by the U.N. and most other scholars/agencies. This is rather conservative, given that the average life expectancy for men and women in Japan in 2009 was 83.0 years.[11] Some recent research indicates that there might be a significant improvement in mortality in the first half of this century because of biomedical breakthroughs and better personal health practices, such as healthy diets, smoking cessation and exercise, etc. (see, e.g., Shekell et al. 2005). Therefore, we made another more optimistic *low mortality scenario*, namely, life expectancy for both sexes combined projected to approach 84.8 in 2050, and 88 years in 2080 (see Table 7.2). This optimistic mortality scenario is subject to uncertainty, but we believe that it is not impossible.[12] Despite the uncertainty, the medium and low mortality scenarios bracket an informative range of possibilities in China during the decades of this century.

[11] Data released by Ministry of Health, Labor, and Welfare of Japan: http://www.mhlw.go.jp/english/database/db-hw/lifetb09/1.html—accessed 03/24/2011.

[12] The Chinese life expectancy at birth was 71.4 in 2000 and 73 in 2005. Assuming the same annual rate of increase as that in 2000–2005 continues into the future, the life expectancy at birth in China would be 87.4 years old in 2050 and 97 years old in 2080. Therefore, assuming a life expectancy at birth of 84.8 in 2050 and 88 in 2080 in China in the low mortality scenario may not be too optimistic.

7.3.2.3 Assumptions About the Retirement Age

As discussed earlier, unlike in the U.S. and some other Western countries where people have relatively more freedom to choose at what age to retire based on their health conditions and preferences, China effectively implements a government-determined compulsory age at retirement, which has been on average 60 for men and 52.2 for women. Note that there are compensative effects between government-allowed (or requested) later retirements for exceptionally skilful professionals and high-ranking officers and the early retirement scheme due to disability certified by a medical doctor. Publicly available representative data for estimating the actual average age at retirement are not available. Therefore, we simply assume that the average age at retirement is approximately equal to the compulsory retirement age.

We assume that the average age at retirement gradually increases from 60 for men and 52.2 for women in 2000 to age 65 for both men and women in 2050, and remains unchanged after 2050. The male and female retirement age between the years 2000 and 2050 are derived through liner interpolations. In another scenario for the comparative analysis, we assume that the current very low ages at retirement will remain unchanged: age 60 for men and 52.2 for women.

7.3.2.4 Assumptions on the Pension Program Parameters

According to published statistics, the overall average replacement rate ($B(t)$) in 2000 was 0.8789, the overall average contribution rate ($P(t)$) in 2000 was 0.1821, and the annual pension deficit rate as proportion of the total wage ($n(t)$) in 2000 was 0.069 (SSB 2001). Because reliable data to reasonably and consistently estimate retirement rates ($r(t)$) and pension program participation rates ($e(t)$) are not available due to the current complications of the Chinese pension schemes between various sectors as discussed earlier, we estimated the ratio of the retirement rate to the pension participation rate ($c(t)$), using Eq. 7.2 and the census data for $d_2(t)$, and officially published average $B(t)$, $P(t)$, and $n(t)$ in 2000. The estimated $c(t)$ in China in 2000 was 1.275, which indicates that the proportion of retirees who received pensions among the elderly persons was 27.5 % higher than the proportion of persons of labor force age who participated in the pension programs. This is mainly due to the fact that the majority of the urban elderly retirees are entitled and currently covered by the old DB pension plan supported by the state, according to the grandfather clause. However, a substantial portion of the urban young and middle-aged workers, who are not entitled to the old DB plan, did not participate in the newly established, mostly voluntary pension plan, which is a combination of DC and DB. In the Chinese rural areas, participation rates in pension program in 2000 for both old and young persons are extremely low.

We expect that the pension participation rate will increase faster than the retirement rate in China in the next few decades because people tend to be more concerned about their old age support in the context of rapid population aging and there will be a large reduction in the traditional family support system due to tremendously reduced fertility. Moreover, to face the serious challenges of population aging, the government is trying to increase the pension participation rate by further increasing the subsidies, especially in rural areas since 2009 as reviewed earlier; and may even pass a new law for compulsory participation in the state subsidized/managed pension programs in the future in both rural and urban areas. Therefore, in all of our scenarios, we assume that $c(t)$ will be reduced linearly from 1.275 in 2000 to 1.0 in 2040 and remain unchanged thereafter; this implies that the equilibrium of pension program participation between older and younger generations ($c(t) = 1.0$) would be reached in and after 2040 in all scenarios.

Given economic growth accompanied with rapid increases in wages, most of the Chinese experts in the field promote gradually reducing the current high replacement rate (0.8789 in 2000), which was defined in Sect. 7.2, to a goal of about 0.6, which is often regarded as the international standard[13] (e.g., He 1998, 2001; Zhang 2007). In early December 2005, the State Council of China released an official notice on "Consummating the Basic Old Age Insurance System for Enterprise Workers[14]" and set up an ultimate goal of a contribution rate of 0.28 (the employer and governmental subsidy contribute 0.20 and the employee contributes 0.08), which is intended to materialize gradually. Therefore, we assume that the replacement rate will be gradually reduced from 0.8789 in 2000 to 0.60 in 2040, and the contribution rate will be gradually increased from 0.1821 in 2000 to 0.28 in 2040, in all of the eight ($= 2 \times 2 \times 2$) scenarios with different combinations of assumptions on fertility, mortality, and retirement age.[15] The rates between 2000 and 2040 are estimated by linear interpolation. The rates after 2040 are assumed to be unchanged. These assumptions about the pension program policy parameters are subject to uncertainty, but imposing them serves well the main purpose of our illustrative application.

In all of the eight scenarios, rural–urban migration is included, assuming that the proportion of the population that is urban will increase from 36 % in 2000 to 61.7 % in 2030, 75 % in 2050, and 90 % in 2080. The net international migration is assumed to be zero in all scenarios, as it is currently negligible (relative to China's huge population size) and there are no reasonable data to predict the future Chinese net international migration.

[13] For example, the average replacement rate in countries of the Organization for Co-operation and Development (OECD) was 0.569 in 2005 (Whiteford and Whitehouse 2006) and 0.587 in 2007 (OECD Statistics 2007).

[14] See website "www.people.com.cn" for details; accessed March 18, 2011.

[15] In fact, we also tried two other sets of scenarios which assume that the anticipated goals of increasing the contribution rate to 0.28, reducing the replacement rate to 0.60, and reducing the $c(t)$ to 1.0 are reached in the year 2030 and 2050 (instead of 2040), respectively. The general patterns and conclusions of these scenarios, which are not presented here due to space limitations, are the same as the scenarios presented.

Table 7.3 Projected annual pension deficit rate as percentage of total wages under different scenarios of fertility, mortality, and retirement age

	Increasing retirement age				Constant retirement age			
	Medium mortality		Low mortality		Medium mortality		Low mortality	
Year	M. Fertility	L. Fertility	M. Fertility	L. Fertility	M. Fertility	L. Fertility	M. Fertility	L. Fertility
2000	6.9	6.9	6.9	6.9	6.9	6.9	6.9	6.9
2010	4.6	4.6	4.6	4.6	7.5	7.5	7.6	7.6
2020	1.9	1.9	2.2	2.2	11.2	11.2	11.5	11.5
2030	3.3	3.9	4.2	4.7	17.0	18.0	18.2	19.1
2040	−1.5	0.2	0.4	2.1	14.4	17.5	16.8	19.8
2050	−1.7	1.2	1.7	4.9	22	29.4	26.9	34.5
2060	2.5	9.2	7.6	15.1	22.3	35.9	28.9	43.7
2070	1.9	12.5	7.3	19.5	23.8	42.7	30.6	52.3
2080	3.7	17.1	9.0	25.1	25.4	49.1	32.3	59.9

7.3.3 The Results

7.3.3.1 Remarkable Impacts of Gradual Increase in Retirement Age

The results show that the annual pension deficit will be largely reduced in the scenarios of gradual increase in retirement age as compared to the scenarios of constant retirement age under different combinations of demographic regimes. In the scenario of the two-child policy and medium mortality with a gradual increase in the retirement age, the annual pension deficit rate will steadily decrease from 6.9 % in 2000 to 3.3 % in 2030, reach negative values of −1.5 % and −1.7 % (pension surplus) in 2040 and 2050, and remain at a very low level (2.5–3.7 %) with some fluctuations after 2050. In the other three scenarios of different combinations of fertility and mortality, gradual increases in retirement age will also lead to steady decreases in annual pension deficit, reaching a close-to-zero rate in 2040–2050, then increasing after 2050, up to 15.1–25.1 % in 2060–2080 under the low fertility (current rigid fertility policy unchanged) plus low mortality scenario, but still remaining at a relatively low level (7.6–9.0 % in 2060–2080) under the two-child policy and low mortality scenario (Table 7.3 and Fig. 7.1).

However, if China keeps its current very low retirement age unchanged, the annual pension deficit will steadily and quickly increase in all cases of different combinations of fertility and mortality levels, despite the assumed substantially increased contribution rate, reduced replacement rate, and equilibrium of pension program participation between older and young generations. For example, under the demographic regime of medium mortality and the two-child policy, a constant retirement age would lead to increasing annual pension deficit rate from 6.9 % in 2000 to 17.0 % in 2030, 22.0 % in 2050, and 25.4 % in 2080; this is in contrast to pension fund surpluses in 2040–2050 and pension deficit rate around 2–3 % thereafter in the scenario with a gradual increase in the retirement age with exactly the same demographic and pension program parameters. In the scenarios of constant retirement age with current fertility policy unchanged, the pension deficit rate would be

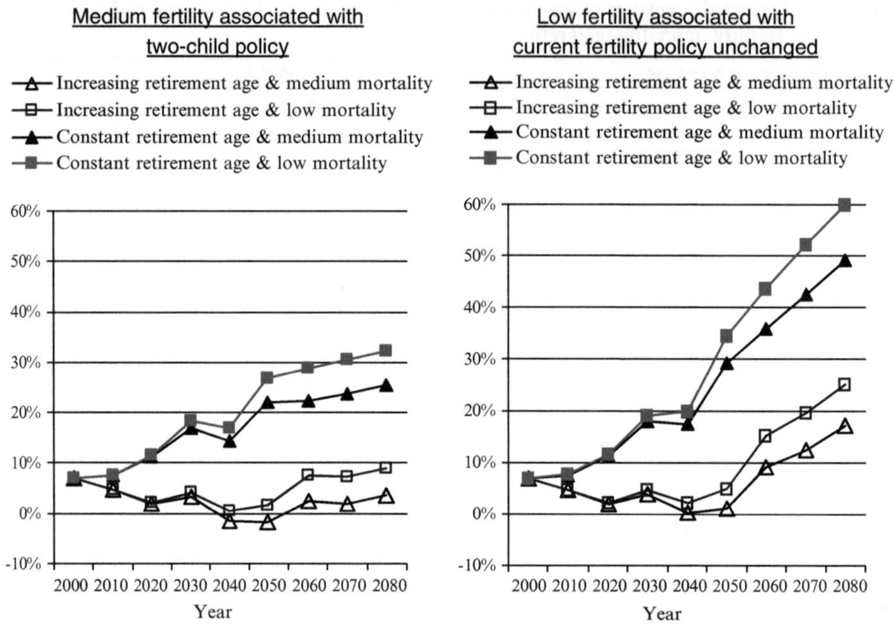

Fig. 7.1 A comparison of projected annual pension deficit rates as a percentage of total wages under different scenarios

even much higher: 18.0–19.1 % of total wages in 2030, 29.4–34.5 % in 2050, and 49.1–59.9 % in 2080 (Table 7.3 and Fig. 7.1). The pension deficit rate under the gradually increasing retirement age scenario would be lower than that under the constant retirement age scenario by 13.7–14.4 percentage points in 2030, 23.7–29.6 percentage points in 2050 and 21.7–34.8 percentage points in 2080 (see panel (III) of Table 7.4).

7.3.3.2 Large Impact of Possible Changes in Fertility Policy

Under either retirement age policy scenario, differences in pension deficit rate between the two-child policy and current fertility policy unchanged will become increasingly large after 2030. In the scenarios of medium or low mortality and gradually increasing retirement age, the annual pension deficit rate under the two-child policy scenario would be lower than that in the current fertility policy unchanged scenario by 0.5–0.6 percentage points in 2030, 2.9–3.2 percentage points in 2050, and 13.4–16.1 percentage points in 2080 (see panel (I) of Table 7.4). These percentage point absolute differences represent very large and quickly increasing relative differences: the pension deficit rate under the current fertility policy unchanged is higher than that under the two-child policy by 11.9–18.2 % in 2030, 170.6–188.2 % in 2050, and 178.9–362.2 % in 2080 (see Fig. 7.1). Under the assumptions of a constant retirement age and medium or low mortality, the pension

Table 7.4 Impacts of possible changes in fertility policy (Two-child policy vs. current policy unchanged), retirement age (increasing vs. constant), and mortality (low vs. medium) on pension deficits, as percentage point differences in projected annual pension deficit rate

	(I) Impacts of fertility policy: two-child policy vs. current policy unchanged				*(II) Impacts of mortality change*: low mortality vs. medium mortality				*(III) Impacts of retirement age change*: increasing retirement age vs. constant retirement age			
	Increasing retirement age		Constant retirement age		Increasing retirement age		Constant retirement age		Medium mortality		Low mortality	
Year	M. mort	L. mort	M. mort	L. mort	M. fert.	L. fert.	M. fert.	L. fert.	M. fert.	L. fert.	M. fert.	L. fert.
2000	0	0	0	0	0	0	0	0	0	0	0	0
2010	0	0	0	0	0	0	0.1	0.1	−2.9	−2.9	−3.0	−3.0
2020	0	0	0	0	0.3	0.3	0.3	0.3	−9.3	−9.3	−9.3	−9.3
2030	−0.6	−0.5	−1.0	−0.9	0.9	0.8	1.2	1.1	−13.7	−14.1	−14.0	−14.4
2040	−1.7	−1.7	−3.1	−3.0	1.9	1.9	2.4	2.3	−15.9	−17.3	−16.4	−17.7
2050	−2.9	−3.2	−7.4	−7.6	3.4	3.7	4.9	5.1	−23.7	−28.2	−25.2	−29.6
2060	−6.7	−7.5	−13.6	−14.8	5.1	5.9	6.6	7.8	−19.8	−26.7	−21.3	−28.6
2070	−10.6	−12.2	−18.9	−21.7	5.4	7.0	6.8	9.6	−21.9	−30.2	−23.3	−32.8
2080	−13.4	−16.1	−23.7	−27.6	5.3	8.0	6.9	10.8	−21.7	−32.0	−23.3	−34.8

deficit rate in the scenario of current fertility policy unchanged would be 18.0–19.1, 29.4–34.5 and 49.1–59.9 percentage points in 2030, 2050 and 2080, respectively (see Table 7.3), which are about 4.9–5.9 %, 28.3–33.6 % and 85.4–93.3 % higher than that under the two-child policy with everything else being equal (see Fig. 7.1).

It is clear that, as compared to the two-child policy, the negative and long-term impacts of keeping the current rigid fertility policy unchanged on pension deficits would be increasingly serious after 2030; the differences between the two fertility policy options are extremely large, especially after 2040 (see Fig. 7.1 and panel (I) of Table 7.4).

7.3.3.3 Impact of Possible Changes in Mortality

The estimates presented in Tables 7.3 and 7.4 and Fig. 7.1 show that the low mortality scenario would result in higher pension deficit rate as compared to that under the medium mortality scenario, with everything else being equal. The possible lower mortality may start to have a sizable impact after 2020, and may increase the pension deficit rate by 0.8–1.2, 3.4–5.1 and 5.3–10.8 percentage points in 2030, 2050, and 2080, respectively (see panel (II) of Table 7.4). In general, the impacts of possible changes in mortality on the future pension deficit are relatively less dramatic as compared to the impacts of possible changes in Chinese fertility and retirement age policies (Table 7.4).

7.4 Discussion

It is not surprising that this illustrative application shows that, everything else being equal, the annual pension deficit in the scenarios with medium and low fertility levels will not differ significantly before 2030 because it takes a couple decades for children born in the early part of the twenty-first century to reach working age and contribute to the pension program. However, under the assumptions of either increasing or constant retirement age and either medium or low mortality, the annual pension deficit after 2030 would be much more serious in the current fertility policy unchanged scenario than that in the two-child policy. From the point of view of pension fund balance, our analysis suggests that China needs to transition as soon as possible from the current rigid fertility policy to a universal two-child policy. This is an additional reason, on top of many other reasons such as potential shortages of family support resources and unbalanced gender structure between men and women of marriage ages, why China should have a fertility policy transition towards a two-child policy (Bongaarts and Greenhalgh 1985; Hesketh et al. 2005; Johnson 1994; Wang 2005; Zeng 2006, 2009; Zeng and Vaupel 1989). The key points in this chapter which go beyond what was discussed in the existing literature are that the magnitudes and timing of the impacts of all possible combinations of two-child policy, current fertility policy unchanged, increasing and constant retirement age, and medium and low mortality on future pension deficits are analyzed innovatively with a comparative approach (see Tables 7.3 and 7.4 and Fig. 7.1).

Note that a premise supporting the impact of fertility policy transition is that relaxing the current rigid fertility policy would translate into an increase in the rural–urban combined TFR from 1.63 in 2000 to 1.89–1.83 in 2015–2030 with an universal two-child plus an appropriate and voluntary spacing policy, and to 1.96–1.92 in 2035–2050 without any policy restriction on reproduction. One may question whether such a policy transition will result in the assumed TFR increase compared to the TFR of neighboring regions/countries such as Hong Kong, Taiwan, Japan, and Korea, which has been well below 1.8 for many years despite the lack of restrictive policies on childbearing. However, China is much more diverse than Hong Kong, Taiwan, Japan, and Korea, which are highly industrialized; about half of the total population of China still currently lives in mostly poor rural areas where the reproductive potential is relatively high once the rigid birth control policy restriction is relaxed.

The results of this study demonstrate that if the average age at retirement in China gradually increases from its very low level to the current international standard (age 65 for both men and women) in 2050, a rather conservative assumption, the annual pension deficit would be largely reduced or eliminated up to the middle of this century, under various demographic regimes. The currently very low age of retirement in China actually provides a good opportunity to face the challenges of population aging, because a gradual increase in the retirement age plus a smooth transition to the two-child policy will help China avoid the serious problem of annual pension deficit, as shown by the analysis presented in this chapter. Clearly, the excuse that future serious pension deficit is unavoidable, used by some policy makers and scholars to support the stagnation and decline of the rural pension program in the period of 1999–2008 (as reviewed earlier), was not supported in this study.

Moreover, working until age 65 would help individuals to continue to lead active and healthy lives. Society would also benefit from older adults' experience and skills. On the other hand, many policy makers and scholars in China argue that delaying the age at retirement might reduce job opportunities for the younger individuals, and thus they strongly oppose increasing the compulsory age at retirement. However, their arguments may be incorrect because the problems of young people's job opportunities can be resolved or reduced by developing more labor-intensive and job-creating businesses, especially the service industries (currently under-developed in China), and by prolonging and supporting the education and professional training period of young people. Another innovative new policy option would be to allow everyone (both young and old) to work fewer hours per week but more years over the whole course of life Vaupel (2010). This policy may not only play a critical role in resolving the conflicts between healthy elderly and young people's job opportunities, but also be useful in improving health. Fewer hours spent working per week would reduce the stress of a heavy workload and give people more spare time. Such an innovative new policy may also stimulate service industries of tourism and other leisure/social activities which will create more jobs and develop the economy, and perhaps increase reproduction as well.

7.5 Concluding Remarks

This chapter presents a simple approach associated with our Profamy model and software for projecting the annual pension deficit rate based on demographic measurements of fertility, mortality, and migrations, retirement age, and a few measurable and predictable pension program parameters. The simple method can also be used to estimate the replacement rate and contribution rate required to achieve a zero annual pension deficit rate in the projection years given the expected demographic conditions, while assuming an equal or a specified differential relative change in contributions/benefits of workers and retirees (see the last part of Appendix 1).

It is easy to understand that changes in fertility, mortality, and retirement age would impact the future pension deficit rate. What is interesting and useful to policy analysis is that the simple method for projecting future pension deficit rate proposed and illustrated in this chapter can be used to predict when the impacts of possible changes in the demographic parameters and retirement age on pension deficit would become apparent, and what would be the sizes of the impacts under various assumptions. Our simple method can also address the relative magnitude of the impacts due to various combinations of changes in the demographic parameters and retirement age. For example, the results presented in Table 7.4 indicate that the impacts of a gradual increase in the retirement age on future pension deficit in China would start much earlier and be much larger compared to impacts of changes in fertility policy and a faster mortality decline. Under either increasing or constant retirement age scenarios, a faster mortality decline starts affecting the pension deficit earlier than an increase in fertility, and the magnitude of the impact of declining mortality on the pension balance is larger through 2030 and more or less the same in 2040–2050, as compared to the impacts of adopting the two-child policy; but the changes in fertility policy have a much greater impact than that due to mortality change after 2050 (see Table 7.4).

Note that in the illustrative application we assumed zero international migration, fixed linearly changing trajectories to reach the targets of anticipated replacement rate, contribution rate and equilibrium of the pension program participation rates between older and younger generations, while altering the parameters of fertility, mortality and average age at retirement in different scenarios. This design is useful in the current Chinese context to investigate the impacts of possible changes in current fertility and retirement age policies and potentially faster mortality decline on the future pension deficit, which is the purpose of the illustrative analysis of this study. However, for different research purposes in the context of other countries, one may predict or assume the size and age/gender structure of international migration and/or pension program parameters (*P(t), B(t), r(t) and e(t)* or *c(t)*), while fixing the fertility, mortality and retirement age parameters, to explore the impacts of alternative policies concerning international migration and/or pension program reform on future pension deficits rate.

As compared to the actuarial approach, the analysis presented in this chapter is relatively simple and requires only commonly available demographic data and a few measurable and predictable pension program policy parameters. This is indeed useful for straightforward overall annual pension deficit projections or simulations which inform policy analysis as a "sentinel" warning system for governments that manage the pension programs. It is especially applicable and useful when detailed age-sex-sector-specific data on labor force participation rates, unemployment rates, earnings profiles, contributors and pensioners, years of employment at retirement, etc., which are required for predicting pension deficits using actuarial models, are not available or not accessible. However, it must be emphasized that investigations using the simple method and limited available data in this article are a "what if" type of demographic and policy analysis exercise rather than any kind of accurate pension revenue and expenditure forecasting. Such analyses cannot be used to distinguish the differentials in input and output of the pension systems between DB and DC schemes, between rural and urban sectors, and between various groups classified by age, gender, and other socioeconomic characteristics. It cannot be used to project the details of a pension system's expenditure and revenues. If an analyst is interested in investigating such details and the required detailed data are available, actuarial models such as the PROST should be employed.

Appendix 1: Derivation of the Simple Method

Let $W(t)$ denote the total number of workers in year t; the term "workers" here refers to those who reside/work in rural or urban areas and participate in the DB or DC pension program.

$R(t)$, the total number of retirees in year t; the term "retirees" here refers to those who are retired and receive pension benefits, regardless of whether it is a DB or DC program and rural or urban residence.

$d(t)$, the retiree-worker ratio in year t, namely, the ratio of total number of retirees to total number of workers in year *t*, $d(t) = \frac{R(t)}{W(t)}$

A(t), the average wage per worker in year *t*;

P(*t*) (the contribution rate in year *t*) and *B*(*t*) (the replacement rate in year *t*) are as defined in the text.

If the total amount of the pension fund premium contributions by workers and employers is equal to the total pension payment to the retirees in year *t*, the following equation holds: $P(t)[A(t)W(t)] = [B(t)A(t)]R(t)$,

Dividing both sides by *A(t) W(t)*, we get:

$$P(t) = B(t) \cdot d(t) \tag{7.3}$$

The above equation is the classic basic equilibrium equation which expresses the annual balanced pension funds (e.g., Becker and Paltsev 2001: 19; Hamayon and Legros 2001; Sin 2005). If *d(t)* could be reasonably projected into the future years, one could simply use the classic basic equilibrium equation to project the annual pension deficits. However, it is extremely difficult to do so, because *d(t)*, the retiree-worker ratio in year *t*, mixes the impacts of demographic parameters (such as fertility, mortality, and migration) and changes in the prevalence of pension program coverage among the elderly and participation among workers. Therefore, Zeng (2011) was motivated to decompose *d(t)* into a couple of demographic parameters and pension program variables that are reasonably predictable.

Let *n(t)* denote the annual pension deficit rate, which is defined and discussed in the text. The following equation holds in any case of pension fund deficit, balance or surplus:

$$P(t)\ [A(t)\ W(t)] + n(t)\ A(t)\ W(t) = [B(t)\ A(t)]\ R(t)$$

Dividing both sides of the above equation by *A(t) W(t)*, we get:

$$\begin{aligned} P(t) + n(t) &= B(t) \cdot d(t) \\ n(t) &= B(t)\ d(t) - P(t) \end{aligned} \tag{7.4}$$

We may rewrite Eq. 7.4 as:

$n(t) = B(t)d_2(t)\frac{d(t)}{d_2(t)} - P(t)$, where $d_2(t)$ is the dependency ratio of elderly (defined in the text) which is easily predictable as it can be derived from commonly available population forecasting based on parameters of fertility, mortality, and migration.

$$\frac{d(t)}{d_2(t)} = \frac{\text{total number of retirees/total number of workers}}{\text{total number of persons over average age at retirement/total number of persons of working age}}$$

We rearrange the components of $\frac{d(t)}{d_2(t)}$ to make it easier for interpretation,

$$\frac{d(t)}{d_2(t)} = \frac{\text{total \# of retirees/total \# of persons over average age at retirement}}{\text{total \# of workers who participate in the pension program/total \# of persons of working age}}$$

the numerator and denominator of the right side of the above equation is *r(t)*, the retirement rate, and *e(t)*, the pension program participation rate (*r(t)* and *e(t)* are defined in the text). Thus,

$$n(t) = B(t) \cdot d_2(t)\frac{r(t)}{e(t)} - P(t) \tag{7.5}$$

As discussed in the text, when reliable data for estimating/projecting *r*(*t*) *and e*(*t*) separately are not available, one may simply project (or assume) the *c(t)* (= *r(t)*/*e* (*t*)). Thus,

$$n(t) = B(t)d_2(t)c(t) - P(t) \tag{7.6}$$

Based on Eq. 7.5 or Eq. 7.6, one may investigate another interesting policy question: How should the contribution rate *P*(*t*), replacement rate *B*(*t*), and/or average age at retirement be set so that the annual pension deficit rate is zero in future years? In such a policy analysis exercise, one may formulate a set of simultaneous equations: one equation is Eq. 7.5 or Eq. 7.6 with *n(t)* being set to zero; another one or two equations present the constraints assumed by the analyst, such as simultaneously adjusting the replacement rate and the contribution rate in opposite directions, while ensuring that relative changes in wages for workers and retirement benefits for retirees are the same or different, depending on either DB or DC participants.[16] The estimators for the adjustment indices to adjust the replacement rate and the contribution rate can be obtained by resolving the simultaneous equations, or numerical trailing/simulation.

While *n(t)* (as percent of the total wages) is a valid indicator of annual pension deficit, one may go one step further to estimate *m(t)*, the index of annual pension deficit as a percentage of GDP. One can compute *m(t)* by multiplying *n(t)* by *S(t)*, the total wages as a percentage of GDP, which is relatively easy to predict, and the time series data are usually available.

$$m(t) = n(t)S(t) = \left[B(t)d_2(t)\frac{r(t)}{e(t)} - P(t)\right]S(t) \tag{7.7}$$

Or

$$m(t) = n(t)S(t) = [B(t)d_2(t)c(t) - P(t)]S(t) \tag{7.8}$$

[16] For example, if the DC program is not substantially subsidized by the state, one wouldn't be free simply to adjust every retiree's benefits, because the benefits of the retirees who participate in DC program are promised by the program according to their contributions in the past. In this case, the government may have to mainly adjust the benefits for those retirees who are eligible for the DB program, but adjust less or do not adjust the benefits of the DC retirees.

Part II
Applications in the United States

Chapter 8
U.S. Family Household Momentum and Dynamics: Projections at the National Level

8.1 Data and Estimates

General data requirements and U.S. data sources for household projections using the ProFamy method are presented and listed in Table 3.1 in Chap. 3. Because race classifications are included for projections in the U.S. application, the age-specific standard schedules described in (a) through (e) of section (2) of Table 3.1 and the demographic summary measures described in section (3) of Table 3.1 are all race-specific. We follow the Census Bureau's most recent classification to distinguish four racial/ethnic groups in the U.S. household projections: (1) White non-Hispanic, (2) Black non-Hispanic, (3) Hispanic, and (4) Asian and other non-Hispanic (Hollmann et al. 2000).[1]

The standard schedules of mortality, net migration, and summary demographic measures (TFR, life expectancy at birth, and number of net migrants) are taken from the Bureau of the Census 1999–2100 population projection (Hollmann et al. 2000). The base population and standard schedules of net rate of leaving the parental home are derived from census micro data files – all are very straightforward (see Sect. 3.2 in Chap. 3). General rates of marriage/union formation and dissolution are derived from pooled survey data in combination with vital statistics (see Sect. 3.3 and Appendix 4 in Chap. 3). The standard schedules of race-sex-age-specific occurrence/exposure (o/e) rates of marriage/union formation and dissolution and race-age-parity specific o/e rates of marital and non-marital fertility are estimated based on pooled survey data – these deserve special attention.

Previous empirical research has shown that combining data from multiple surveys can provide enhanced estimates by increasing the sample size, extending coverage, and increasing the accuracy of measurements derived for smaller population groups and smaller areas (see Sect. 3.3 of Chap. 3). With this rationale and

[1] We do not consider multiple or mixed races in this study. If a person reported his or her race as White or Black plus another race in the 2000 census, we consider him or her as White or Black only.

Y. Zeng et al., *Household and Living Arrangement Projections*, The Springer Series on Demographic Methods and Population Analysis 36,
DOI 10.1007/978-90-481-8906-9_8,

justification, we estimated race-sex-age-specific o/e rates of marital/union status transitions and age-parity-marital/union status-specific o/e rates of fertility needed for U.S. household projections based on retrospective event history data from the following four national surveys (Zeng et al. 2012b):

(a) National Survey of Families and Households (NSFH) conducted in 1987–1988, 1992–1994, and 2002;
(b) National Survey of Family Growth (NSFG) conducted in 1983, 1988, 1995, and 2002;
(c) Current Population Surveys (CPS) conducted in 1980, 1985, 1990, and 1995;
(d) Survey of Income and Program Participation (SIPP) conducted in 1996.

There are, in total, 97,778 men (aged 15–95) and 304,536 women (aged 15–98) in the pooled sample. Note three points about this pooled sample. First, our pooled dataset reduces problems of small sample sizes for minority racial/ethnic groups. For estimates without race or for the majority group only, one large sample survey data set is sufficient. However, when the race-sex-age-status-specific o/e rates are estimated for different race groups, the sub-sample sizes for the minority groups are likely too small if estimated from only one survey. This problem is particularly serious for male minorities. Second, vital registration (VR) data, which have the advantage of large sample size, are not an option given our focus on union regimes, including cohabitation, and racial differentials. Specifically, VR numerators are obtained from marriage, divorce, and birth registrations, whose design can vary from state to state; denominators are obtained from the census and population projections. The VR and the census forms often do not ask questions in an identical manner. Thus, the race-specific numerators and denominators used in computing the age-specific rates by race are not fully compatible. Morgan et al. (1999) presented evidence that the race-specific fertility estimates based on VR data can be seriously flawed. The pooled retrospective data obtained from the NSFH, NSFG, CPS and SIPP surveys do not have such inconsistencies because numerators and denominators within surveys are calculated using precisely the same definitions of racial/ethnic categories. Moreover, the VR data do not contain cohabitation information, which is of major interest in U.S. projections. Third, the concepts and definitions of age, sex, race, marital/union status, births, and dates of marital/union status changes, the only measurements needed to estimate the race-sex-age-specific o/e rates, are similar in the NSFH, NSFG, CPS and SIPP surveys.

The CPS, SIPP, NSFH, and NSFG datasets all contain detailed event histories of marriage formations and dissolutions that provide large sample sizes for reliable estimates of race-sex-age-specific o/e rates for first marriage, divorce, and remarriage of divorced and widowed persons. Information on the timing and age of current and previous cohabitations was only collected in the NSFH and NSFG; thus, sample sizes for these o/e rates are much smaller in comparison. To increase comparability and reliability of the estimates using all available data on marital/union status transitions, we employed a straightforward demographic estimation procedure (presented in Appendix 1) to adjust the race-sex-age-specific o/e rates of marital/union (including cohabitation) status transitions based on the NSFH and NSFG data. The adjustments make these rates consistent with the race-sex-age-specific o/e rates of first marriage, divorce, and remarriage of divorced and widowed persons based on all of the data from all four surveys (Zeng et al. 2012b).

The survey data used in this study include marriage histories for the most recent, first, and second marriages (up to three marriages for each respondent). This limitation is not problematic because the number of respondents with more than three marriages is very small.

The period sex-age-specific occurrence/exposure (o/e) rate is defined as the number of events that occurred (occurrence) divided by the number of person-years lived at risk of experiencing the event (exposure). We employed the method of event history analysis (Allison 1995) to estimate the race-sex-age-specific o/e rates of marriage/union formation and dissolution and age-parity-marital/union status-specific o/e rates of fertility for the four racial/ethnic groups in the 1990s. In our event history analysis models, age (in 5-year groups) and race are treated as covariates in each of the events. To account for the sampling design so that each survey could maintain the representativeness of its target population, the original sampling weights of each survey are applied on relevant observations in the pooled dataset. We have examined all sets of the o/e rates by gender, race/ethnic group, and the relevant statuses. The smoothness and plausibility of these estimates are confirmed by graphical analyses (not shown).

Based on estimates of the sex-age-specific and marital status (never-married, married, widowed, divorced) specific death rates for all races combined in the 1990s used in Schoen and Standish (2001) and the race-sex-age-specific death rates released by the NCHS (Arias 2004), we estimated race-sex-age-marital status-specific death rates in the 1990s. Given that death rates for cohabiting persons are not available and the literature considers cohabitation as mostly a transitional stage before marriage in the U.S. (e.g., Goldstein and Kenney 2001), we assume that the race-age-specific death rates of never-married and cohabiting men/women are equal to the average of the corresponding death rates of never-married and married men/women; the race-age-specific death rates of widowed and cohabiting men/women are equal to the average of the corresponding death rates of widowed and married men/women; and the race-age-specific death rates of divorced and cohabiting men/women are equal to the average of the corresponding death rates of divorced and married men/women.

To evaluate and validate our approach of pooling the relevant data from the four national surveys, we compare the summary measures from multistate marital status life tables (without information on cohabitation) for all races combined as presented by Schoen and Standish (2001) with our estimates of the corresponding measures for all races combined (excluding cohabitation) based on the pooled data. Comparing our estimates with theirs, the lifetime proportions of first marriage, divorce, and remarriage are generally consistent – among eight pairs of summary measures, four have discrepancy rates of 1–2 %, and four have discrepancy rates of 7–10 % (see Appendix 2). Furthermore, as demonstrated in Zeng et al. (2012b), the levels, trends and racial differentials derived from multistate life table analysis based on the pooled survey data are generally consistent with other previous studies using totally different approaches. Thus, we are confident in using our 1990–1996 estimates of race-sex-age-specific o/e rates of marriage/union formation and dissolution, and race-age-parity specific o/e rates of births for married, cohabiting, and not-married/not-cohabiting women, as standard schedules for the U.S. household projection by race.[2]

8.2 Medium Projections

Our medium projections use the time-varying race-specific medium mortality (e_0), medium fertility (TFR), and medium international net migration projections adopted by the Census Bureau's population projection (Hollmann et al. 2000).

The race-specific general rates of marriage, divorce, cohabitation, and cohabitation-union dissolution, and the race-specific mean age at first marriage and at births (of all orders combined), are derived based on observations in 1990–1996 from vital statistics and pooled survey data (see Sect. 8.1) and are assumed to be constant from 2001 to 2050 (see Appendix 3). The race-sex specific proportions of people eventually leaving the parental home, race-sex-age-specific proportions of people living in institutional households, and race-sex-age-marital-status specific proportions of elderly living with children are derived from the 2000 census micro file, and are also assumed to remain unchanged.

Note that one common approach in population projection is to hold some of the current demographic rates constant throughout the projection horizon (e.g., Day 1996; Treadway 1997). Smith et al. (2001: 83–84) argued that holding some of the rates and proportions constant in the demographic projections can be justified on either of two grounds. One is that future rates and proportions are not likely to differ much from the current level. Another justification for holding the rates and proportions constant is the belief that neither the direction nor the magnitude of future changes can be predicted accurately. The argument here is not so much that the current rates will remain constant, but rather that scientific theories and past history do not provide a reliable basis for predicting how those rates will change. If upward or downward movements are equally likely, the current rates provide a reasonable forecast of future rates. In addition, we have calculated low and high bounds of the rates and medium household projections (to be discussed in Sect. 8.3); these projections, together with the medium variant, yield an array of possible changes in the rates and households in the

[2] Using the pooled survey data, we tried to estimate the race-age-specific o/e rates of marriage, union break, and fertility by parity for never-married and cohabiting, widowed and cohabiting, and divorced and cohabiting persons separately. But the results were not satisfactory due to problems of subsample sizes for minority groups that are too small. We thus combine relevant data and use the same race-age-specific o/e rates of marriage, union break, and fertility by parity for three different kinds of cohabiting people. The race-age-specific o/e rates of cohabitation union formation and fertility by parity for never-married, widowed, and divorced persons were estimated separately with general satisfaction.

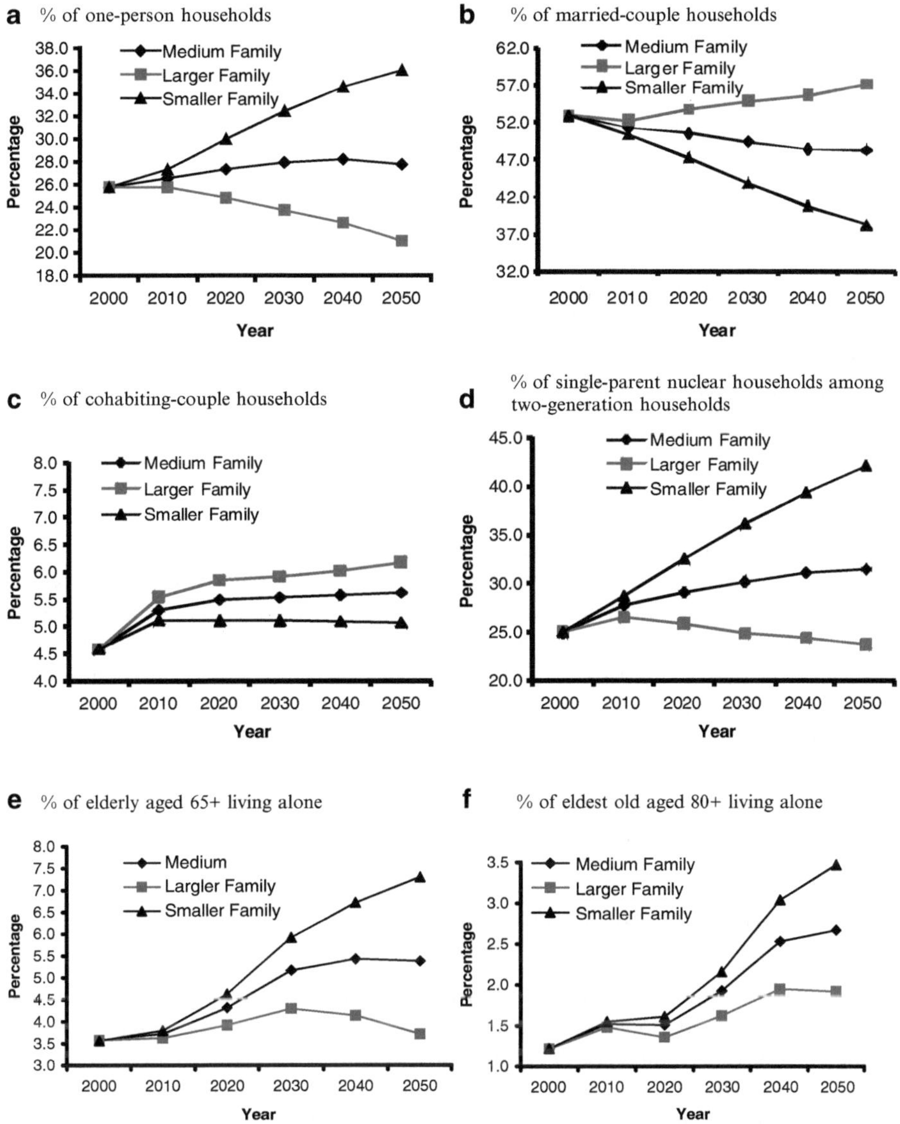

Fig. 8.1 Medium, low, and high bounds of household and living arrangement projection

future. This facilitates an assessment of the impacts of any assumed constant rates and proportions.

Under the medium projections, as shown in Fig. 8.1, the average household size would decrease from 2.59 in 2000 to 2.41 in 2020, and remain stable afterwards; one-person and two-person households would increase from 25.8 % and 32.5 % in 2000 to 28.5 % and 35.6 % in 2020, and to 29.2 % and 35.1 % in 2050. Husband-wife households would decrease from 54.0 % in 2000 to 49.9 % in 2020, a relative decline of 9.1 %, and become 47.6 % in 2050. Cohabiting-couple households would increase from 4.7 % in 2000 to 5.4 % in 2020, a relative increase of 14.9 %, and become 5.5 % in 2050. Among two-generation households, single-parent households would increase from 24.6 % in 2000 to 28.6 % in 2020, and to 31.1 % in 2050.

The medium projections also indicate that there would be 14.6 and 22.7 million elderly aged 65 and older living alone in the years 2020 and 2050 respectively, in contrast to 10.1 million in 2000. The proportion of elderly aged 65+ living alone within the total population will increase by 23.2 and 53.4 % in 2020 and 2050, respectively, as compared with 2000. The number of oldest-old persons aged 80+ living alone would be 3.4, 5.1, and 11.3 million in 2000, 2020, and 2050, respectively. Although the age-sex specific proportions of the elderly living in institutions are assumed to remain unchanged, the number of those elders aged 65+ living in institutions in 2050 would be 2.5 times as large as in 2000.

The large increase in the number and percentage of the elderly living alone among the total population is due to the mixed effects of the increase in the proportion of the elderly population in general and changes in marital status and living arrangements across cohorts and periods. Changes in the relative percentage distributions of marital status and living arrangement within age groups of the elderly primarily reflect changes in marriage/union formation and dissolution across cohorts and periods. The percentage of those who are divorced and living alone among the elderly aged 65+ would be 4.8, 8.9, and 10.5 in 2000, 2020, and 2050; the percentage of those who are cohabiting with a partner among the elderly aged 65+ would be 1.3, 3.5, and 3.7 in 2000, 2020, and 2050, respectively. The percentage of the elderly who are widowed and not cohabiting is projected to decrease steadily, mainly because more widowed elderly will tend to cohabit with partners.

8.3 Family Household Momentum

As summarized above and depicted in Fig. 8.1, under the medium projections, the proportional distributions of household types and sizes and elderly living arrangements will change considerably from 2000 to 2020 and remain more or less stable after 2020; the percentage of the oldest-old living alone will continue to increase substantially after 2020. As discussed earlier, however, the medium projections assume that from 2000 to 2050, general rates of marriage/union formation and dissolution will remain constant. Even under a constant scenario with everything (marriage union formation and dissolution, fertility, mortality, migration, etc.) after 2000 assumed to remain the same as in 2000, the proportional distributions of household types/size and living arrangements of the elderly will change considerably until 2020 or so and remain stable afterwards. Why would distributions of households and elderly living arrangements change considerably from 2000 to 2020 while the demographic parameters remain constant in the same period? Our explanation is that family household momentum plays an important role.

Cohorts who were younger in 2000 experienced, and will continue to experience, stabilized (or constant) higher rates of marriage/union disruption and lower rates of marriage/union formation than cohorts who were older in 2000 and had already completed most of their family life course. Profiles of households and elderly living arrangements in 2000 represent the mixed cumulative life course experiences of younger and older cohorts in the past few decades. Although the marriage/union formation and dissolution rates are assumed to remain constant during the period 2000–2050, the distributions of households and elderly living arrangements would change considerably because older cohorts, who had more traditional family patterns, will be replaced by younger cohorts with more current family patterns. Such family household momentum is similar to the well-known population momentum (Keyfitz 1971), in which population size could continue to increase after fertility is equal to or even below the replacement level. The medium projections using the ProFamy method/program and constant rate scenario provide both a tool to investigate and empirical evidence of family household momentum.

8.4 Low and High Bounds of Household and Living Arrangement Projections

We next explore low and high bounds for the projection of household and living arrangements by examining smaller and larger family scenarios. The smaller family scenario assumes that, as compared to the medium projections, the general rate of divorce and general rate of cohabiting-union dissolution are higher by 15 % in 2020 and 25 % in 2050 and that the general rates of marriage and of cohabitation are lower by 15 % in 2020 and 25 % in 2050. General rates between 2000, 2020, and 2050 are derived through linear interpolation. This scenario employs the low fertility, low mortality, and low international net migration assumptions adopted in the Census Bureau's latest population projection (Hollmann et al. 2000). The smaller family scenario assumes increasing marriage/union dissolution, decreasing marriage/union formation, decreasing fertility and mortality,[3] and receipt of fewer international immigrants. We expect that such a combination of demographic rates will result in the low bounds for household size and percentages of married- or cohabiting-couple households and high bounds for percentages of one-person households, single-parent households, and so on.

[3] Low mortality may (1) reduce the U.S. average household size through increasing number of elderly households that are mostly small (one or two persons); and (2) increase the size of some households by increasing the survivorship of adults and children in these larger households. The effects of (2) may be smaller than those of (1) because a further decrease in adult and child mortality in the U.S. is limited, but the prolongation of elderly life span may have larger relative impact.

The larger family projection assumes that, as compared to the medium projections, the general rates of divorce and of cohabiting-union dissolution are lower by 15 % in 2020 and 25 % in 2050 and general rates of marriage and cohabitation are higher by 15 % in 2020 and 25 % in 2050. General rates between 2000, 2020, and 2050 are derived through linear interpolation. This scenario employs the high fertility, high mortality, and high net international migration assumptions adopted by the Census Bureau's latest population projection. The larger family scenario assumes that the family will regain its traditional values with decreasing marriage/union dissolution, increasing marriage/union formation, and increasing fertility, accompanied by high mortality and a larger number of international immigrants. The combination of demographic rates in the larger family scenario may result in high bounds for household size and percentage of married- or cohabiting-couple households, and low bounds for percentages of one-person households, single-parent households, and so on.

The assumptions that there will be 15 % and 25 % increases (or decreases) in general rates of marriage, divorce, cohabitation, and union dissolution in 2020 and 2050 constitute educated guesses about the largest possible changes in marriage/union formation and dissolution in the next few decades. Although we made these guesses with reference to time series data of the general rates from 1970 to 2002, they are largely arbitrary because of uncertainties about future trends. Nevertheless, similar to conventional deterministic population projections of low and high variants that formulate the possible bounds of population growth, our smaller and larger family scenarios formulate the possible low and high bounds of future household and living arrangement distributions.

As shown in Fig. 8.1a, the average U.S. household sizes are projected to be 2.3–2.5 and 2.2–2.7 in 2020 and 2050 respectively. The projections of the percentage of one-person households have ranges of 26.0–31.1 and 22.5–37.3 in 2020 and 2050 (Fig. 8.1b). In 2020, the projected percentages of married-couple and cohabiting-couple households would be 46.8–53.1 and 5.5–6.4, respectively; the corresponding figures in 2050 would be 38.1–55.9 and 5.3–6.8, respectively (Fig. 8.1c, d). The possible range of the percentage of single-parent households among two-generation households would be quite large: 25.4–31.9 in 2020 and 23.8–41.0 in 2050 (Fig. 8.1e).

Figure 8.1f, g show that the ranges of the projected percentage of elderly aged 65 + living alone among the total population are 4.1–4.5 in 2020 and 4.2–6.8 in 2050; the ranges of the projected percentage of oldest-old aged 80+ living alone among the total population are 1.4–1.6 in 2020 and 2.2–3.2 in 2050. It is clear that the effects of demographic rates on future elderly living arrangements are substantial.

Projected numbers of households in future years are practically useful in market trend analysis and socioeconomic planning for the consumption of housing, energy, automobile, and other household related goods and services. Table 8.1 shows the projected possible ranges of the number of households by type as well as the total projected numbers of elderly living alone, based on the smaller and larger family scenarios. We present only the summary results in this chapter due to space limitations. As discussed in Sect. 4.2.5 of Chap. 4, however, the number of household and living arrangement projections are classified by household type/size, race, age,

Table 8.1 Projected possible ranges of the number of households by type as well as total number of the elderly living alone (unit: millions)

	Number of households					Elderly living alone	
Year	Total	One-person	Single-parent	Cohabiting-couple	Married-couple	Ages 65+	Ages 80+
2000	105.2	27.1	11.3	5.2	57.8	3.6	1.2
2010	120.9–122.3	32.6–34.2	13.2–13.7	6.6–7.3	61.0–64.0	3.7–3.8	1.5–1.5
2020	133.0–137.4	35.7–41.4	14.2–15.4	7.2–8.7	62.3–73.0	4.1–4.5	1.4–1.6
2030	142.8–153.2	38.3–48.1	15.8–17.1	7.7–10.0	61.9–82.6	4.7–5.7	1.8–2.1
2040	149.5–171.7	41.2–53.6	18.1–18.5	8.1–11.5	60.4–93.6	4.6–6.3	2.2–2.8
2050	152.8–192.0	43.2–57.0	19.6–20.5	8.1–13.1	58.2–107.3	4.2–6.8	2.2–3.2

sex, and marital status in the ProFamy household projection output files, which can be used for the purposes of market analysis and socioeconomic planning.

8.5 Racial Differentials in Dynamics of Households and Living Arrangements

Figure 8.2 presents racial/ethnic differentials in the projected dynamics of households and living arrangements from 2000 to 2050 under medium projections. Figure 8.2a shows that White non-Hispanics have the smallest average household size; Hispanics and Asian/other non-Hispanics have much larger average household size than White and Black categories throughout the first half of the century. Black average household size is considerably larger than the White average up to 2030, but the difference becomes smaller by 2050. Black non-Hispanics and White non-Hispanics have much higher percentages of one-person households than the Hispanics and the Asian/other non-Hispanics, and the difference tends to grow larger with time – the Black and White percentages will be about twice as large as those of Hispanics and Asian/other non-Hispanics in 2040–2050. The Black and White percentages of one-person households are pretty close to each other up to 2010, with the Black percentage becoming larger than the White afterwards.

The Black non-Hispanic population category has the lowest percentage of married-couple households (either with or without children) throughout the first half of this century; the White non-Hispanic category has the second lowest and the Asian/other non-Hispanic category has the highest, showing very large racial differentials in the percentage of married-couple households (see Fig. 8.2c). Figure 8.2d shows that the proportion of cohabiting-couple households (either with or without children) is the highest within the Hispanic population category, and the lowest among White non-Hispanics. Throughout the first half of this century, the Black non-Hispanic percentage of single-parent households among two-generation households is projected to be dramatically higher than that of any other race groups (see Fig. 8.2e).

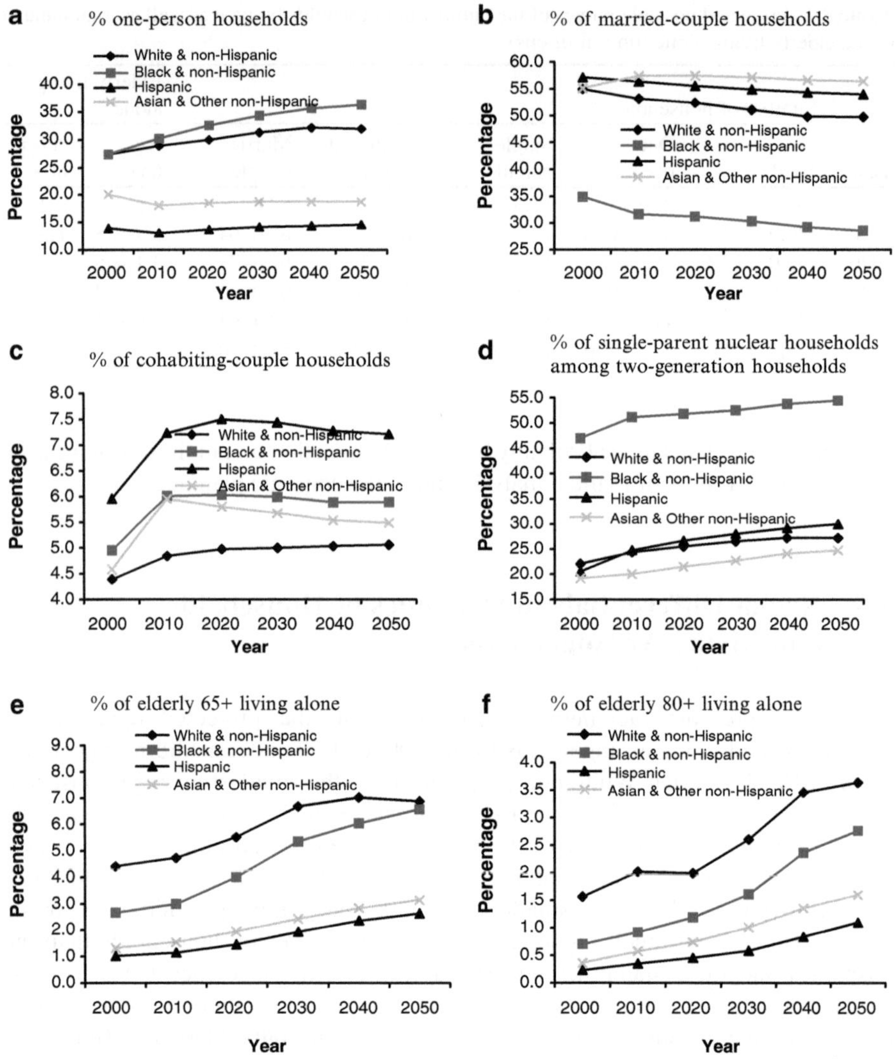

Fig. 8.2 Racial differentials of households and living arrangements based on the medium projection

The White non-Hispanic group is projected to have much higher percentages of elderly aged 65+ living alone and oldest-old aged 80+ living alone than any other racial/ethnic group; the percentage of elderly (including the oldest-old) living alone among Black non-Hispanics is substantially lower than that among the White non-Hispanics, but much higher than that among Hispanic and Asian/other non-Hispanics.

8.6 Concluding Remarks

The foregoing U.S. household projection scenarios have shown that, even assuming constant demographic rates, the distributions of households and elderly living arrangements will continue to change considerably in the next couple of decades. We have named this demographic trend "family household momentum." It will occur because older cohorts, with more traditional family patterns, will be replaced by younger cohorts with contemporary family patterns even if current age-sex-specific demographic rates remain unchanged. Our household projections also show large expected racial/ethnic differentials in household types/sizes and living arrangements in the first half of this century, due to racial differentials in demographic rates in the past that may continue for the next few decades. Our smaller and larger family scenarios generate low and high bounds for the future household and living arrangements distributions. This is useful because it provides informative ranges of possibilities of future trends rather than one set of figures involving many uncertainties.

To our knowledge, the household projections we report using the ProFamy method/program and demographic rates as input were the first to have found empirical evidence of family household momentum. These projections also provide informative low and high bounds of various indices of projected future household and living arrangement distributions based on possible changes in demographic parameters.

The ProFamy model and its associated computer program produce a large number of output tables and graphics by household type and size for each of the projection years (see Table 4.5 in Chap. 4). ProFamy also projects the entire population cross-classified by race, sex, age, marital status (including cohabitation), whether living with one or two parents or not living with parents, and by number of co-residing children. Due to space limitations, however, we have presented only the main output of the general trends and the racial differentials in this chapter. More detailed output of households and living arrangements, which could be useful in household consumption projections for business market analysis and socioeconomic planning, can be obtained. For example, the numbers of households by type, size, and age of the reference person projected by ProFamy can be combined with distributions of average home ownership rates classified by the number of bedrooms and household types, sizes, and age distributions (observed from the survey or census) to forecast housing consumption (see Smith et al. 2008, 2012, and Chap. 15 of this book for details). Similar forecasts of household and consumption using the ProFamy method can be conducted for home-based consumptions of vehicles (see, Feng et al. 2011; Prskawetz et al. 2004; and Chap. 11 of this book for details), energy (Dalton et al. 2008), durable goods, home-based services, and the like.

It is important to distinguish between using the ProFamy model for forecasting and using it for simulation. Forecasts with less than 20 years of time horizon may be used for business and governmental planning, but any results beyond that should be considered to be simulations only, due to large uncertainties after more than 20 years. In this chapter, we follow the expert opinion approach for projecting national future demographic summary measures; this serves reasonably well for the purpose of demographic simulation, but cannot be regarded as accurate forecasts, especially the projection results of more than 20 years after the starting year. Within the ProFamy model framework, however, more accurate forecasts with less than 20 years of time horizon can be performed. For example, future demographic summary measures can be forecasted using time series analysis. Further research may also need to include time series data for other related socioeconomic covariates in the forecasting of demographic summary measures. It should always be noted that the accuracy of the household forecasts relies heavily on the validity of assumptions regarding the demographic summary measures. Erroneous assumptions for too many covariates can quickly lead to forecasts that are far off the mark.

Appendix 1: A Procedure to Adjust the o/e Rates of Marital/Cohabiting Union Status Transitions Based on the NSFH and NSFG Data to Be Consistent with the o/e Rates of Marital Status Transitions Based on the CPS, SIPP, NSFH, and NSFG Data

Although we perform adjustments for each of the race groups, men and women, respectively, we omit race and sex dimensions in the formulas for simplicity of presentation.

Let $m4_{ij}(x)$ denote the o/e rate of transition from marital status i to marital status j between age x and $x + 1$ based on the CPS, SIPP, NSFH, and NSFG data, using a classic 4 marital status model ($i,j = 1,2,3,4$, represent never-married, married, widowed, and divorced, respectively, excluding cohabitation).

$m^*_{ij}(x)$, observed and unadjusted age-specific o/e rates of transitions from marital/union status i to j ($i,j = 1,2,3,4,5,6,7$, including cohabitation, see the definitions and flow chart in Fig. 2.2 of Chap. 2), based on NSFH and NSFG data;

$m_{ij}(x)$, the final adjusted age-specific o/e rates of transitions from marital/union status i to j ($i,j = 1,2,3,4,5,6,7$, including cohabitation) based on pooled survey data and adjusted to be consistent with $m4_{ij}(x)$;

$P_{ij}(x)$, age-specific probabilities of transitions from marital/union status i to j; $m_{ij}(x)$ can be analytically transferred into $P_{ij}(x)$ using the standard formula in multistate demography (see, e.g., Willekens et al. 1982; Schoen 1988; Preston et al. 2001).

$l_i(x)$, life table number of persons aged x with marital/union status i;

$L_i(x)$, number of person-years lived in marital/union status i between ages x and $x + 1$;

$$l_i(x+1) = \sum_k l_k(x) P_{ki}(x), \quad k = 1, 2, \ldots, 7 \tag{8.1}$$

$$L_i(x) = 0.5[l_i(x) + l_i(x+1)] \tag{8.2}$$

The goal of the adjustment is to make the average number of marriages (including first and re-marriages) (*AM7*) and average number of divorces (*AD7*) in the life time in the 7-status life table (including cohabitation) based on NSFH and NSFG data equal to the corresponding average numbers (*AM4* and *AD4*) in the 4-status life table (excluding cohabitation) based on all of the data from CPS, SIPP, NSFH, and NSFG.

We use the $m4_{ij}(x)$ to compute $P4_{ij}(x)$, age-specific probabilities of marital status transitions based on CPS, SIPP, NSFH, and NSFG data, using the standard formula. Based on $P4_{ij}(x)$, we construct a multi-state life table to get $L4_i(x)$, using formulas (8.1) and (8.2). We then use $m4_{ij}(x)$ and $L4_i(x)$ to compute the *AM4* and *AD4* in the 4 marital status model based on CPS, SIPP, NSFH, and NSFG data.

$$AM4 = \frac{\sum_{x=\partial}^{\omega} [L4_i(x)m4_{i2}(x)]}{100,000}, \quad i = 1,3,4$$

$$AD4 = \frac{\sum_{x=\partial}^{\omega} [L4_2(x)m4_{24}(x)]}{100,000}$$

We then employ the following two-step procedure to adjust the observed o/e rates of first marriage, divorce, and remarriages ($m^*_{12}(x)$, $m^*_{52}(x)$, $m^*_{24}(x)$, $m^*_{32}(x)$, $m^*_{62}(x)$, $m^*_{42}(x)$ and $m^*_{72}(x)$), but do not need to adjust the observed o/e rates of cohabitation union formation and dissolution ($m^*_{15}(x)$, $m^*_{36}(x)$, $m^*_{47}(x)$, $m^*_{51}(x)$, $m^*_{63}(x)$, $m^*_{74}(x)$) based on NSFH and NSFG.

Step 1. Adjustment for the o/e rates of first marriage, remarriage, and divorce
We use the unadjusted survey-based $m^*_{ij}(x)$ to compute $P^*_{ij}(x)$, and we then use $P^*_{ij}(x)$ to construct an initial multi-state life table and get the initial $L^*_i(x)$ using formulas (8.1) and (8.2); we then use $m^*_{ij}(x)$ and $L^*_i(x)$ to compute the initial *AM7** and *AD7** in the 7 marital/union status model based on the NSFH and NSFG data.

$$AM7^* = \frac{\sum_{x=\partial}^{\omega} [L^*_i(x)m^*_{i2}(x)]}{100,000}, \quad i = 1,3,4,5,6,7 \tag{8.3}$$

$$AD7^* = \frac{\sum_{x=\alpha}^{\omega} [L^*_2(x)m^*_{24}(x)]}{100,000} \tag{8.4}$$

We use *AM4*/*AM7**, *AD4*/*AD7** as adjustment factors (not age-specific) to adjust the corresponding age-specific o/e rates of first marriage, remarriage, and divorce for not-cohabiting and cohabiting persons at ages x ($x = \alpha$ to ω).

$$m'_{12}(x) = m^*_{12}(x)AM4/AM7^* \tag{8.5}$$

$$m'_{52}(x) = m^*_{52}(x)\ AM4/AM7^* \tag{8.6}$$

$$m'_{32}(x) = m^*_{32}(x)\ AM4/AM7^* \tag{8.7}$$

$$m'_{62}(x) = m^*_{62}(x)\ AM4/AM7^* \tag{8.8}$$

$$m'_{42}(x) = m^*_{42}(x)\ AM4/AM7^* \quad (8.9)$$

$$m'_{72}(x) = m^*_{72}(x)\ AM4/AM7^* \quad (8.10)$$

$$m'_{24}(x) = m^*_{24}(x)\ AM4/AM7^* \quad (8.11)$$

Step 2. Check whether the goal of the adjustment is achieved

We use the first adjusted $m'_{ij}(x)$ to compute the first adjusted $P'_{ij}(x)$, and use $m'_{ij}(x)$ to replace $m^*_{ij}(x)$ in the formulas (8.3) and (8.4) to get the first adjusted *AM7'* and *AD7'*. If the absolute values of the relative difference between *AM7'* and *AM4* and between *AD7'* and *AD4* are all less than a selected criterion (e.g., 0.5 %), we have completed Step 2 and have the final estimates of the o/e rates ($m_{ij}(x)$). Otherwise, we will have to use the first adjusted *AM7'* and *AD7'* to replace *AM7** and *AD7** in formulas (8.5), 8.6, 8.7, 8.8, 8.9, 8.10 and (8.11) to repeat the iterative procedures described in Step 1 and Step 2 until the selected criterion is achieved.

Appendix 2: Comparisons of Summary Measures of Marital Status Life Tables (Excluding Cohabitation) Between Our Estimates Based on the Pooled Survey Data and Schoen's Estimates Based on Vital Statistics, All Races Combined

Data set	Schoen	Surveys		
Period	1995	1990s	Abs. Diff	% Diff
Women				
Lifetime proportion of first marriage	0.887	0.878	−0.009	−1.0
Lifetime proportion of divorce	0.425	0.430	0.005	1.2
Lifetime proportion of remarriage of widows	0.048	0.049	0.001	2.1
Lifetime proportion of remarriage of divorcees	0.687	0.619	−0.068	−9.9
Men				
Lifetime proportion of first marriage	0.831	0.841	0.01	1.2
Lifetime proportion of divorce	0.437	0.392	−0.045	−10.3
Lifetime proportion of remarriage of widows	0.123	0.132	0.009	7.3
Lifetime proportion of remarriage of divorcee	0.781	0.723	−0.058	−7.4

Sources: "Schoen" refers to: Schoen, R. and N. Standish (2001); "Surveys" refer to our estimates based on pooled data sets of the CPS, SIPP, NSFH and NSFG

Appendix 3: Major Parameters of Medium Forecasts, Smaller and Larger Family Household Scenarios

	White and non-Hispanic			Black and non-Hispanic			Hispanic			Asian and other non-Hispanic		
	2000	2025	2050	2000	2025	2050	2000	2025	2050	2000	2025	2050
Medium mortality												
Male e_0	74.9	77.8	81.1	68.7	73.6	78.5	77.4	80.0	83.0	79.7	81.9	84.6
Female e_0	80.3	83.6	86.4	75.4	80.5	84.6	83.8	86.1	88.4	85.9	87.6	89.7
Low mortality												
Male e_0	74.9	79.2	83.5	68.8	75.3	81.3	77.4	81.5	85.5	79.7	83.4	86.9
Female e_0	80.3	84.5	88.0	75.5	81.7	86.5	83.9	87.1	90.0	86.0	88.6	91.2
High mortality												
Male e_0	74.8	76.9	79.5	68.5	72.4	76.6	77.2	79.0	81.4	79.6	81.0	83.0
Female e_0	80.2	82.6	84.8	75.3	79.3	82.7	83.8	85.1	86.8	85.8	86.6	88.0
Medium fertility												
TFR-all births	1.84	2.03	2.04	2.08	2.12	2.11	2.97	2.68	2.56	2.26	2.18	2.16
TFR(1)-1st birth	0.80	0.87	0.88	0.82	0.77	0.77	0.95	0.94	0.90	0.96	0.91	0.91
TFR(2)-2nd birth	0.66	0.71	0.71	0.67	0.66	0.66	0.90	0.81	0.77	0.73	0.70	0.69
TFR(3)-3rd birth	0.26	0.30	0.31	0.38	0.39	0.39	0.55	0.50	0.48	0.37	0.35	0.35
TFR(4)-4th birth	0.09	0.10	0.10	0.14	0.18	0.18	0.28	0.25	0.24	0.12	0.13	0.13
TFR(5)-5+ birth	0.03	0.05	0.05	0.07	0.12	0.12	0.20	0.18	0.18	0.08	0.08	0.08
Low fertility												
TFR-all births	1.82	1.73	1.67	2.06	1.80	1.73	2.96	2.28	2.09	2.23	1.86	1.76
TFR(1)-1st birth	0.79	0.74	0.72	0.81	0.66	0.63	0.94	0.80	0.73	0.95	0.78	0.74
TFR(2)-2nd birth	0.65	0.60	0.58	0.67	0.56	0.54	0.89	0.69	0.63	0.72	0.59	0.57
TFR(3)-3rd birth	0.26	0.26	0.25	0.37	0.33	0.32	0.55	0.42	0.39	0.37	0.30	0.29
TFR(4)-4th birth	0.09	0.09	0.08	0.14	0.15	0.14	0.28	0.21	0.20	0.12	0.11	0.11
TFR(5)-5+ birth	0.03	0.04	0.04	0.07	0.11	0.10	0.20	0.16	0.14	0.08	0.07	0.07

(continued)

	White and non-Hispanic			Black and non-Hispanic			Hispanic			Asian and other non-Hispanic		
	2000	2025	2050	2000	2025	2050	2000	2025	2050	2000	2025	2050
High fertility												
TFR-all births	1.86	2.33	2.42	2.10	2.44	2.50	2.99	3.08	3.03	2.28	2.51	2.56
TFR(1)-1st birth	0.81	0.95	0.95	0.82	0.89	0.91	0.95	0.95	0.95	0.96	0.95	0.95
TFR(2)-2nd birth	0.67	0.84	0.90	0.68	0.76	0.78	0.90	0.90	0.90	0.74	0.86	0.89
TFR(3)-3rd birth	0.26	0.36	0.38	0.38	0.45	0.46	0.56	0.66	0.63	0.38	0.43	0.45
TFR(4)-4th birth	0.09	0.12	0.13	0.14	0.20	0.21	0.28	0.33	0.32	0.13	0.16	0.17
TFR(5)-5+ birth	0.03	0.05	0.06	0.07	0.14	0.15	0.20	0.24	0.23	0.08	0.10	0.11
Medium marriage/union formation and dissolution												
General marriage rate	0.0704	0.0704	0.0704	0.0362	0.0362	0.0362	0.0593	0.0593	0.0593	0.0676	0.0676	0.0676
General divorce rate	0.0292	0.0292	0.0292	0.0308	0.0308	0.0308	0.0184	0.0184	0.0184	0.0214	0.0214	0.0214
General cohabiting rate	0.1094	0.1094	0.1094	0.0775	0.0775	0.0775	0.0996	0.0996	0.0996	0.1187	0.1187	0.1187
General union break rate	0.2992	0.2992	0.2992	0.3612	0.3612	0.3612	0.2013	0.2013	0.2013	0.3341	0.3341	0.3341
Male mean age first mar.	27.59	28.45	28.45	29.99	30.57	30.57	27.38	28.14	28.14	30.54	31.12	31.12
Female mean age first mar.	25.45	26.16	26.16	28.57	29.18	29.18	25.83	26.68	26.68	28.25	28.92	28.92
Mean age at births	27.56	28.07	28.07	25.74	26.19	26.19	26.88	27.61	27.61	28.77	29.22	29.22

Note: The medium/low/high mortality and medium/low/high fertility parameters are those assumed in the Census Bureau's latest population projection (Hollmann et al. 2000)

Chapter 9
Household and Living Arrangement Projections for the 50 States, Washington DC, and Relatively Large Counties in the U.S.

9.1 Data and Parameter Assumptions

The data sources for the projections are listed in the last column of Table 3.1 in Chap. 3. As discussed in Core Idea 4 in Sect. 2.2.4 of Chap. 2, we employ model standard schedules of race-sex-age-specific demographic rates (except domestic migration rates) estimated from national datasets for household and living arrangement projections for each of the 50 states, DC, SC, and M-S. Based on the 2000 census 5 % micro dataset, we also estimated race-sex-age-specific probabilities of domestic out-migration from each of the states, DC, counties, and area to the rest of the country and race-sex-age-specific frequencies of in-migration from the rest of the country to each of the states, DC, counties, and area (see Appendix 1 for some details).

The state-race-sex-specific life expectancies at birth and the race-parity-specific TFRs from 2000 to 2050 used in these projections are estimated/projected from regional data with the medium assumptions of the Census Bureau population projections (Hollmann et al. 2000; U.S. Census Bureau 2008). The state-specific numbers of domestic in-migrants and out-migrants and international net migrants are estimated from the combined data of the ACS from 2000 to 2006; the migration parameters are assumed to be constant after 2006.

The conceptual ideas for estimating the standardized general rates of marriage/union formation and dissolution were discussed in Sect. 3.4 of Chap. 3 and the associated technical procedures for estimating the standardized general rates in 2000 at the sub-national level are presented in Appendix 5 of Chap. 3. Instead of constant assumptions, time-dependent changes were specified for some of the parameters for the period 2000–2010, so that the projected values for 2010 were consistent with the corresponding 2010 census results. These parameters include: the race-specific general rates of marriage/union formation and dissolution, race-age-sex-specific proportion of persons who live in group quarters (PGQ), race-sex-specific proportion of those aged 45–49 who do not live with parents (PNP), and race-household size-specific average number of other relatives (other than spouse/partner, parents, and children) and non-relatives living in the same household (ORNR). This

Y. Zeng et al., *Household and Living Arrangement Projections*,
The Springer Series on Demographic Methods and Population Analysis 36,
DOI 10.1007/978-90-481-8906-9_9, © Springer Science+Business Media Dordrecht 2014

is similar to the practice adopted by other demographic projections in which an earlier census year is the starting point of the projection and the most recent census year is within the projection period and the main results of the most recent census are published but detailed micro data to derive the base population of the projections are not yet available (U.S. Census Bureau 2008). The race-specific general rates of marriage/union formation and dissolutions, PGQ, PNP, and ORNR from 2010 to 2050 are simply assumed to be constant at the 2010 level in our medium projection.[1]

The number of racial groups represented by state depends on the state's racial composition. Four race groups (White non-Hispanic, Black non-Hispanic, Hispanic, other minorities) defined by the Census Bureau are distinguished in 14 states where each of the four racial groups has a sufficiently large population size for the projection. Due to the small population sizes of the minority groups, three racial groups (White non-Hispanic, Black non-Hispanic, Hispanic plus other minorities) are distinguished in 13 states, two racial groups (White non-Hispanic and all other races combined) are distinguished in the other 13 states. All race groups are combined in 11 states, because the population size of all non-white groups combined in these states is not sufficiently large for distinct projections.

9.2 Low and High Bounds of Household and Living Arrangement Projections

In order to explore the possible low and high bounds of household and living arrangement projections, we examined small and large family scenarios for each of the 50 U.S. states and DC and the other sub-regional areas. The small family scenario assumes that, as compared to the medium projections, the general rate of divorce and general rate of cohabitation union dissolution are higher by 15 % in 2025 and 25 % in 2050 and the general rates of marriage and of cohabitation are lower by 15 % in 2025 and 25 % in 2050. To obtain the TFR, e_0, and number of international migrants in 2025 and 2050 for the small family scenario, we multiplied the medium fertility (TFR), medium mortality (e_0), and medium international net migration in 2025 and 2050 adopted by the Census Bureau's latest population projection released in 2008 by the ratios of the low TFR, low mortality (high e_0), and low international net migration to the corresponding medium variants adopted by the Census Bureau's population projection in 2000 (Hollmann et al. 2000). This small family scenario assumes increasing marriage/union

[1] One common approach in population projection is to hold some of the current demographic rates constant throughout the projection horizon (e.g., Day 1996; Treadway 1997). Smith et al. (2001: 83–84) argued that neither the direction nor the magnitude of future changes can be predicted accurately, and thus if upward or downward movements are more or less equally likely, constant demographic rates provide a reasonable forecast of future rates.

dissolution, decreasing marriage/union formation, decreasing fertility and mortality, and receipt of fewer international immigrants. We expect that such a combination of demographic rates will result in low bounds of household size and percentages of married- or cohabiting-couple households, high bounds of percentages of one-person households, single-parent households, and so on.

The large family scenario assumes that, as compared to the medium projections, the general rates of divorce and of cohabiting-union dissolution are lower by 15 % in 2025 and 25 % in 2050 and the general rates of marriage and of cohabitation are higher by 15 % in 2025 and 25 % in 2050. To obtain the TFR, e_0, and number of international migrants in 2025 and 2050 for the large family scenario, we multiplied the medium fertility (TFR), medium mortality (e_0), and medium international net migration in 2025 and 2050 adopted by the Census Bureau's latest population projection released in 2008 by the ratios of the high TFR, high mortality (low e_0), and high international net migration to the corresponding medium variants adopted by the Census Bureau's population projection in 2000 (Hollmann et al. 2000). This large family scenario assumes that the family will regain its traditional values, reflected in decreasing marriage/union dissolution, increasing marriage/union formation, and increasing fertility, accompanied by a larger number of international immigrants and relatively higher mortality. The combination of demographic rates in the large family scenario will produce high bounds for projections of household size and percentages of married- or cohabiting-couple households, low bounds of percentages of one-person households, single-parent households, and so on.

The general rates of marriage, divorce, cohabitation, and union dissolution, TFR, e_0, and number of international migrants for all individual years between 2010, 2025, and 2050 in all scenarios are linearly interpolated. The assumptions that there will be 15 % and 25 % increases (or decreases) in general rates of marriage, divorce, cohabitation, and union dissolution in 2025 and 2050 are based on our educated guesses about the largest possible changes in marriage/union formation and dissolution in the next few decades. Although we made these guesses with reference to the available time series data of the general rates, they are largely arbitrary because of uncertainties about future trends. Nevertheless, similar to conventional population projections of low and high variants that formulate possible bounds of population growth, our small and large family scenarios generate possible low and high bounds of future household and living arrangement distributions.[2]

[2] The race-sex-specific demographic parameters (TFR is parity-specific) (see parameters (a) to (h) in panel (3) of Table 3.1) in the medium, small, and large family scenarios in selected years from 2000 to 2050 for each of the 50 states, DC, each of the M-S Area and six countries of SC require one large table occupying about one full page of space. To include them in this book would require 58 pages, which is not feasible and not necessary, and thus they are not presented.

9.3 Summary of Projection Outcomes

Relatively detailed numerical outcomes for the main indeices of the medium projections in 2010, 2020, 2030, 2040, and 2050, and the low and high bounds after 2010, are presented in Tables 9.1–9.8 of Appendix 2. We also include available census observations in Tables 9.1, 9.2, 9.3 and 9.4 to compare projections to observations in 2010. We summarize here the insights from the main indices of the decennial projections but could not present the details due to space limitations.

As shown in Table 9.1, the average household size would decrease moderately and pervasively in almost all states and DC in the period of 2000–2020, from an overall mean of 2.58 persons for the whole country in 2000 to 2.52 (SI: 2.48, 2.55) in 2020 (SI is the abbreviation for the scenarios interval between the low and high bounds; hereafter we will present the medium variant followed by the SI). The average household size may continue to decline after 2020 in 46 states, but at a slightly slower rate as compared to the period 2000–2020, from an average of 2.44 (SI: 2.40, 2.48) in 2020 to an average of 2.36 (SI: 2.14, 2.61) in 2050. In four other states (California, Colorado, Maryland, New Jersey) and DC, the trend after 2020 is different: the average household size may slightly increase from 2.54 (SI: 2.50, 2.57) in 2020 to 2.60 (SI: 2.36, 2.86) in 2050.

The proportion of one-person households is projected to increase substantially in almost all states in the first two decades. As compared to 2000, the proportion of one-person households in 2020 in 8, 17, 22 and 4 states would increase by <5.0 %, 5.0–9.99 %, 10.0–14.99 % and ≥ 15.0 %, respectively. But in the period 2020–2050, this increasing trend would slow down considerably in 44 states: from 0.275 (SI: 0.265, 0.284) in 2020 to 0.301 (SI: 0.238, 0.361). The proportion of one-person households in 2020–2050 in the other six states (California, Maryland, New Jersey, Arizona, Hawaii, Nevada) would decline slightly from 0.275 (SI: 0.265, 0.284) in 2020 to 0.266 (SI: 0.210, 0.317) in 2050 (see Table 9.2). The pattern of change in average proportions of one-person households in the six counties of Southern California in the first half of this century is totally different as compared to the overall trends in most of the states: it declines from 0.227 in 2000 to 0.211 (SI: 0.204, 0.217) in 2020 and 0.187 (SI: 0.157, 0.221) in 2050 (see eighth line from the bottom of Table 9.2).

The projected declines in the proportion of the one-person households in the six counties of SC during the first half of this century and in the six states after 2020 are likely due to the large racial/ethnic differentials and changes in racial compositions. While the proportions of one-person households will increase substantially in all racial groups in the next few decades, the Hispanic population, which has the lowest proportion of one-person households and largest average household size, will compose a substantially higher percentage of the total future population in the six states and the six counties of SC. For example, the proportion of one-person households in the Hispanic group was 0.099 in 2000 in the six counties of SC, but the corresponding figures in the White non-Hispanic, Black non-Hispanic, and Asian and other non-Hispanic were 0.30, 0.29, and 0.19. The Hispanic group, which consisted of 40.6 % of the total population in 2000, will become the majority in 2020 (53.5 %) and 2050 (65.8 %) in the six counties of SC. Consequently, these racial composition changes will result in a decline in the proportion of one-person households for all races combined.

Husband-wife households would decrease moderately in almost all states (with a few exceptions of slight increases). Under the medium scenario, as compared to 2000, the proportion of married-couple households among all households in 6, 15, 14 and 13 states in 2020 would decrease by <5.0 %, 5.0–9.99 %, 10.0–14.99 % and ≥15.0 %, respectively. The decrease in the proportion of married-couple households would considerably slow down in the period 2020–2050: the proportion in 14, 19, 9 and 6 states in 2050 would decrease by <5.0 %, 5.0–9.99 %, 10.0–14.99 % and ≥15.0 %, respectively, as compared to 2020 (see Table 9.3). The proportion of cohabiting-couple households among all households would increase dramatically in the first two decades of this century: the proportion in 3, 7, 11, 18 and 11 states in 2020 would be higher than that in 2000 by <10 %, 10.0–29.0 %, 30.0–49.9 %, 50.0–69.9 % and 70.0–89.9 %, respectively, under the medium scenario (see Table 9.4). The proportion of cohabiting households would remain relatively stable after 2020 in almost all states.

Directions of changes in the percentage of single-parent households in the first half of this century are diversified, increasing moderately in some states but decreasing moderately or remaining more or less unchanged in other states. The average percentage of single-parent households across all states and DC was 30.9 in 2000 and is projected to be 30.8 (SI: 29.3, 32.2) in 2020 and 33.1 (SI: 24.7, 44.4) in 2050 (see Table 9.5). Such patterns may be explained by the opposite effects of moderate declines in marriages and substantial increases in cohabitation, plus stable divorce and union dissolution rates.

The aging trends shown in the household and living arrangement projections presented in Tables 9.6, 9.7 and 9.8 are striking. Under the medium scenario, as compared to 2000, the proportion of elderly households (with householder aged 65+) in 11, 17, 15, 4 and 3 states in 2020 would increase by <10 %, 10–19.9 %, 20–29.9 %, 30–39.9 % and ≥40 %. In the period 2020–2050, household aging will accelerate further. More specifically, as compared to 2020, the proportion of elderly households in 2, 24, 20 and 4 states in 2050 will increase by <20.0 %, 20–29.9 %, 30–39.9 % and ≥40 %, respectively; only DC, which attracts a lot of young in-migrants, is an exception (see Table 9.6). As compared to 2000, elderly households will slightly more than double in Hawaii and New Hampshire and nearly triple in Alaska by the middle of this century. Similar to the general pattern of increase in elderly households, the proportion of elderly aged 65+ living alone will increase dramatically and pervasively across all states (see Table 9.7). Table 9.8 demonstrates that the oldest-old aged 80+ living alone will even more dramatically increase in the next a few decades across all states. The average percentage of the oldest-old aged 80+ living alone across all states and DC in 2020 and 2050 would be 1.48 (SI: 1.44, 1.51) and 2.41 (SI: 1.85, 2.96), representing a 23.8 % increase in 2020 and slightly more than doubled by 2050 as compared to 2000, under the medium assumption. As compared to 2000, the percentage of oldest-old living alone in 2050 would increase by 44.5–79.9 % and 80–99.9 % in 7 and 14 states; more than double (but less than triple) in 24 states, and more than triple in 5 states (Louisiana, South Carolina, Hawaii, New Hampshire, and Alaska), under the medium scenario.

9.4 Discussion and Concluding Remarks

Applying the ProFamy extended cohort-component method and national model standard schedules of age-sex-race-specific demographic rates based on commonly available survey and census data, we have demonstrated that comprehensive and simultaneous projections of households, living arrangements, and population at the sub-national level require a census micro data file and projected (or assumed) demographic summary parameters.

For the purpose of illustrative applications, we calculated household and living arrangement projections from 2000 to 2050 with medium, small, and large family scenarios, for each of the 50 U.S. states, DC, six counties of SC, and the M-S Area. Among many interesting numerical outcomes of household and living arrangements projections with medium, low and high bounds, the aging of American households over the next few decades across all states/areas is particularly striking. To our knowledge, these were the first comprehensive household and living arrangement projections by race, age of the householders, and various household types/sizes using conventional demographic rates as input for each of the states, DC, and other counties/area in the United States (Zeng et al. 2013a).

Limitations of the present study and potentials for further investigations should also be noted. First, our projections of the demographic summary parameters are based on trend extrapolations and expert opinions. Thus far, we have not included other socioeconomic factors relevant to changes in demographic parameters; this can be done in future research. Second, we discussed main results for all races combined due to space limitations, although we have made race-specific projections for most of the 50 U.S. states, DC, six counties of SC, and the M-S Area; analysis of many other detailed state-race-specific projection outcomes may be an interesting topic for further research.

We also note that the ProFamy model cannot be directly applied to project household and living arrangements for small areas that do not have adequate data to estimate the needed demographic summary parameters. However, as discussed and illustrated in Chap. 6, it is possible to project household and living arrangements for small areas, employing the well-established ratio method and the projection of the small area's parental region (a state or province or large county/city) produced by the ProFamy approach.

Appendix 1: Information About Population Sizes of the 50 States, DC, the Six Counties of SC, and the M-S Area

Among the 50 states and DC, the state of Wyoming had the smallest total population size of 0.49 million and California had the largest population size of 33.9 million in 2000.

Among the six counties of Southern California, Imperial county had the smallest total population size of 0.14 million and Los Angeles county had the largest population size of 9.51 million in 2000. The total population size of the Minneapolis-St. Paul Area (including 7 counties of Anoka, Carver Dakota, Hennepin, Ramsey, Scott, Washington, and Minnesota) in 2000 was 2.64 million. For such relatively large counties, cities, or areas with adequate data to estimate the summary demographic parameters, we can apply the ProFamy approach to project household and living arrangements, using demographic race-age-sex-specific model standard schedules at the national level. Employing the ProFamy approach, Wang (2009a, b, 2011a, b) has successfully conducted household and living arrangement projections for each of the six counties of SC and the M-S Area, upon requests from the Southern California Association of Governments (SCAG) and the Minnesota Twin City Municipality Government. The main results for the six counties of SC and M-S Area presented in Tables 9.1, 9.2, 9.3, 9.4, 9.5, 9.6, 9.7 and 9.8 in Appendix 2 are cited from these projects' final reports by Wang (2009a, b, 2011a, b). We are very grateful to Simon Choi and Todd Graham for their comments on these reports.

Appendix 2: Output of Household and Living Arrangement Projections, U.S. Sub-National Level

Table 9.1 Projections of average household size for the 50 states, DC, and some counties, 2000–2050

		2010		2020	2030	2040	2050
	2000	Obs.	Proj.	Med (Low-High)	Med (Low-High)	Med (Low-High)	Med (Low-High)
United States	2.58	2.58	2.56	2.52 (2.48–2.55)	2.49 (2.38–2.62)	2.49 (2.30–2.70)	2.49 (2.26–2.77)
Alabama	2.49	2.48	2.46	2.39 (2.35–2.43)	2.34 (2.24–2.45)	2.32 (2.16–2.50)	2.32 (2.12–2.55)
Alaska	2.73	2.65	2.62	2.59 (2.54–2.62)	2.55 (2.41–2.66)	2.53 (2.31–2.71)	2.52 (2.24–2.73)
Arizona	2.62	2.63	2.61	2.64 (2.60–2.67)	2.63(2.51–2.77)	2.63 (2.44–2.86)	2.62 (2.37–2.91)
Arkansas	2.49	2.47	2.48	2.42 (2.39–2.46)	2.37 (2.26–2.48)	2.33 (2.16–2.51)	2.30 (2.09–2.53)
California	2.84	2.90	2.88	2.86 (2.82–2.90)	2.87 (2.74–3.01)	2.88 (2.69–3.12)	2.89 (2.65–3.19)
Colorado	2.52	2.49	2.48	2.48 (2.45–2.52)	2.49 (2.37–2.62)	2.51 (2.32–2.73)	2.52 (2.29–2.80)
Connecticut	2.53	2.52	2.52	2.47 (2.40–2.54)	2.45 (2.30–2.62)	2.45 (2.25–2.66)	2.44 (2.25–2.66)
Delaware	2.54	2.55	2.54	2.49 (2.46–2.53)	2.47 (2.37–2.59)	2.46 (2.29–2.65)	2.46 (2.24–2.70)
District of Columbia	2.15	2.11	2.08	2.11 (2.04–2.14)	2.15 (2.00–2.22)	2.19 (1.99–2.29)	2.22 (2.00–2.37)
Florida	2.45	2.48	2.49	2.46 (2.44–2.48)	2.43 (2.37–2.51)	2.43 (2.31–2.57)	2.45 (2.26–2.66)
Georgia	2.64	2.63	2.67	2.63 (2.58–2.68)	2.61 (2.47–2.75)	2.61 (2.41–2.83)	2.62 (2.38–2.90)
Hawaii	2.88	2.89	2.84	2.80 (2.76–2.85)	2.81 (2.69–2.94)	2.80 (2.61–3.00)	2.77 (2.54–3.00)
Idaho	2.68	2.66	2.63	2.57 (2.54–2.60)	2.50 (2.38–2.61)	2.45 (2.28–2.63)	2.44 (2.22–2.66)
Illinois	2.62	2.59	2.58	2.51 (2.48–2.55)	2.48 (2.36–2.61)	2.48 (2.29–2.71)	2.49 (2.25–2.79)
Indiana	2.52	2.52	2.49	2.44 (2.39–2.48)	2.41 (2.28–2.55)	2.41 (2.22–2.63)	2.43 (2.19–2.70)
Iowa	2.45	2.41	2.41	2.33 (2.29–2.36)	2.25 (2.15–2.36)	2.21 (2.05–2.41)	2.21 (1.99–2.47)
Kansas	2.51	2.49	2.50	2.44 (2.41–2.48)	2.40 (2.28–2.52)	2.38 (2.20–2.58)	2.38 (2.15–2.63)
Kentucky	2.47	2.45	2.45	2.39 (2.36–2.42)	2.33 (2.23–2.44)	2.31 (2.14–2.49)	2.30 (2.08–2.54)
Louisiana	2.61	2.55	2.56	2.46 (2.43–2.48)	2.39 (2.29–2.49)	2.35 (2.20–2.53)	2.34 (2.14–2.56)
Maine	2.39	2.32	2.32	2.25 (2.23–2.28)	2.19 (2.10–2.28)	2.17 (2.02–2.34)	2.19 (1.98–2.41)
Maryland	2.60	2.61	2.59	2.59 (2.56–2.63)	2.61 (2.50–2.75)	2.64 (2.45–2.86)	2.65 (2.41–2.93)
Massachusetts	2.51	2.48	2.49	2.42 (2.39–2.45)	2.39 (2.29–2.51)	2.38 (2.20–2.58)	2.37 (2.15–2.63)
Michigan	2.55	2.49	2.48	2.40 (2.37–2.43)	2.34 (2.24–2.46)	2.32 (2.16–2.51)	2.32 (2.11–2.56)
Minnesota	2.52	2.48	2.46	2.36 (2.32–2.41)	2.31 (2.20–2.44)	2.28 (2.11–2.48)	2.27 (2.07–2.54)
Mississippi	2.62	2.58	2.56	2.48 (2.45–2.50)	2.41 (2.32–2.51)	2.39 (2.23–2.56)	2.38 (2.17–2.60)
Missouri	2.47	2.45	2.42	2.35 (2.31–2.38)	2.28 (2.18–2.40)	2.26 (2.09–2.44)	2.26 (2.04–2.50)
Montana	2.44	2.35	2.36	2.25 (2.22–2.28)	2.16 (2.07–2.26)	2.11 (1.96–2.27)	2.09 (1.90–2.30)
Nebraska	2.49	2.46	2.43	2.33 (2.30–2.37)	2.27 (2.16–2.38)	2.24 (2.07–2.43)	2.22 (2.01–2.48)
Nevada	2.60	2.65	2.62	2.64 (2.57–2.70)	2.64 (2.49–2.80)	2.64 (2.43–2.88)	2.64 (2.38–2.94)
New Hampshire	2.53	2.46	2.48	2.37 (2.34–2.41)	2.2 (2.19–2.40)	2.24 (2.07–2.42)	2.22 (2.00–2.45)
New Jersey	2.67	2.68	2.67	2.64 (2.61–2.67)	2.66 (2.55–2.79)	2.69 (2.49–2.92)	2.70 (2.46–3.00)
New Mexico	2.62	2.55	2.51	2.46 (2.42–2.49)	2.41 (2.29–2.54)	2.37(2.19–2.58)	2.35 (2.12–2.62)
New York	2.59	2.57	2.55	2.47 (2.44–2.51)	2.45 (2.33–2.60)	2.44 (2.24–2.70)	2.44 (2.21–2.78)
North Carolina	2.48	2.48	2.47	2.43 (2.40–2.47)	2.41 (2.30–2.51)	2.41 (2.24–2.57)	2.42 (2.21–2.63)
North Dakota	2.40	2.30	2.33	2.18 (2.11–2.26)	2.08(1.95–2.22)	2.03 (1.89–2.21)	2.02 (1.86–2.20)
Ohio	2.48	2.44	2.41	2.34 (2.31–2.38)	2.29 (2.19–2.41)	2.27 (2.10–2.47)	2.27 (2.05–2.54)
Oklahoma	2.48	2.49	2.46	2.38 (2.35–2.42)	2.33 (2.22–2.44)	2.29(2.13–2.48)	2.28 (2.06–2.52)
Oregon	2.50	2.47	2.46	2.41 (2.37–2.45)	2.38 (2.26–2.51)	2.37 (2.19–2.58)	2.38 (2.15–2.65)
Pennsylvania	2.47	2.45	2.43	2.37 (2.34–2.41)	2.35 (2.24–2.46)	2.34 (2.17–2.52)	2.34 (2.12–2.57)
Rhode Island	2.47	2.44	2.47	2.36 (2.33–2.40)	2.28 (2.18–2.39)	2.21 (2.06–2.39)	2.14 (1.96–2.39)
South Carolina	2.52	2.49	2.48	2.38 (2.35–2.41)	2.32 (2.22–2.42)	2.28 (2.13–2.45)	2.26 (2.08–2.47)
South Dakota	2.49	2.42	2.40	2.31 (2.27–2.34)	2.22 (2.11–2.34)	2.18 (2.01–2.36)	2.17 (1.96–2.40)
Tennessee	2.48	2.48	2.46	2.42 (2.38–2.45)	2.38 (2.28–2.50)	2.37 (2.20–2.56)	2.37 (2.16–2.62)
Texas	2.73	2.75	2.72	2.67 (2.63–2.72)	2.64 (2.51–2.80)	2.64 (2.42–2.88)	2.64 (2.36–2.95)
Utah	3.10	3.10	3.03	2.93 (2.89–2.97)	2.86 (2.72–3.01)	2.81 (2.59–3.04)	2.76 (2.47–3.05)
Vermont	2.44	2.34	2.36	2.25 (2.22–2.29)	2.19 (2.08–2.30)	2.15 (1.98–2.33)	2.12 (1.91–2.34)
Virginia	2.53	2.54	2.52	2.47 (2.44–2.51)	2.45 (2.34–2.57)	2.44 (2.27–2.64)	2.44 (2.23–2.69)
Washington	2.53	2.51	2.51	2.45 (2.41–2.49)	2.42 (2.30–2.55)	2.41 (2.23–2.62)	2.42 (2.19–2.67)
West Virginia	2.40	2.36	2.38	2.33 (2.30–2.36)	2.28 (2.18–2.38)	2.24 (2.09–2.40)	2.22 (2.03–2.41)

(continued)

Table 9.1 (continued)

		2010		2020	2030	2040	2050
	2000	Obs.	Proj.	Med (Low-High)	Med (Low-High)	Med (Low-High)	Med (Low-High)
Wisconsin	2.49	2.43	2.43	2.36 (2.33–2.40)	2.32 (2.21–2.44)	2.31 (2.14–2.51)	2.33 (2.11–2.60)
Wyoming	2.47	2.42	2.40	2.32 (2.29–2.36)	2.26 (2.16–2.36)	2.23 (2.07–2.38)	2.22 (2.03–2.40)
6-county region, SC	2.97	3.03	3.00	2.96 (2.92–3.01)	2.96 (2.84–3.09)	2.95 (2.77–3.13)	2.94 (2.73–3.17)
Imperial county	3.31	3.34	3.32	3.38 (3.33–3.42)	3.35 (3.24–3.44)	3.28 (3.16–3.41)	3.28 (3.08–3.38)
Los Angeles county	2.95	2.98	2.93	2.85 (2.81–2.89)	2.84 (2.73–2.96)	2.84 (2.67–3.00)	2.83 (2.63–3.03)
Orange county	2.96	2.99	3.03	3.02 (2.97–3.07)	3.06 (2.91–3.22)	3.08 (2.87–3.33)	3.11 (2.85–3.41)
Riverside county	2.96	3.14	3.09	3.09 (3.05–3.13)	3.04 (2.94–3.17)	2.98 (2.81–3.18)	2.95 (2.73–3.19)
S Bernardino county	3.13	3.26	3.26	3.20 (3.15–3.26)	3.15 (3.02–3.29)	3.10 (2.91–3.28)	3.08 (2.86–3.30)
Ventura county	3.01	3.04	2.99	2.97 (2.92–3.02)	2.96 (2.83–3.11)	2.95 (2.75–3.17)	2.95 (2.70–3.22)
M-S 7-county region	2.52	2.50	2.53	2.48 (2.48–2.52)	2.46 (2.46–2.61)	2.45 (2.45–2.69)	2.45 (2.45–2.75)

Table 9.2 Projections of % one-person households for the 50 states, DC, and some counties, 2000–2050

		2010		2020	2030	2040	2050
	2000	Obs.	Proj.	Med (Low-High)	Med (Low-High)	Med (Low-High)	Med (Low-High)
United States	25.8	26.7	26.9	27.9 (26.9–28.7)	28.7 (25.8–31.3)	29.1 (24.3–33.3)	29.1 (22.8–34.8)
Alabama	26.1	27.4	27.4	29.4 (28.4–30.2)	31.0 (28.1–33.4)	31.9 (27.2–35.9)	32.3 (26.0–37.9)
Alaska	23.5	25.6	25.2	25.3 (24.8–27.0)	26.1 (24.2–30.1)	26.2 (22.7–32.2)	26.2 (21.5–33.9)
Arizona	24.8	26.1	25.9	25.4 (24.5–26.3)	25.5 (22.8–28.0)	25.2 (20.9–29.5)	25.1(19.4–31.0)
Arkansas	25.6	27.1	26.7	28.1 (27.1–29.1)	29.9 (26.8–32.7)	31.3 (26.5–36.0)	32.4 (26.0–38.9)
California	23.5	23.3	23.6	23.8 (23.1–24.5)	23.6 (21.3–25.5)	22.9 (19.3–25.7)	22.3 (17.7–26.1)
Colorado	26.3	27.9	28.3	29.6 (28.6–30.5)	30.7 (27.7–33.4)	31.0 (26.0–35.4)	31.1 (24.5–36.9)
Connecticut	26.4	27.3	27.5	28.8 (27.5–29.8)	29.5 (26.0–32.4)	29.8 (24.7–34.4)	29.9 (23.9–35.9)
Delaware	25.0	25.6	25.7	26.6 (25.7–27.5)	27.4 (24.5–29.9)	27.9 (23.4–32.0)	28.1 (22.2–33.9)
District of Columbia	43.8	44.0	44.5	44.0 (42.5–45.2)	42.9 (39.2–45.8)	41.8 (36.4–45.9)	40.8 (34.0–45.8)
Florida	26.6	27.2	27.5	28.7 (28.0–29.5)	29.8 (27.5–31.9)	30.5 (26.5–33.9)	30.7 (25.0–35.6)
Georgia	23.6	25.4	25.7	26.7 (25.7–27.4)	27.5 (24.6–29.9)	27.7 (23.2–31.7)	27.5 (21.7–32.8)
Hawaii	21.9	23.3	23.0	22.4 (21.2–23.4)	20.8 (17.8–23.4)	20.8 (16.8–24.9)	21.8 (17.2–27.0)
Idaho	22.4	23.8	23.5	24.6 (23.6–25.5)	25.9 (23.1–28.8)	26.9 (22.4–31.6)	27.2 (21.4–33.8)
Illinois	26.8	27.8	27.7	28.9 (27.9–29.8)	29.7 (26.6–32.5)	30.0 (24.9–34.6)	29.9 (23.1–36.2)
Indiana	25.9	26.9	27.3	28.9 (27.8–30.0)	30.5 (27.1–33.4)	31.3 (26.0–35.9)	31.5 (24.6–37.6)
Iowa	27.2	28.4	28.4	30.4 (29.3–31.3)	32.8 (29.4–35.6)	34.7 (29.0–39.3)	35.7 (27.8–42.2)
Kansas	27.0	27.8	28.1	28.1 (27.0–29.2)	28.9 (25.7–31.9)	29.4 (24.5–34.2)	29.4 (23.2–36.1)
Kentucky	26.0	27.5	27.4	29.6 (28.6–30.5)	31.6 (28.5–34.3)	32.9 (27.8–37.2)	33.4 (26.5–39.5)
Louisiana	25.3	26.9	26.8	27.8 (27.4–28.2)	28.7 (26.1–30.9)	29.3 (24.9–33.2)	29.5 (23.7–35.1)
Maine	27.0	28.6	28.7	30.3 (29.1–31.4)	32.3 (28.9–35.5)	33.5 (28.0–38.7)	33.8 (26.2–40.9)
Maryland	25.0	26.1	26.3	26.5 (25.6–27.3)	26.6 (23.8–29.0)	26.4 (22.0–30.3)	26.2 (20.5–31.3)
Massachusetts	28.0	28.7	29.0	29.5 (28.6–30.4)	30.1 (27.1–32.7)	30.6 (25.6–34.9)	30.9 (24.3–36.7)
Michigan	26.2	27.9	28.2	28.9 (28.0–29.8)	29.9 (26.8–32.7)	30.5 (25.5–35.1)	30.6 (24.0–37.0)
Minnesota	26.9	28.0	28.3	29.6 (28.5–30.4)	31.1 (27.9–33.7)	32.5 (27.1–37.0)	32.9 (25.5–39.5)
Mississippi	24.6	26.3	26.3	27.8 (27.4–28.1)	29.1 (26.7–31.2)	29.8 (25.7–33.6)	30.1 (24.6–35.5)
Missouri	27.3	28.3	28.2	30.2 (29.2–31.2)	32.1 (28.9–34.9)	33.4 (28.2–37.9)	34.0 (26.8–40.2)
Montana	27.4	29.7	29.2	32.0 (31.0–32.9)	34.8 (31.8–37.6)	36.9 (31.7–41.5)	38.2 (31.0–44.8)
Nebraska	27.6	28.7	29.0	30.4 (29.3–31.4)	32.2 (28.8–35.3)	33.6 (28.1–38.6)	34.3 (27.0–41.3)
Nevada	24.9	25.7	25.5	25.6 (24.5–26.6)	25.8 (22.8–28.5)	25.7 (21.2–30.0)	25.6 (19.8–31.2)
New Hampshire	24.4	25.6	25.8	27.1 (26.0–28.0)	28.4 (25.4–31.4)	29.7 (24.7–34.6)	30.4 (23.7–37.4)
New Jersey	24.5	25.2	24.9	25.0 (24.3–25.7)	24.9 (22.2–27.3)	24.5 (20.1–28.5)	24.2 (18.6–29.4)
New Mexico	25.4	28.0	28.2	29.3 (28.2–30.2)	30.4 (27.3–33.1)	31.2 (26.2–35.7)	31.8 (24.9–38.1)
New York	28.1	29.1	29.3	30.9 (29.9–31.8)	31.7 (28.3–34.4)	32.0 (26.4–36.3)	32.0 (24.5–37.5)
North Carolina	25.4	27.0	27.3	28.6 (27.6–29.4)	29.6 (26.8–32.2)	30.1 (25.7–34.1)	30.0 (24.3–35.6)
North Dakota	29.3	31.5	31.0	34.5 (32.7–35.7)	37.4 (33.4–40.6)	39.2 (33.3–44.0)	40.1 (32.5–46.5)

(continued)

Table 9.2 (continued)

		2010		2020	2030	2040	2050
	2000	Obs.	Proj.	Med (Low-High)	Med (Low-High)	Med (Low-High)	Med (Low-High)
Ohio	27.3	28.9	28.9	30.6 (29.5–31.6)	32.3 (29.0–35.2)	33.3 (28.0–38.1)	33.8 (26.5–40.3)
Oklahoma	26.7	27.5	27.9	29.9 (28.9–30.8)	31.8 (28.7–34.4)	33.2 (28.1–37.5)	34.0 (27.1–40.1)
Oregon	26.1	27.4	27.5	29.1 (28.0–30.1)	30.3 (27.1–33.3)	31.0 (25.8–35.8)	31.1 (24.2–37.6)
Pennsylvania	27.7	28.6	28.7	28.9 (27.8–29.9)	29.4 (26.2–32.4)	29.7 (24.8–34.5)	29.5 (23.2–35.9)
Rhode Island	28.6	29.6	29.8	30.8 (29.7–31.7)	32.5 (29.2–35.1)	35.2(29.6–39.3)	39.1 (30.0–44.6)
South Carolina	25.0	26.5	26.6	28.2 (27.3–29.1)	29.4 (26.6–31.9)	30.2 (25.7–34.3)	30.6 (24.7–36.4)
South Dakota	27.6	29.4	29.5	31.5 (30.3–32.6)	33.3 (29.8–36.5)	34.5 (29.1–39.6)	34.9 (27.8–41.8)
Tennessee	25.8	26.9	27.0	28.2 (27.2–29.1)	29.3 (26.3–31.9)	29.9 (25.2–34.3)	30.0 (23.8–36.0)
Texas	23.7	24.2	24.6	26.1 (25.2–26.9)	27.0 (24.2–29.6)	27.3 (22.8–31.5)	27.2 (21.3–32.9)
Utah	17.8	18.7	19.0	19.9 (19.0–20.7)	21.5 (19.1–24.1)	23.6 (19.6–28.1)	25.5 (20.1–32.4)
Vermont	26.2	28.2	27.7	30.5 (29.2–31.5)	32.7 (29.2–36.2)	34.7 (28.9–40.1)	36.0 (29.3–43.5)
Virginia	25.1	26.0	25.7	26.9 (25.9–27.8)	27.7 (24.7–30.3)	28.0 (23.3–32.3)	28.0 (21.9–33.8)
Washington	26.2	27.2	27.0	28.2 (27.1–29.2)	29.2 (25.9–32.1)	29.7 (24.5–34.3)	29.8 (22.9–35.9)
West Virginia	27.1	28.4	28.4	27.5 (26.5–28.5)	27.7 (24.8–30.6)	27.9 (23.6–32.6)	27.9 (22.5–34.5)
Wisconsin	26.8	28.2	27.9	29.3 (28.2–30.4)	30.9 (27.5–33.9)	31.5 (26.2–36.5)	31.4 (24.5–38.1)
Wyoming	26.3	28.0	27.9	29.3 (28.2–30.4)	30.7 (27.5–33.8)	31.8 (26.9–36.7)	32.2 (26.2–38.8)
6-county region, SC	22.7	22.1	22.0	21.1 (20.4–21.7)	20.0 (18.1–21.7)	19.2 (16.9–21.8)	18.7 (15.7–22.1)
Imperial county	17.1	17.0	16.8	15.4 (14.9–15.8)	14.5 (13.6–15.4)	14.5 (13.1–14.4)	13.2 (13.1–14.7)
Los Angeles county	24.6	24.2	23.9	23.5 (22.8–24.1)	22.5 (20.7–24.2)	21.8 (19.7–24.2)	21.4 (18.6–24.4)
Orange county	21.1	20.9	21.0	19.7 (19.0–20.3)	18.5 (16.5–20.2)	17.8 (14.8–20.5)	17.2 (13.7–20.6)
Riverside county	20.7	19.3	19.3	18.4 (17.6–19.1)	17.8 (15.9–19.8)	17.6 (14.7–20.7)	17.4 (13.9–21.6)
S Bernardino county	18.4	17.7	17.9	16.7 (16.1–17.3)	15.8 (14.1–17.4)	15.4 (13.1–17.9)	15.0 (12.2–18.2)
Ventura county	18.9	19.9	20.3	19.1 (18.4–19.7)	18.3 (16.5–20.1)	17.9 (15.1–20.7)	17.3 (13.9–21.2)
M-S 7-county region	27.5	28.5	28.4	28.9 (27.9–28.9)	29.5 (26.4–29.5)	30.2 (25.5–30.2)	30.7 (24.5–30.7)

Table 9.3 Projections of % married-couple households for the 50 states, DC and some counties, 2000–2050

		2010		2020	2030	2040	2050
	2000	Obs.	Proj.	Med (Low-High)	Med (Low-High)	Med (Low-High)	Med (Low-High)
United States	51.4	48.4	48.4	46.0 (44.9–47.2)	44.5 (40.7–48.5)	43.6 (37.3–50.2)	43.2 (34.3–52.1)
Alabama	52.1	47.9	47.6	44.6(43.4–45.9)	42.5 (38.7–46.5)	41.0 (34.6–47.6)	40.0 (31.1–49.2)
Alaska	50.7	49.4	49.4	49.7 (48.3–51.4)	49.2 (44.8–54.2)	49.1 (41.5–56.6)	49.2 (38.7–58.5)
Arizona	51.9	48.1	49.4	49.2 (48.1–50.5)	48.9 (45.1–52.8)	48.8 (42.3–55.0)	48.8 (39.6–57.0)
Arkansas	52.5	49.5	50.4	48.3 (46.8–49.7)	46.3 (41.9–50.8)	44.7 (37.6–51.6)	43.3 (33.5–52.6)
California	50.8	49.4	49.6	47.6 (46.5–48.7)	46.7 (43.2–50.5)	46.6 (40.7–53.0)	46.8 (38.2–55.3)
Colorado	52.1	49.2	49.6	47.1 (45.9–48.4)	45.0 (41.0–49.1)	44.2 (37.5–51.0)	43.9 (34.5–52.8)
Connecticut	48.1	49.0	48.8	47.0 (45.9–48.1)	45.6 (42.1–49.0)	44.8 (38.9–50.0)	44.6 (36.3–51.2)
Delaware	51.3	48.3	48.1	46.5 (45.3–47.9)	45.2 (41.2–49.3)	44.4 (37.7–50.8)	43.9 (34.6–52.1)
District of Columbia	22.4	22.0	22.0	21.0 (20.2–21.8)	20.6 (17.9–23.3)	20.4 (16.1–24.9)	20.5 (14.8–26.4)
Florida	50.3	46.6	46.5	42.4 (41.4–43.3)	39.8 (36.7–43.1)	38.0 (32.6–44.0)	36.8 (29.0–45.6)
Georgia	51.0	47.8	47.6	45.6 (44.7–46.5)	43.5 (40.0–47.3)	42.3 (36.2–48.6)	41.7 (33.2–50.2)
Hawaii	51.8	50.5	50.6	49.6 (48.3–51.1)	52.1 (47.7–56.6)	53.3 (46.5–59.3)	53.1 (44.4–59.8)
Idaho	58.8	55.3	55.6	53.3 (51.9–54.8)	51.9 (47.2–56.6)	51.9 (44.0–59.2)	52.6 (41.6–61.7)
Illinois	51.5	48.2	48.5	46.0 (44.9–47.2)	45.0(40.9–49.1)	44.7 (37.7–51.7)	45.0 (35.0–54.3)
Indiana	53.5	49.6	49.6	45.0 (43.9–46.3)	41.6 (37.8–45.7)	38.9 (32.5–45.7)	36.8 (28.0–46.2)
Iowa	56.4	51.2	51.0	46.1 (45.0–47.3)	42.2 (38.6–46.0)	39.0 (33.1–45.6)	36.7 (28.3–46.3)
Kansas	55.5	51.1	51.1	51.5 (50.1–53.0)	50.9 (46.5–55.1)	50.3 (43.3–56.4)	50.1 (40.4–57.7)
Kentucky	54.0	49.3	49.2	45.4 (44.2–46.7)	42.5 (38.7–46.6)	40.4 (34.0–47.3)	39.1 (30.0–48.7)
Louisiana	43.7	44.4	44.7	44.0 (43.6–44.5)	42.8 (39.5–46.3)	41.9 (36.0–48.0)	41.4 (33.0–49.5)
Maine	53.6	48.5	48.3	44.8 (43.5–46.1)	42.2 (38.3–46.4)	40.6 (34.0–47.7)	40.1 (30.8–49.9)
Maryland	49.8	47.6	47.7	47.2 (45.9–48.5)	46.8 (42.7–51.0)	46.5 (39.7–53.1)	46.4 (37.0–54.8)
Massachusetts	45.8	46.3	46.4	44.4 (43.2–45.5)	42.8 (39.2–46.2)	41.7 (35.6–47.2)	41.1 (32.5–48.4)
Michigan	51.8	48.0	47.7	47.3 (46.1–48.4)	46.7 (42.7–50.9)	46.3 (39.5–53.1)	46.3 (36.6–55.1)

(continued)

Table 9.3 (continued)

		2010		2020	2030	2040	2050
	2000	Obs.	Proj.	Med (Low-High)	Med (Low-High)	Med (Low-High)	Med (Low-High)
Minnesota	55.7	50.8	50.8	48.5 (47.7–49.3)	46.5 (43.2–49.9)	44.5 (38.5–50.7)	43.5 (34.5–52.4)
Mississippi	48.7	45.4	45.6	44.0 (43.6–44.3)	42.6 (39.4–45.9)	41.6 (35.8–47.4)	41.0 (32.7–49.1)
Missouri	52.5	48.4	48.4	44.9 (43.6–46.2)	42.5 (38.5–46.7)	40.8 (34.2–48.0)	39.9 (30.5–50.0)
Montana	55.0	49.2	49.5	44.7 (43.5–45.9)	41.0 (37.2–45.0)	38.4 (32.0–45.3)	36.6 (27.7–46.6)
Nebraska	55.9	50.8	50.8	47.8 (46.4–49.2)	45.3 (41.2–49.6)	43.3 (36.5–50.1)	41.9 (32.3–50.9)
Nevada	49.5	46.0	46.1	43.2 (41.7–44.7)	41.7 (37.1–46.6)	40.9 (33.4–48.6)	40.3 (30.3–50.5)
New Hampshire	50.0	52.1	51.9	49.5 (48.3–50.8)	47.6 (43.5–51.7)	46.1 (39.1–53.0)	45.4 (35.3–54.7)
New Jersey	53.2	51.1	51.1	50.4 (49.4–51.3)	50.3 (46.6–53.8)	50.4 (44.0–56.2)	50.8 (41.7–58.0)
New Mexico	50.6	45.3	45.5	42.7 (41.4–44.1)	41.0 (36.9–45.4)	40.0 (33.3–47.2)	39.4 (30.1–49.3)
New York	47.0	43.6	43.9	40.1 (38.9–41.2)	38.4 (34.8–42.2)	37.7 (31.6–44.0)	37.6 (29.2–46.0)
North Carolina	52.2	48.4	48.0	45.4 (44.2–46.7)	43.9 (40.0–47.9)	43.0 (36.5–49.4)	42.6 (33.6–51.2)
North Dakota	55.2	48.6	48.6	42.1 (40.9–43.4)	38.4 (34.5–42.6)	36.0 (29.6–43.2)	34.6 (25.8–44.8)
Ohio	51.4	47.2	47.1	43.9 (42.6–45.3)	41.9 (37.7–46.3)	40.6 (33.6–47.8)	39.9 (30.1–49.6)
Oklahoma	54.1	49.5	49.3	45.7 (44.5–47.0)	42.9 (38.9–47.2)	40.8 (34.1–47.8)	39.1 (29.9–49.1)
Oregon	52.6	48.3	48.1	43.6 (42.3–45.0)	41.4 (37.0–45.8)	40.0 (32.8–47.4)	39.4 (29.4–49.5)
Pennsylvania	52.1	48.2	48.3	47.2 (45.8–48.6)	46.4 (42.0–50.7)	45.8 (38.8–52.3)	45.9 (36.2–54.1)
Rhode Island	40.2	44.5	44.2	40.8 (39.7–42.0)	36.6 (33.1–40.3)	32.5 (27.2–38.8)	27.6 (20.6–37.9)
South Carolina	47.0	47.2	46.9	45.2 (44.1–46.4)	44.1 (40.4–47.9)	43.5 (37.3–49.7)	43.3 (34.5–51.6)
South Dakota	56.0	50.1	49.9	45.9 (44.4–47.4)	43.8 (39.2–48.6)	42.6 (35.2–50.1)	42.0 (31.8–51.8)
Tennessee	52.4	48.7	48.8	47.0 (45.8–48.3)	45.8 (41.6–50.1)	45.0 (38.0–52.0)	44.6 (34.9–54.0)
Texas	53.8	50.6	50.2	47.7 (46.4–49.0)	46.4 (42.2–50.8)	45.8 (38.8–52.9)	45.5 (35.8–55.0)
Utah	64.1	61.0	60.6	59.3 (57.7–60.9)	58.3 (53.5–62.7)	57.7 (50.0–64.0)	57.4 (46.6–65.3)
Vermont	53.7	48.5	48.2	42.6 (41.2–44.2)	39.8 (34.9–44.7)	37.6 (29.9–45.4)	36.1 (25.8–45.5)
Virginia	52.5	50.2	50.0	48.1 (46.8–49.5)	47.1 (43.0–51.4)	46.7 (39.8–53.5)	46.6 (37.1–55.5)
Washington	52.0	49.2	49.1	45.6 (44.3–47.0)	43.7 (39.4–48.2)	42.6(35.6–49.9)	42.1 (32.5–51.9)
West Virginia	49.9	49.8	50.1	52.9 (51.5–54.4)	53.2 (48.9–57.4)	53.4 (46.5–59.3)	53.7 (44.2–61.2)
Wisconsin	54.9	49.6	49.7	45.7 (44.3–47.1)	43.4 (39.0–47.9)	42.1 (35.0–49.3)	41.7 (31.9–51.1)
Wyoming	54.6	50.9	50.8	48.2 (46.7–49.7)	46.6 (41.9–51.3)	45.6 (38.0–52.9)	45.1 (34.6–54.1)
6-county region, SC	43.0	49.6	50.0	50.5 (49.4–51.5)	50.9 (47.5–54.1)	51.2 (45.8–55.6)	51.7 (44.3–57.2)
Imperial county	37.6	53.8	54.1	56.0 (55.1–56.6)	56.2 (54.0–57.5)	56.0 (53.8–58.0)	57.4 (51.8–58.3)
Los Angeles county	42.8	45.7	46.7	46.5 (45.5–47.5)	46.6 (43.3–50.0)	46.4 (40.9–51.1)	46.2 (38.6–52.6)
Orange county	45.7	54.2	53.9	53.9 (52.9–54.9)	53.1 (49.8–56.2)	52.7 (47.1–57.6)	52.5 (44.8–58.8)
Riverside county	42.1	54.8	54.8	55.9 (54.7–57.1)	56.3 (52.5–59.5)	56.7 (51.0–61.1)	57.1 (49.6–62.3)
S Bernardino county	39.8	53.5	53.1	54.1 (52.9–55.2)	54.9 (51.4–58.1)	55.2 (49.9–59.3)	55.8 (48.7–60.5)
Ventura county	44.5	56.4	56.4	57.6 (56.5–58.6)	57.2 (54.0–59.9)	57.2 (52.0–61.2)	57.5 (50.1–62.4)
M-S 7-county region	45.0	48.6	48.8	47.4 (47.4–48.3)	45.2 (45.2–48.3)	43.5 (43.5–48.3)	42.3 (42.3–48.6)

Table 9.4 Projections of % cohabiting-couple households for the 50 states, DC and some counties, 2000–2050

		2010		2020	2030	2040	2050
	2000	Obs.	Proj.	Med (Low-High)	Med (Low-High)	Med (Low-High)	Med (Low-High)
United States	5.0	6.6	6.6	7.4 (7.0–7.7)	7.6 (6.8–8.2)	7.7 (6.8–8.3)	7.7 (6.8–8.4)
Alabama	3.0	4.7	4.7	5.0 (4.7–5.2)	5.1 (4.6–5.4)	5.2 (4.6–5.4)	5.2 (4.6–5.4)
Alaska	7.8	8.6	8.5	8.7 (7.4–8.0)	8.5 (6.1–6.9)	8.4 (5.9–6.6)	8.4 (5.8–6.6)
Arizona	5.8	7.8	7.7	8.2 (7.8–8.6)	8.2 (7.5–8.8)	8.1 (7.4–8.8)	8.1 (7.3–8.9)
Arkansas	3.3	5.7	5.6	6.3(6.0–6.5)	6.4 (5.9–6.7)	6.4 (6.0–6.7)	6.5 (6.0–6.6)
California	5.7	7.2	7.2	7.8 (7.4–8.2)	8.1 (7.3–8.9)	8.6 (7.7–9.4)	8.7 (7.8–9.4)
Colorado	5.3	6.5	6.4	6.7 (6.4–7.0)	6.8 (6.2–7.4)	6.8 (6.1–7.4)	6.9 (6.1–7.4)
Connecticut	6.5	6.6	6.6	6.9 (6.5–7.3)	7.0 (6.1–7.8)	6.9 (6.0–8.2)	6.9 (5.9–8.4)
Delaware	5.7	7.3	7.3	7.5 (7.1–7.9)	7.5 (6.8–8.2)	7.5 (6.7–8.1)	7.5 (6.6–8.2)
District of Columbia	5.1	7.7	8.0	8.1 (7.4–8.8)	8.0 (6.5–9.7)	8.0 (6.1–10.0)	8.0 (5.8–10.4)
Florida	5.4	7.3	7.3	8.7 (8.2–9.2)	9.1 (8.0–10.2)	9.3 (7.9–10.5)	9.5 (7.7–10.7)
Georgia	4.4	5.9	6.0	6.2 (5.9–6.4)	6.4 (5.7–7.1)	6.6 (5.8–7.3)	6.7 (5.8–7.5)
Hawaii	5.5	7.3	7.3	9.6 (9.0–10.1)	9.8 (9.2–10.3)	8.2 (7.7–8.7)	6.6 (6.0–7.5)
Idaho	4.8	6.3	6.4	7.1 (6.9–7.4)	7.0 (6.6–7.2)	6.2 (6.0–6.1)	5.3 (5.2–5.2)
Illinois	4.5	6.3	6.4	7.3 (6.9–7.6)	7.5 (6.8–8.2)	7.7 (6.9–8.3)	7.8 (6.9–8.4)
Indiana	5.0	6.9	6.9	8.1 (7.6–8.6)	8.6 (7.6–9.6)	9.1 (7.8–10.3)	9.6 (7.9–10.9)
Iowa	4.5	6.7	6.8	8.4 (7.8–8.9)	8.9 (7.8–9.9)	9.3 (7.9–10.3)	9.6 (7.8–10.5)
Kansas	3.7	5.8	5.8	6.0 (5.8–6.2)	6.0 (5.6–6.3)	5.9 (5.6–6.3)	5.9 (5.7–6.5)
Kentucky	4.1	6.4	6.4	7.0 (6.6–7.4)	7.3 (6.5–7.8)	7.5 (6.6–8.0)	7.6 (6.5–8.1)
Louisiana	6.3	6.8	6.8	6.7 (6.4–6.9)	6.6 (5.9–7.2)	6.5 (5.8–7.1)	6.5 (5.7–7.1)
Maine	6.9	9.4	9.2	10.9 (10.3–11.4)	11.2 (10.1–12.1)	11.3 (10.1–12.0)	11.3 (9.9–11.8)
Maryland	5.2	6.4	6.2	6.4 (6.1–6.7)	6.5 (5.9–6.9)	6.5 (5.9–6.9)	6.5 (5.9–6.9)
Massachusetts	6.7	7.0	7.0	7.5 (7.1–8.0)	7.5 (6.6–8.5)	7.5 (6.5–8.7)	7.5 (6.3–8.9)
Michigan	5.1	6.4	6.4	6.5 (6.2–6.8)	6.4 (5.9–6.8)	6.3 (5.8–6.7)	6.3 (5.8–6.6)
Minnesota	5.0	6.9	6.9	7.4 (7.0–7.8)	7.5 (6.7–8.3)	7.6 (6.7–8.4)	7.7 (6.7–8.4)
Mississippi	4.3	5.7	5.5	5.6 (5.4–5.8)	5.6 (4.9–6.2)	5.6 (4.9–6.2)	5.6 (4.9–6.2)
Missouri	4.6	6.7	6.7	7.6 (7.2–8.0)	7.8 (7.0–8.5)	8.0 (7.0–8.6)	8.2 (7.0–8.7)
Montana	4.6	6.7	6.8	8.2 (7.7–8.6)	8.6 (7.6–9.4)	8.9 (7.7–9.7)	9.2 (7.7–9.8)
Nebraska	4.0	6.0	6.1	6.9 (6.5–7.2)	7.1 (6.4–7.6)	7.1 (6.4–7.6)	7.2 (6.4–7.6)
Nevada	6.6	8.7	8.5	10.1 (9.7–10.6)	10.4 (9.5–11.2)	10.6 (9.5–11.3)	10.7 (9.4–11.3)
New Hampshire	8.3	8.3	8.4	9.5 (9.1–10.0)	9.6 (8.8–10.2)	9.6 (8.8–10.1)	9.7 (8.8–10.0)
New Jersey	4.6	5.9	6.1	6.6 (6.4–6.9)	6.7 (6.1–7.3)	6.6 (6.0–7.3)	6.5 (5.9–7.3)
New Mexico	5.4	8.2	8.2	9.5 (9.0–9.9)	9.6 (8.8–10.3)	9.6 (8.6–10.3)	9.6 (8.4–10.1)
New York	5.1	6.8	6.8	8.0 (7.5–8.4)	8.1 (7.2–9.2)	8.2 (7.1–9.4)	8.2 (6.9–9.5)
North Carolina	4.3	5.9	5.9	6.5 (6.1–6.8)	6.6 (6.0–7.1)	6.7 (6.0–7.1)	6.8 (6.0–7.1)
North Dakota	4.4	6.4	6.5	7.8 (7.3–8.2)	8.2 (7.2–9.1)	8.5 (7.3–9.4)	8.8 (7.3–9.5)
Ohio	4.7	6.8	6.7	7.6 (7.2–8.0)	7.8 (7.1–8.5)	8.0 (7.1–8.6)	8.1 (7.0–8.7)
Oklahoma	3.5	5.9	6.0	6.7 (6.3–7.0)	6.9 (6.2–7.5)	7.1 (6.2–7.6)	7.3 (6.2–7.7)
Oregon	5.8	8.0	7.9	9.6 (9.1–10.1)	10.0 (9.0–10.8)	10.3 (9.1–10.9)	10. 4 (9.0–10.9)
Pennsylvania	4.6	6.6	6.6	7.6 (7.3–7.9)	7.7 (7.1–8.2)	7.7 (7.1–8.3)	7.7 (7.0–8.4)
Rhode Island	8.7	7.6	7.5	7.5(7.1–8.0)	7.4 (6.5–8.5)	7.3 (6.1–8.5)	7.0 (5.5–8.6)
South Carolina	5.8	6.0	6.0	6.2 (5.8–6.5)	6.1 (5.5–6.6)	6.0 (5.4–6.5)	6.0 (5.3–6.5)
South Dakota	4.4	6.5	6.5	7.5 (7.1–7.8)	7.8 (7.1–8.2)	7.9 (7.2–8.2)	8.0 (7.2–8.1)
Tennessee	3.8	5.8	5.8	6.2 (5.9–6.5)	6.2 (5.7–6.7)	6.3 (5.7–6.6)	6.4 (5.7–6.6)
Texas	4.0	6.0	6.0	6.6 (6.3–6.9)	6.7 (6.1–7.1)	6.8 (6.2–7.1)	6.8 (6.1–7.0)
Utah	3.2	4.6	4.6	4.8 (4.7–5.0)	4.8 (4.6–5.0)	4.7 (4.7–4.8)	4.7 (4.8–4.7)
Vermont	7.6	9.2	9.3	11.5 (11.1–12.0)	12.1 (11.1–12.7)	12.3 (11.2–12.8)	12.5 (11.1–12.5)
Virginia	4.4	5.7	5.7	6.1 (5.8–6.4)	6.2 (5.6–6.7)	6.3 (5.6–6.7)	6.3 (5.6–6.7)
Washington	6.0	7.7	7.7	8.9 (8.4–9.3)	9.1 (8.2–9.7)	9.2 (8.2–9.7)	9.2 (8.0–9.6)
West Virginia	5.6	6.6	6.7	6.3 (6.1–6.5)	6.1 (5.8–6.3)	5.9 (5.7–6.1)	5.8 (5.7–5.9)
Wisconsin	5.2	7.3	7.4	8.9 (8.5–9.4)	9.3 (8.4–10.0)	9.4 (8.4–10.1)	9.5 (8.3–10.2)
Wyoming	5.5	7.1	6.9	7.7 (7.4–8.0)	7.8 (7.3–8.1)	7.9 (7.4–8.0)	7.9 (7.4–8.0)
6-county region, SC	7.9	7.0	7.1	7.2 (6.8–7.5)	7.0 (6.3–7.7)	6.9 (6.1–7.7)	6.8 (5.9–7.9)
Imperial county	10.7	6.1	6.3	5.8 (5.4–6.3)	5.7 (4.9–6.9)	5.6 (4.7–7.2)	5.7 (4.6–7.6)
Los Angeles county	7.3	7.2	7.5	7.4 (7.0–7.7)	7.1 (6.4–7.7)	7.0 (6.2–7.5)	6.9 (6.0–7.6)
Orange county	7.7	5.7	5.8	6.5 (6.1–6.9)	7.0 (6.3–7.8)	7.1 (6.3–8.1)	7.2 (6.3–8.5)
Riverside county	9.7	7.5	7.3	7.2 (6.9–7.6)	6.9 (6.2–7.7)	6.6 (5.9–7.7)	6.4 (5.6–7.7)
S Bernardino county	9.8	7.6	7.6	7.7 (7.4–8.1)	7.4 (6.7–8.2)	7.3 (6.4–8.5)	7.1 (6.1–8.7)
Ventura county	9.4	6.1	6.1	6.0 (5.7–6.3)	5.9 (5.4–6.6)	5.8 (5.2–6.8)	5.8 (5.1–7.1)
M-S 7-county region	7.1	6.9	6.6	6.5 (6.5–6.8)	6.5 (6.5–7.2)	6.6 (6.6–7.4)	6.6 (6.6–7.7)

Table 9.5 Projections of % single-parent households for the 50 states, DC, and some counties, 2000–2050

		2010	2020	2030	2040	2050
	2000	Proj.	Med (Low-High	Med (Low-High)	Med (Low-High)	Med (Low-High)
United States	30.9	28.7	30.8 (29.3–32.2)	32.1 (27.6–37.0)	32.8 (26.1–40.9)	33.1 (24.7–44.4)
Alabama	34.8	33.0	35.8 (34.5–37.3)	37.7 (33.6–42.4)	39.2 (32.7–47.0)	40.1 (31.5–51.1)
Alaska	28.1	25.5	24.8 (23.8–26.7)	24.9 (21.7–30.6)	25.1 (20.2–34.1)	25.1 (19.1–37.3)
Arizona	30.1	27.1	27.2 (25.8–28.7)	27.6 (23.4–32.0)	28.2 (21.9–35.3)	28.4 (20.4–38.0)
Arkansas	32.9	28.4	29.5 (27.9–31.2)	30.6 (25.9–35.8)	31.5 (24.5–40.2)	32.2 (23.2–44.1)
California	30.0	27.2	29.6 (28.1–31.1)	30.6 (26.0–35.5)	30.8 (23.8–39.0)	31.2 (22.3–42.6)
Colorado	26.8	25.2	27.9 (26.4–29.5)	30.2 (25.8–35.2)	30.8 (24.4–39.1)	30.8 (23.0–42.3)
Connecticut	35.6	30.3	31.5 (30.1–32.8)	32.8 (29.1–37.0)	33.6 (28.3–40.3)	33.8 (27.4–43.0)
Delaware	31.6	29.4	31.2 (29.5–32.9)	32.5 (27.9–37.8)	33.3 (26.7–41.8)	33.8 (26.0–45.2)
District of Columbia	61.0	57.1	60.5 (58.9–62.4)	62.4 (57.6–67.9)	63.5 (56.3–71.6)	64.0 (54.9–74.0)
Florida	33.9	31.2	34.9 (33.3–36.5)	37.5 (32.5–42.8)	39.2 (31.0–48.1)	40.2 (29.4–52.6)
Georgia	33.9	32.5	34.5 (33.6–35.5)	36.7 (32.5–41.0)	38.1 (31.4–45.4)	38.8 (30.1–48.9)
Hawaii	31.7	29.1	28.4 (26.8–30.0)	26.9 (22.9–31.2)	27.7 (22.3–34.1)	28.7 (22.3–37.4)
Idaho	23.0	20.3	21.7 (20.2–23.3)	22.7 (18.4–27.6)	22.9 (16.9–30.9)	22.7 (15.7–33.2)
Illinois	29.8	27.8	29.5 (28.2–30.8)	30.0 (25.5–34.9)	29.6 (22.9–38.1)	28.8 (20.7–40.8)
Indiana	28.4	26.2	30.3 (28.9–31.8)	33.6 (29.1–38.8)	36.2 (28.8–45.0)	38.1 (28.0–50.4)
Iowa	22.0	22.1	26.0 (24.5–27.6)	29.3 (25.0–34.8)	31.3 (24.6–40.8)	32.4 (23.8–45.7)
Kansas	25.3	24.0	23.8 (22.3–25.5)	24.4 (20.5–29.3)	25.1 (20.0–32.8)	25.4 (19.7–35.6)
Kentucky	29.6	28.0	30.6 (29.2–32.1)	32.8 (28.6–37.5)	34.4 (27.8–42.6)	35.6 (26.8–47.7)
Louisiana	44.8	35.3	36.1 (35.5–36.8)	38.0 (33.9–42.3)	39.4 (32.9–46.8)	40.4 (31.9–50.3)
Maine	24.6	24.7	26.9 (25.2–28.6)	29.0 (24.2–34.2)	30.4 (22.8–39.1)	30.4 (21.0–42.5)
Maryland	33.3	31.6	31.8 (30.3–33.3)	31.7 (27.5–36.6)	31.8 (25.8–39.9)	32.0 (24.8–43.0)
Massachusetts	34.4	30.4	33.4(31.9–34.9)	35.6 (31.5–40.4)	36.6 (30.7–44.7)	37.2 (29.8–48.6)
Michigan	30.4	29.1	29.5 (28.3–30.8)	29.8 (25.6–34.4)	29.7 (23.4–37.7)	29.5 (21.8–40.6)
Minnesota	21.7	22.7	25.5 (24.5–26.7)	27.2 (23.6–31.9)	28.3 (22.7–36.5)	28.8 (21.8–40.4)
Mississippi	39.2	36.1	37.5 (36.9–38.1)	39.0 (34.9–43.3)	40.1 (33.4–47.5)	40.6 (31.8–50.9)
Missouri	29.6	27.5	30.0 (28.3–31.7)	31.7 (26.8–37.0)	32.5 (25.2–41.7)	32.7 (23.4–45.8)
Montana	24.9	24.5	27.5 (25.8–29.2)	29.9 (24.7–35.8)	31.2 (23.2–41.5)	31.4 (21.2–45.7)
Nebraska	23.2	23.6	26.3 (24.7–28.0)	27.8 (23.6–33.2)	28.6 (22.5–37.6)	29.2 (22.0–41.8)
Nevada	31.7	28.2	30.1 (28.1–32.2)	31.5 (25.8–37.8)	32.5 (24.2–42.2)	33.1 (23.0–46.1)
New Hampshire	32.9	23.1	24.4 (22.8–26.1)	26.5 (21.9–31.9)	27.7 (20.8–36.9)	28.0 (19.4–40.8)
New Jersey	28.6	27.4	28.0 (27.1–28.9)	28.2 (24.8–32.2)	28.0 (23.1–34.9)	27.9 (22.0–37.6)
New Mexico	32.8	29.3	31.1 (29.3–33.0)	32.4 (27.3–38.1)	33.0 (25.5–42.1)	33.3 (23.9–45.7)
New York	33.6	33.4	37.1 (35.5–38.7)	38.6 (33.9–43.9)	39.1 (32.4–47.7)	39.2 (31.1–50.8)
North Carolina	32.5	30.0	32.2 (30.7–33.7)	33.4 (29.2–38.3)	33.9 (27.8–42.1)	34.0 (26.3–45.2)
North Dakota	21.1	24.5	30.9 (29.8–32.8)	34.2 (29.1–40.9)	35.9 (27.4–46.9)	36.1 (24.8–50.9)
Ohio	30.8	28.6	30.7 (29.0–32.4)	32.0 (27.1–37.6)	32.7 (25.3–42.0)	32.9 (23.9–45.7)
Oklahoma	29.9	26.6	29.1 (27.5–30.8)	30.9 (26.2–36.4)	32.2 (25.1–41.4)	33.2 (24.0–46.2)
Oregon	27.2	26.2	29.6 (27.8–31.5)	31.4 (26.1–37.4)	32.2 (24.3–42.0)	32.6 (22.8–46.1)
Pennsylvania	29.3	28.4	29.7 (28.2–31.2)	30.5 (26.0–35.7)	30.9 (24.5–39.2)	31.0 (23.5–42.1)
Rhode Island	44.8	35.0	39.6 (37.9–41.3)	44.4 (39.6–49.7)	48.2 (40.8–56.5)	52.4 (41.3–63.8)
South Carolina	41.8	33.9	35.0 (33.5–36.5)	35.9 (31.7–40.5)	36.4 (29.9–44.0)	36.4 (28.1–46.9)
South Dakota	22.6	25.0	28.9 (27.0–30.8)	30.8 (25.6–36.7)	31.4 (23.9–40.9)	31.6 (22.4–44.6)
Tennessee	33.0	29.0	30.1 (28.6–31.5)	30.7 (26.3–35.6)	31.0 (24.1–39.2)	31.1 (22.5–42.5)
Texas	30.0	28.2	30.2 (28.7–31.7)	31.2 (26.7–36.2)	31.7 (24.9–39.9)	32.0 (23.5–43.5)
Utah	19.9	19.6	20.5 (19.1–22.0)	21.3 (17.4–26.0)	21.9 (16.6–29.3)	22.2 (15.8–32.5)
Vermont	23.3	24.3	26.9 (24.8–29.2)	27.9 (22.1–35.0)	28.3 (20.8–39.4)	28.1 (20.1–42.8)
Virginia	30.5	28.0	29.4 (27.9–31.0)	30.1 (25.9–34.9)	30.4 (24.2–38.3)	30.4 (22.9–41.2)
Washington	26.9	25.0	27.9 (26.3–29.7)	29.6 (24.9–35.2)	30.5 (23.8–39.6)	30.8 (22.5–43.1)
West Virginia	36.3	26.1	23.2 (21.8–24.6)	23.4 (19.8–27.8)	23.9 (18.7–30.9)	23.9 (17.7–33.3)
Wisconsin	24.2	25.3	28.5 (26.7–30.3)	30.3 (25.1–36.1)	31.2 (23.6–40.4)	31.5 (22.3–43.7)
Wyoming	25.7	23.8	25.8 (24.1–27.5)	27.0 (22.4–32.4)	27.1(20.5–36.0)	26.8 (19.1–39.5)
6-county region, SC	40.9	28.8	29.7 (28.3–31.1)	30.7 (26.9–35.0)	31.3 (25.9–38.0)	31.5 (24.8–40.4)
Imperial county	48.4	26.9	26.2 (25.3–27.3)	26.9 (24.7–29.9)	27.5 (24.4–32.5)	27.8 (23.9–35.5)
Los Angeles county	38.4	31.4	33.5 (32.1–34.9)	35.2 (31.0–39.8)	36.7 (30.2–44.1)	37.4 (29.1–47.7)
Orange county	40.0	26.2	27.3 (26.1–28.7)	29.2 (25.5–33.4)	30.2 (24.6–37.2)	30.8 (23.6–40.1)
Riverside county	46.8	24.4	23.6 (22.3–25.0)	24.2 (20.8–28.2)	24.7 (20.3–30.5)	25.1 (20.1–32.4)
S Bernardino county	49.2	26.6	26.4 (25.0–27.9)	26.6 (23.0–30.9)	26.9 (22.3–33.2)	27.0 (21.5–35.1)
Ventura county	42.5	24.0	24.1 (22.9–25.4)	25.5 (22.4–29.5)	26.3 (21.8–32.5)	26.8 (21.0–35.1)
M-S 7-county region	37.0	27.5	30.0 (29.3–30.0)	32.5 (29.6–32.5)	33.8 (29.5–33.8)	34.7 (29.3–34.7)

Table 9.6 Projections of % elderly households with householders aged 65+ for the 50 states, DC, and some counties, 2000–2050

	2000	2010 Proj.	2020 Med (Low-High)	2030 Med (Low-High)	2040 Med (Low-High)	2050 Med (Low-High)
United States	17.8	17.6	20.5 (20.3–20.6)	23.5 (22.4–24.1)	24.5 (22.3–25.8)	25.3 (22.2–27.3)
Alabama	18.1	18.2	21.0 (20.8–21.1)	24.2 (23.2–24.5)	25.1 (23.5–26.0)	26.1 (23.6–27.4)
Alaska	6.8	11.3	15.4 (15.3–15.4)	17.7 (17.4–17.9)	18.5 (17.9–18.7)	19.3 (18.5–19.8)
Arizona	18.8	20.3	23.2 (23.1–23.2)	25.7 (24.7–26.3)	27.1 (25.2–28.2)	28.2 (25.4–29.7)
Arkansas	20.9	19.9	22.4 (22.4–22.4)	24.8 (24.4–24.9)	25.8 (24.9–26.1)	26.6 (25.2–27.1)
California	15.4	15.5	18.5 (18.3–18.6)	21.5 (19.7–22.4)	22.9 (19.5–25.5)	24.0 (19.3–27.6)
Colorado	13.5	14.3	18.3 (18.2–18.4)	21.8 (20.7–22.2)	23.1 (21.1–24.0)	24.2 (21.4–25.6)
Connecticut	18.1	20.0	22.6 (22.3–22.7)	25.7 (24.0–26.7)	26.6 (23.2–28.8)	26.9 (22.3–29.8)
Delaware	18.4	18.9	21.4 (21.3–21.5)	24.0 (23.3–24.4)	25.0 (23.6–25.8)	26.0 (24.0–27.3)
District of Columbia	15.3	14.2	13.5 (13.2–13.6)	12.9 (12.1–13.2)	12.9 (11.8–13.0)	13.3 (12.2–13.5)
Florida	23.8	23.6	25.6 (25.5–25.7)	28.0 (26.6–28.8)	28.8 (26.0–30.5)	29.5 (25.5–31.9)
Georgia	13.2	15.2	18.4 (18.2–18.5)	21.4 (20.6–21.9)	23.0 (21.4–24.0)	24.1 (21.7–25.7)
Hawaii	15.7	18.4	22.7 (22.4–22.8)	26.3 (24.6–27.0)	28.5 (25.5–29.7)	30.1 (26.0–31.7)
Idaho	16.6	17.8	20.7 (20.6–20.7)	23.2 (22.9–23.4)	24.3 (23.5–24.6)	25.9 (24.6–26.6)
Illinois	18.1	16.9	19.3 (19.2–19.4)	22.2(21.0–22.8)	23.1 (20.6–24.3)	23.6 (20.1–25.3)
Indiana	17.9	17.1	19.0 (18.9–19.0)	20.5 (19.9–20.8)	20.3 (19.2–21.0)	20.5 (18.9–21.5)
Iowa	22.0	20.1	23.3 (23.2–23.3)	26.5 (25.7–26.9)	27.0 (25.3–28.1)	28.2 (25.7–30.1)
Kansas	19.5	18.1	21.6 (21.6–21.6)	25.4 (24.6–25.9)	26.2 (24.1–27.4)	27.0 (23.6–29.0)
Kentucky	17.6	18.5	21.5 (21.5–21.5)	24.4 (23.9–24.6)	25.1(24.1–25.8)	25.8 (24.1–26.8)
Louisiana	15.6	16.2	20.0 (19.8–20.1)	24.4 (23.6–24.8)	25.8 (24.2–26.7)	27.0 (24.5–28.4)
Maine	20.7	20.9	25.6 (25.6–25.5)	30.1 (29.8–30.2)	31.9 (31.0–32.2)	33.0 (31.5–33.6)
Maryland	15.6	16.1	18.9 (18.6–19.0)	21.6 (20.0–22.3)	22.4 (19.7–23.8)	23.3 (19.7–25.2)
Massachusetts	17.9	18.1	21.3 (21.1–21.4)	25.0 (23.3–25.8)	26.2 (22.8–27.8)	26.5 (21.8–28.8)
Michigan	17.9	17.6	21.2 (21.0–21.3)	25.3 (24.2–25.8)	26.8 (24.8–27.7)	27.7 (24.8–29.2)
Minnesota	18.4	17.3	22.1 (21.5–22.5)	27.9 (26.5–28.7)	28.8 (26.6–30.1)	29.6 (26.8–31.3)
Mississippi	17.1	17.0	19.8 (19.7–19.9)	23.5 (22.7–23.9)	25.3 (23.8–26.1)	26.9 (24.7–28.2)
Missouri	19.2	18.3	21.4 (21.3–21.4)	24.7 (24.1–25.1)	25.3 (23.9–26.0)	25.7 (23.7–27.0)
Montana	19.2	19.2	23.5 (23.5–23.5)	26.9 (26.7–27.0)	27.8 (27.3–28.1)	29.3 (28.4–30.0)
Nebraska	20.5	19.4	22.9 (22.9–22.9)	26.3 (25.6–26.6)	27.4 (25.8–28.2)	28.5 (26.2–29.6)
Nevada	14.9	16.3	18.2 (17.9–18.3)	19.6(18.6–20.1)	20.8 (19.0–21.8)	21.9 (19.6–23.6)
New Hampshire	15.1	17.2	23.7 (23.7–23.8)	29.3 (28.6–29.5)	31.0 (29.6–31.5)	32.0 (29.8–32.6)
New Jersey	19.5	18.4	20.4 (20.1–20.6)	23.4 (21.3–24.4)	24.6 (20.7–26.3)	25.2 (20.6–27.5)
New Mexico	16.4	17.3	21.3 (21.2–21.4)	24.8 (23.9–25.2)	25.7 (23.7–26.7)	27.6 (24.3–29.3)
New York	19.0	17.8	20.3 (20.0–20.4)	23.5 (21.1–24.8)	24.6 (19.9–27.2)	25.1 (19.2–28.5)
North Carolina	16.6	16.8	19.7 (19.6–19.8)	22.7 (22.6–23.2)	24.1 (24.0–25.2)	24.8 (24.4–26.7)
North Dakota	21.8	20.1	23.2 (23.0–23.3)	26.8 (26.3–27.1)	27.6 (26.5–28.4)	28.9 (27.3–30.1)
Ohio	18.9	18.3	21.6 (21.5–21.7)	25.0 (24.3–25.3)	25.7 (24.1–26.4)	26.2 (23.7–27.6)
Oklahoma	19.2	18.8	21.6 (21.5–21.6)	24.0 (23.4–24.2)	24.2 (23.1–24.9)	24.8 (23.3–26.0)
Oregon	18.2	17.8	21.4 (21.3–21.4)	23.6 (22.7–23.9)	24.1 (22.4–25.0)	25.0 (22.6–26.5)
Pennsylvania	22.8	20.7	23.8 (23.7–23.8)	27.7 (26.8–28.1)	28.2 (26.2–29.2)	28.7 (25.6–30.4)
Rhode Island	19.6	18.1	19.8 (19.7–19.9)	20.7 (19.8–21.0)	21.0 (19.1–21.7)	22.9 (20.2–23.8)
South Carolina	15.8	17.1	20.8 (20.7–20.8)	24.4 (23.6–24.7)	26.1 (24.6–26.7)	27.4 (25.3–28.3)
South Dakota	21.3	19.8	22.5 (22.5–22.5)	26.0 (25.8–26.1)	27.0 (26.3–27.5)	28.2 (26.7–29.1)
Tennessee	16.8	17.6	21.0 (20.9–21.1)	23.9 (23.2–24.2)	24.9 (23.6–25.5)	25.6 (23.7–26.6)
Texas	14.4	14.3	16.7 (16.5–16.8)	19.2 (18.2–19.8)	20.2 (18.4–21.8)	21.3 (18.7–23.9)
Utah	13.4	13.1	15.2 (15.1–15.2)	17.2 (16.8–17.5)	18.5 (17.7–19.1)	20.1 (18.7–21.1)
Vermont	18.9	19.0	24.0 (24.0–24.1)	28.1 (27.8–28.2)	29.3 (29.2–29.3)	31.1 (30.5–31.0)
Virginia	15.3	16.6	19.7 (19.5–19.8)	22.4 (21.2–22.9)	23.3 (21.1–24.4)	24.3 (21.2–25.8)
Washington	16.0	16.2	19.5 (19.3–19.5)	22.0 (21.0–22.4)	22.8 (21.0–23.8)	23.9 (21.3–25.1)
West Virginia	20.5	19.2	24.0 (23.9–24.0)	27.4 (27.0–27.6)	28.5 (27.5–29.0)	29.7 (28.3–30.4)
Wisconsin	19.9	18.5	22.1 (22.1–22.1)	25.9 (25.4–26.0)	26.7 (25.5–27.2)	27.2 (25.3–28.3)
Wyoming	16.8	16.7	19.8 (19.7–19.8)	22.0 (21.7–22.1)	22.5 (21.9–22.7)	23.5 (22.4–23.8)
6-county region, SC	13.8	14.7	17.9 (17.4–17.7)	22.6 (20.2–21.7)	26.3 (21.6–24.6)	28.3 (21.9–26.0)
Imperial county	12.6	16.0	16.5 (15.7–16.2)	18.5 (15.7–17.5)	19.5 (15.0–17.4)	20.6 (15.4–18.6)
Los Angeles county	13.1	14.1	17.8 (17.3–17.6)	24.2 (21.2–23.0)	29.6 (23.5–27.4)	31.9 (23.4–28.9)
Orange county	13.8	15.0	19.9 (19.1–19.6)	25.1 (21.4–23.5)	29.5 (22.2–26.7)	31.1 (21.4–27.4)
Riverside county	20.3	17.2	16.9 (16.7–16.8)	19.0 (18.3–18.8)	21.0 (19.6–20.6)	23.8 (21.5–23.1)
S Bernardino county	12.4	13.3	15.6 (15.3–15.5)	18.2 (17.0–17.8)	20.8 (17.8–19.6)	23.6 (19.3–21.9)
Ventura county	13.6	16.5	20.7 (20.1–20.5)	24.6 (21.8–23.3)	26.6 (21.2–24.5)	27.9 (20.7–25.2)
M-S 7-county region	12.5	14.6	18.2 (17.9–18.2)	21.4 (19.9–21.4)	22.0 (19.5–22.0)	21.8 (18.7–21.8)

Table 9.7 Projections of % elderly aged 65+ living alone for the 50 states, DC, and some counties, 2000–2050

	2000	2010 Proj.	2020 Med (Low-High)	2030 Med (Low-High)	2040 Med (Low-High)	2050 Med (Low-High)
United States	3.53	3.78	4.22 (4.13–4.30)	4.82 (4.35–5.23)	4.82 (4.23–5.93)	5.31 (4.05–6.62)
Alabama	3.94	4.30	4.92 (4.84–5.01)	5.71 (5.30–6.08)	5.71 (5.30–6.83)	6.41 (5.22–7.68)
Alaska	1.44	1.94	2.46 (2.43–2.54)	2.91 (2.73–3.22)	2.91 (2.72–3.66)	3.27(2.69–4.16)
Arizona	3.26	3.79	4.19 (4.10–4.28)	4.54 (4.07–4.95)	4.54 (3.85–5.50)	4.89 (3.68–6.08)
Arkansas	4.17	4.22	4.47 (4.40–4.56)	4.90 (4.51–5.27)	4.90 (4.41–5.90)	5.50 (4.41–6.70)
California	2.80	3.02	3.36 (3.28–3.42)	3.79 (3.30–4.18)	3.79 (3.08–4.72)	4.07 (2.86–5.16)
Colorado	2.78	3.09	3.74 (3.65–3.80)	4.40 (3.95–4.80)	4.40 (3.92–5.63)	5.16 (3.84–6.49)
Connecticut	3.51	4.11	4.44 (4.27–4.61)	4.94 (4.30–5.54)	4.94 (4.04–6.21)	5.20 (3.74–6.64)
Delaware	3.64	4.08	4.54 (4.45–4.62)	5.12 (4.68–5.51)	5.12 (4.56–6.19)	5.67 (4.43–6.98)
District of Columbia	4.89	4.26	4.18 (4.10–4.33)	4.12 (3.80–4.55)	4.12 (3.56–4.66)	4.13 (3.43–4.89)
Florida	4.71	5.21	5.64 (5.57–5.72)	6.15 (5.62–6.57)	6.15 (5.36–7.25)	6.68 (5.17–8.11)
Georgia	2.77	2.97	3.43 (3.33–3.49)	3.95 (3.57–4.28)	3.95 (3.53–4.89)	4.41 (3.41–5.49)
Hawaii	2.54	3.10	3.53 (3.45–3.60)	3.77 (3.37–4.14)	3.77 (3.22–4.63)	4.22 (3.16–5.23)
Idaho	3.03	3.45	3.89 (3.82–3.96)	4.46 (4.11–4.79)	4.46 (4.09–5.43)	5.10 (4.03–6.29)
Illinois	3.64	3.69	3.94 (3.85–4.01)	4.42 (3.96–4.82)	4.42 (3.76–5.41)	4.71 (3.51–5.95)
Indiana	3.72	3.76	4.00 (3.91–4.09)	4.39 (4.02–4.75)	4.39 (3.87–5.16)	4.81 (3.85–5.81)
Iowa	4.48	4.48	4.96 (4.88–5.04)	5.84 (5.35–6.28)	5.84 (5.32–7.29)	7.04 (5.35–8.79)
Kansas	4.05	4.15	4.50 (4.41–4.58)	5.16 (4.68–5.60)	5.16 (4.50–6.32)	5.55 (4.17–7.04)
Kentucky	3.93	4.11	4.68 (4.59–4.74)	5.44 (5.05–5.81)	5.44 (5.05–6.59)	6.17 (4.99–7.47)
Louisiana	3.06	4.00	4.69 (4.63–4.75)	5.54 (5.18–5.88)	5.54 (5.10–6.55)	5.99 (4.90–7.19)
Maine	4.34	4.89	5.72 (5.64–5.81)	6.82 (6.38–7.25)	6.82 (6.61–8.37)	7.96 (6.58–9.45)
Maryland	3.24	3.45	3.78 (3.71–3.85)	4.18 (3.75–4.55)	4.18 (3.57–5.03)	4.45 (3.39–5.48)
Massachusetts	3.80	4.21	4.59 (4.49–4.67)	5.25 (4.67–5.74)	5.25 (4.47–6.57)	5.66 (4.13–7.17)
Michigan	3.62	3.96	4.55 (4.43–4.64)	5.37 (4.88–5.79)	5.37 (4.87–6.66)	6.01 (4.65–7.40)
Minnesota	3.46	3.72	4.49 (4.35–4.61)	5.72 (5.16–6.20)	5.72 (5.29–7.41)	6.86 (5.14–8.63)
Mississippi	3.75	3.92	4.50 (4.45–4.54)	5.30 (4.99–5.60)	5.30 (5.06–6.35)	6.06 (5.05–7.18)
Missouri	4.13	4.29	4.93 (4.83–5.00)	5.82 (5.36–6.24)	5.82 (5.32–7.07)	6.51 (5.15–8.00)
Montana	3.97	4.44	5.41 (5.32–5.49)	6.47 (6.07–6.86)	6.47 (6.26–7.93)	7.91 (6.54–9.43)
Nebraska	4.14	4.17	4.60 (4.50–4.68)	5.27 (4.77–5.71)	5.27 (4.70–6.60)	6.07 (4.61–7.62)
Nevada	2.94	3.41	3.70 (3.58–3.80)	4.03 (3.58–4.45)	4.03 (3.51–5.10)	4.61 (3.46–5.82)
New Hampshire	2.76	3.88	4.76 (4.67–4.85)	5.73 (5.29–6.16)	5.73 (5.36–7.07)	6.54 (5.24–7.93)
New Jersey	3.74	3.72	3.80 (3.71–3.87)	4.11 (3.64–4.51)	4.11 (3.39–5.01)	4.26 (3.10–5.33)
New Mexico	3.34	3.91	4.74 (4.63–4.83)	5.61 (5.09–6.10)	5.61 (5.00–6.96)	6.55 (4.97–8.16)
New York	3.89	3.92	4.26 (4.16–4.35)	4.91 (4.22–5.50)	4.91 (3.93–6.39)	5.37 (3.61–7.05)
North Carolina	3.49	3.82	4.30 (4.24–4.38)	4.94 (4.74–5.34)	4.94 (4.87–6.13)	5.52 (4.83–6.90)
North Dakota	4.51	4.54	5.39 (5.18–5.57)	6.62 (6.02–7.23)	6.62 (6.17–8.32)	7.97 (6.35–9.70)
Ohio	4.06	4.22	4.75 (4.66–4.83)	5.54 (5.09–5.97)	5.54 (5.00–6.75)	6.11 (4.78–7.58)
Oklahoma	3.94	4.34	4.86 (4.78–4.96)	5.56 (5.12–5.96)	5.56 (5.05–6.69)	6.30 (5.04–7.70)
Oregon	3.64	3.85	4.48 (4.38–4.58)	5.09 (4.60–5.55)	5.09 (4.49–6.36)	5.79 (4.40–7.31)
Pennsylvania	4.62	4.68	4.99 (4.90–5.08)	5.72 (5.23–6.17)	5.72 (5.02–6.86)	5.96 (4.62–7.44)
Rhode Island	3.88	4.47	4.39 (4.29–4.47)	4.56 (4.12–4.90)	4.56 (3.95–5.36)	5.20 (4.09–6.31)
South Carolina	3.07	4.04	4.76 (4.69–4.84)	5.45 (5.07–5.79)	5.45 (5.02–6.47)	6.04 (4.95–7.20)
South Dakota	4.16	4.27	4.93 (4.83–5.01)	5.91 (5.47–6.33)	5.91 (5.53–7.30)	6.82 (5.46–8.35)
Tennessee	3.73	4.03	4.59 (4.50–4.65)	5.18 (4.78–5.55)	5.18 (4.69–6.18)	5.64 (4.52–6.85)
Texas	2.70	2.91	3.33 (3.23–3.42)	3.88 (3.45–4.25)	3.88 (3.39–4.92)	4.43 (3.31–5.66)
Utah	2.00	2.18	2.38 (2.33–2.44)	2.67 (2.41–2.91)	2.67 (2.36–3.37)	3.09 (2.34–3.99)
Vermont	3.88	4.25	5.23 (5.12–5.33)	6.26 (5.70–6.82)	6.26 (5.68–7.93)	7.26 (5.87–9.19)
Virginia	3.24	3.54	3.96 (3.87–4.06)	4.46 (4.04–4.85)	4.46 (3.88–5.40)	4.80 (3.69–5.97)
Washington	3.23	3.51	4.06 (3.97–4.15)	4.69 (4.20–5.11)	4.69 (4.12–5.83)	5.29 (4.03–6.57)
West Virginia	4.56	5.06	5.36 (5.26–5.44)	5.82 (5.46–6.18)	5.82 (5.25–6.60)	5.91 (4.95–7.02)
Wisconsin	3.72	3.91	4.49 (4.41–4.58)	5.34 (4.87–5.78)	5.34 (4.82–6.62)	5.90 (4.56–7.34)
Wyoming	3.43	3.90	4.58 (4.50–4.66)	5.23 (4.87–5.57)	5.23 (4.88–6.25)	5.91 (4.88–7.10)
6-county region, SC	2.32	3.01	3.17 (3.05–3.26)	3.38 (2.96–3.72)	3.44 (2.75–4.03)	3.45 (2.54–4.25)
Imperial county	2.01	2.35	2.18 (2.06–2.26)	2.09 (1.77–2.33)	1.95 (1.54–2.28)	1.88 (1.44–2.27)
Los Angeles county	2.23	2.84	3.25 (3.15–3.34)	3.77 (3.37–4.13)	4.14 (3.43–4.78)	4.31 (3.31–5.20)
Orange county	2.24	2.93	3.23 (3.09–3.33)	3.51 (3.02–3.93)	3.62 (2.81–4.34)	3.60 (2.56–4.53)
Riverside county	2.88	3.25	2.81 (2.72–2.87)	2.68 (2.43–2.88)	2.67 (2.26–3.01)	2.82 (2.25–3.38)
S Bernardino county	1.82	2.60	2.69 (2.60–2.77)	2.86 (2.58–3.11)	2.89 (2.45–3.33)	2.99 (2.37–3.65)
Ventura county	2.36	3.30	3.61 (3.48–3.71)	3.82 (3.36–4.22)	3.73 (2.99–4.40)	3.58 (2.65–4.51)
M-S 7-county region	2.54	3.44	3.98 (3.83–3.98)	4.62 (4.05–4.62)	4.82 (3.82–4.82)	4.79 (3.48–4.79)

Table 9.8 Projections of oldest-old aged 80+ living alone for the 50 states, DC, and some counties, 2000–2050

		2010	2020	2030	2040	2050
	2000	Proj.	Med (Low-High)	Med (Low-High)	Med (Low-High)	Med (Low-High)
United States	1.19	1.50	1.48 (1.44–1.51)	1.77 (1.58–1.94)	2.19 (1.80–2.55)	2.41 (1.85–2.96)
Alabama	1.25	1.54	1.58 (1.54–1.62)	1.93 (1.78–2.08)	2.41 (2.10–2.72)	2.66 (2.18–3.16)
Alaska	0.31	0.54	0.65 (0.64–0.66)	0.90 (0.83–1.00)	1.21 (1.04–1.42)	1.36 (1.11–1.69)
Arizona	1.05	1.33	1.39 (1.34–1.42)	1.62 (1.44–1.76)	1.88 (1.54–2.18)	2.07 (1.59–2.51)
Arkansas	1.46	1.53	1.44 (1.42–1.48)	1.65 (1.51–1.79)	1.94 (1.67–2.20)	2.11 (1.73–2.53)
California	0.93	1.23	1.23 (1.18–1.26)	1.47 (1.26–1.65)	1.80 (1.37–2.18)	2.01 (1.39–2.54)
Colorado	0.90	1.17	1.20 (1.17–1.22)	1.55 (1.37–1.69)	2.01 (1.63–2.36)	2.28 (1.69–2.82)
Connecticut	1.23	1.79	1.72 (1.64–1.78)	2.01 (1.72–2.28)	2.42 (1.84–2.96)	2.67 (1.87–3.43)
Delaware	1.13	1.53	1.60 (1.56–1.64)	1.90 (1.73–2.05)	2.29 (1.93–2.61)	2.56 (2.03–3.08)
District of Columbia	1.43	1.42	1.19 (1.16–1.24)	1.21 (1.09–1.35)	1.29 (1.11–1.49)	1.35 (1.10–1.59)
Florida	1.62	2.03	2.01 (1.98–2.04)	2.28 (2.07–2.45)	2.63 (2.22–2.98)	2.85 (2.25–3.40)
Georgia	0.85	1.01	1.03 (1.00–1.05)	1.29 (1.15–1.42)	1.60 (1.32–1.87)	1.82 (1.41–2.24)
Hawaii	0.78	1.29	1.42 (1.39–1.45)	1.70 (1.51–1.86)	2.07 (1.68–2.40)	2.39 (1.80–2.90)
Idaho	1.12	1.39	1.39 (1.37–1.43)	1.70 (1.56–1.83)	2.13 (1.84–2.39)	2.32 (1.90–2.76)
Illinois	1.32	1.56	1.46 (1.41–1.49)	1.69 (1.49–1.85)	2.05 (1.65–2.42)	2.22 (1.65–2.76)
Indiana	1.26	1.49	1.42 (1.38–1.45)	1.60 (1.44–1.75)	1.83 (1.54–2.10)	1.91 (1.53–2.29)
Iowa	1.82	2.07	1.97 (1.93–2.00)	2.34 (2.12–2.53)	3.02 (2.53–3.49)	3.30 (2.52–4.07)
Kansas	1.58	1.84	1.75 (1.71–1.78)	2.09 (1.88–2.28)	2.70 (2.23–3.12)	2.88 (2.19–3.55)
Kentucky	1.24	1.45	1.46 (1.42–1.50)	1.81 (1.66–1.96)	2.28 (1.96–2.60)	2.52 (2.04–3.03)
Louisiana	0.87	1.44	1.51 (1.48–1.55)	1.89 (1.74–2.03)	2.45 (2.13–2.77)	2.69 (2.20–3.19)
Maine	1.54	1.98	2.06 (2.05–2.09)	2.65 (2.48–2.79)	3.43 (3.05–3.78)	3.87 (3.28–4.47)
Maryland	1.03	1.36	1.36 (1.32–1.39)	1.61 (1.43–1.76)	1.92 (1.57–2.24)	2.13 (1.62–2.60)
Massachusetts	1.32	1.83	1.72 (1.68–1.76)	2.07 (1.81–2.29)	2.58 (2.03–3.06)	2.87 (2.08–3.60)
Michigan	1.23	1.61	1.58 (1.53–1.62)	1.97 (1.77–2.15)	2.55 (2.14–2.93)	2.82 (2.20–3.40)
Minnesota	1.42	1.66	1.67 (1.56–1.78)	2.22 (1.95–2.47)	3.18 (2.63–3.68)	3.64 (2.83–4.41)
Mississippi	1.19	1.33	1.32 (1.30–1.35)	1.63 (1.52–1.74)	2.08 (1.84–2.32)	2.31 (1.96–2.71)
Missouri	1.47	1.64	1.64 (1.61–1.68)	2.04 (1.86–2.20)	2.58 (2.21–2.94)	2.84 (2.29–3.44)
Montana	1.46	1.74	1.82 (1.78–1.85)	2.32 (2.18–2.48)	2.99 (2.67–3.33)	3.29 (2.78–3.86)
Nebraska	1.72	1.95	1.90 (1.85–1.94)	2.21 (2.01–2.41)	2.75 (2.30–3.16)	2.94 (2.28–3.60)
Nevada	0.70	1.02	1.09 (1.06–1.12)	1.27 (1.12–1.40)	1.51 (1.23–1.77)	1.74 (1.33–2.17)
New Hampshire	0.83	1.43	1.58 (1.55–1.61)	2.10 (1.94–2.26)	2.77 (2.41–3.13)	3.15 (2.58–3.73)
New Jersey	1.29	1.62	1.52 (1.48–1.55)	1.69 (1.46–1.88)	1.97 (1.54–2.33)	2.15 (1.57–2.67)
New Mexico	1.06	1.42	1.59 (1.55–1.63)	1.99 (1.81–2.18)	2.55 (2.11–2.96)	2.85 (2.19–3.53)
New York	1.33	1.59	1.51 (1.46–1.55)	1.77 (1.50–2.00)	2.16 (1.60–2.67)	2.42 (1.61–3.13)
North Carolina	1.08	1.41	1.42 (1.40–1.45)	1.74 (1.66–1.89)	2.15 (2.01–2.50)	2.41 (2.18–2.98)
North Dakota	1.83	1.98	1.93 (1.86–2.02)	2.35 (2.13–2.57)	3.08 (2.65–3.50)	3.38 (2.77–4.03)
Ohio	1.37	1.73	1.68 (1.64–1.71)	2.01 (1.85–2.18)	2.55 (2.17–2.92)	2.75 (2.20–3.35)
Oklahoma	1.37	1.62	1.64 (1.60–1.67)	1.98 (1.80–2.14)	2.42 (2.07–2.77)	2.63 (2.13–3.18)
Oregon	1.35	1.57	1.48 (1.44–1.53)	1.83 (1.66–2.00)	2.28 (1.89–2.63)	2.44 (1.89–3.01)
Pennsylvania	1.60	2.04	1.83 (1.78–1.87)	2.13 (1.96–2.31)	2.72 (2.33–3.10)	2.91 (2.32–3.53)
Rhode Island	1.28	2.08	1.82 (1.78–1.86)	1.90 (1.71–2.05)	2.16 (1.82–2.46)	2.42 (1.96–2.90)
South Carolina	0.80	1.35	1.45 (1.43–1.48)	1.85 (1.71–1.99)	2.29 (1.98–2.57)	2.57 (2.14–3.03)
South Dakota	1.75	1.86	1.84 (1.81–1.87)	2.15 (1.98–2.30)	2.76 (2.40–3.11)	3.06 (2.52–3.64)
Tennessee	1.20	1.43	1.47 (1.43–1.49)	1.81 (1.65–1.96)	2.21 (1.90–2.52)	2.43 (1.98–2.92)
Texas	0.88	1.09	1.11 (1.07–1.14)	1.34 (1.18–1.50)	1.70 (1.37–2.01)	1.93 (1.45–2.43)
Utah	0.78	0.91	0.92 (0.89–0.93)	1.08 (0.96–1.18)	1.33 (1.11–1.55)	1.47 (1.13–1.83)
Vermont	1.37	1.66	1.76 (1.72–1.81)	2.33 (2.15–2.52)	2.96 (2.55–3.41)	3.15 (2.51–3.94)
Virginia	1.00	1.35	1.37 (1.35–1.42)	1.68 (1.50–1.83)	2.01 (1.66–2.33)	2.21 (1.70–2.70)
Washington	1.17	1.46	1.43 (1.40–1.47)	1.76 (1.56–1.92)	2.19 (1.80–2.53)	2.41 (1.85–2.92)
West Virginia	1.38	1.88	1.78 (1.75–1.83)	2.12 (1.99–2.27)	2.60 (2.31–2.88)	2.71 (2.31–3.13)
Wisconsin	1.43	1.74	1.67 (1.64–1.70)	2.01 (1.85–2.18)	2.58 (2.21–2.93)	2.82 (2.25–3.40)
Wyoming	1.15	1.55	1.60 (1.56–1.64)	1.96 (1.82–2.11)	2.46 (2.17–2.75)	2.68 (2.23–3.15)
6-county region, SC	0.69	1.24	1.26 (1.20–1.31)	1.46 (1.25–1.63)	1.77 (1.39–2.09)	2.01 (1.47–2.47)
Imperial county	0.49	0.98	0.91 (0.86–0.97)	0.86 (0.69–1.00)	0.94 (0.72–1.12)	1.02 (0.75–1.20)
Los Angeles county	0.65	1.12	1.20 (1.16–1.24)	1.46 (1.30–1.63)	1.90 (1.55–2.20)	2.35 (1.78–2.82)
Orange county	0.68	1.18	1.26 (1.19–1.32)	1.52 (1.28–1.74)	1.88 (1.43–2.28)	2.19 (1.54–2.76)
Riverside county	0.81	1.43	1.25 (1.19–1.29)	1.18 (1.06–1.28)	1.31 (1.11–1.48)	1.47 (1.19–1.73)
S Bernardino county	0.47	1.05	1.07 (1.04–1.11)	1.20 (1.08–1.33)	1.49 (1.25–1.72)	1.68 (1.34–2.03)
Ventura county	0.72	1.37	1.44 (1.38–1.48)	1.70 (1.47–1.91)	2.03 (1.61–2.41)	2.13 (1.57–2.66)
M-S 7-county region	0.88	1.50	1.48 (1.41–1.48)	1.77 (1.52–1.77)	2.20 (1.71–2.20)	2.37 (1.68–2.37)

Chapter 10
Effects of Changes in Household Structure and Living Arrangements on Future Home-Based Care Costs for Disabled Elders in the United States

10.1 Introduction

It is projected that the number of older Americans aged 65 and older will climb to 89 million in 2050, which is more than two times the number in 2010 (Vincent and Velkoff 2010; also see Chaps. 8 and 9 of this book); this increase is the consequence of a much faster growth rate for the elderly than for the population as a whole. The most significant growth will be among the oldest-old aged 80 or older, who have the highest probabilities of disability and use of home-based and other kinds of long-term care services. The share of the oldest-old among the total population will hit 7.0 % in 2050, almost tripling than that today (U.S. Census Bureau 2008). Thus, this surge will likely produce a similar increase in the demand and costs for care services for disabled elders, i.e., assistance in performing basic activities of daily living (ADL) such as eating, bathing, and dressing (Congressional Budget Office [CBO] 2004).

Numerous studies using data from different surveys in the United States have consistently demonstrated that home-based care utilization and costs differ substantially by household structure and living arrangement among older adults; these differentials have remained rather stable across the last three decades (e.g., Houser et al. 2010; Kaye et al. 2010; Liu et al. 2000; Orsini 2010; Robinson 2007; Weiss et al. 2005). Much empirical research has established that living arrangement is a major determinant of the amount and type of home-based care costs for disabled older adults (e.g., Breeze et al. 1999; Chappell 1991; Freedman 1996; Houser et al. 2010; Liu et al. 2000; Morris et al. 1998; Robinson 2007; Soldo et al. 1990). It follows that models to project disability and home-based care costs should include living arrangements of older adults in addition to other basic demographic characteristics (see introduction of Chap. 5). However, a literature search shows that previous projections of disability and care costs for older adults in the U.S. included age and gender, and occasionally racial differentials, but none included household structure and living arrangements. To fill this research gap, the present study aims to project numbers of disabled elders and their home-based care

Y. Zeng et al., *Household and Living Arrangement Projections*,
The Springer Series on Demographic Methods and Population Analysis 36,
DOI 10.1007/978-90-481-8906-9_10,

costs classified by age, gender, race, and living arrangement, which are the main demographic and familial determinants of the home-based care for disabled elders. We conduct projections and analyses for the United States from 2010 to 2050, and focus on investigating the effects of changes in household structure and living arrangements on future home-based care costs for disabled elders. The remainder of this introduction briefly reviews relevant extant research literature.

The trend of population and household aging, accompanied by declines in informal care resources resulting from reduced marriage prevalence, smaller family sizes, and an increased proportion of women in the labor force (Boaz and Muller 1992), challenges the sustainability of the financing and budgeting for disabled elderly care (Congressional Budget Office 2004). According to CBO projections (2004), the total long-term care (LTC) expenditures for elders (including government and private spending but excluding the value of donated care) in 2000 was about $125.5 billion (1.3 % of GDP), or roughly $15,000 per impaired senior; LTC expenditures are projected to climb to $346 billion (in 2000 dollars) in 2040.

Although growth in the elderly population may have the most powerful impact on future demand for LTC spending, other relevant factors could also play important roles. For example, a decline or increase in the prevalence of ADL disability could slow down or accelerate the growth of LTC spending. Growing evidence has shown that the proportion of older Americans with functional ADL disability has steadily declined since the 1990s, although the number of older adults with ADL disability is increasing (Freedman et al. 2004). Researchers have shown that the prevalence of disability among American elders has fallen 6 % per decade from 1910 to 1980, and 1 % per year since the 1980s (Costa 2000; Manton et al. 2006). However, some studies have projected that this declining trend in the prevalence of disability among elders will reverse in the future (e.g., Lakdawalla et al. 2003). Accordingly, projections of the prevalence of disability in future years, and thus LTC care costs, are subject to a high degree of uncertainty. Despite that uncertainty, the expected increase in the number of elders as the baby boomers age is so large that LTC spending is likely to rise substantially over time because the number of disabled elders will grow even if the prevalence of disability declines. Spending could be even higher if, as some researchers believe, the prevalence of disability may increase in the future (Martin et al. 2010).

Out of the $135 billion spent on LTC for the elderly in the US in 2004, $92.4 billion was for nursing home care and $42.5 billion for home-based services (CBO 2004). The average health care expenditure per Medicare enrolee in 2001 was $10,000. Previous research reveals that the total health care expenditure in the rest of life for a person aged 65 is around $19,000, equal to 60 % of his or her whole lifetime healthcare expenditure; most of these expenditures occur at oldest-old ages (Alemayehu and Warner 2004).

Home-based care costs grew 90.7 % from 1990 to 1995 and 39 % from 1999 to 2004, in contrast to a 33.4 % and 24 % increase for institutional care costs in the corresponding periods (Stallard 2000; Hartman et al. 2008). Clearly, home-based care costs increase much faster than institutional care costs, especially for the oldest-old (Cutler and Meara 1999; Hartman et al. 2008). Lakdawalla and Philipson

(2002) show that the faster increase in the number of male elders will result in relatively reduced need for institutional care because elderly men may be more likely to rely on their spouse and live at home.

Research shows that, on average, disabled elders received 21.6 hours of home-based care per week in 1994 (Liu et al. 2000). Home-based care needs and costs vary by demographic, socioeconomic, and living arrangement statuses. For instance, African-Americans tend to receive more family care than their White counterparts (Mui and Bumette 1994; Chadiha et al. 1995). According to the National Long Term Care Survey (NLTCS), Hispanic and Asian American disabled elders also receive more family care per person than Whites (Pinquart and Sörensen 2005; Weiss et al. 2005). The home-based care costs are likely determined by living arrangements (Bass et al. 1992) which determine the availability of family caregiving resources (Lawton et al. 1992). Care costs are also related to program service availability, requirements, and regions (Kenney and Dubay 1992).

Some actuarial studies have employed the methodology of population projection in combination with assumed age-sex-specific disability rates, while others have used cohort methods combined with regressions to project the quantity and age-sex distributions of disabled elders in the future. For example, Bhaltacharya et al. (2004) used estimated age-specific disability incidence rates to project the elderly population with disability classified by age and gender. Lakdawalla et al. (2003) used regressions to forecast the nursing home population while considering effects of gender, ethnicity, marital status, education, number of surviving children, health practices, and diseases. Some scholars have used structural models with disability as a function of demographic characteristics, lifestyle behaviors, and other risk factors, to forecast the future disability status of the elderly (see, Manton et al. 1993). Some studies attempted to include the possible effects of policy/programs on future expenditures (e.g., Heffler et al. 2005). However, in conducting such multiple-regression-based projections, the analyst needs to first develop the forecasts for the multiple factors affecting disability, which is very difficult and complicated, often not feasible, and may lead to forecast instability (Lee and Miller 2002). Some authors also used stochastic projections for Medicare spending in future years based on probabilistic population projection and estimated Medicare spending (Lee and Miller 2002).

A dynamic microsimulation approach has been recently used to project future estimates of number of disabled elders and long-term care use and costs (Kemper et al. 2005; Johnson et al. 2007). This approach uses transition probabilities to simulate events and statuses, year by year, cohort by cohort, to construct individual life histories. The microsimulation method is able to incorporate as many covariates researchers wish if the data are available. However, its disadvantage is that it requires very large samples to obtain reliable transition probabilities classified by age, gender, race, and detailed statuses. It also needs additional adjustments for crucial indicators such as the demographic rates in order to match external population projection (Kemper et al. 2005).

As noted earlier in this chapter and Chap. 5, almost all of the previously published projection models for elderly disability and home-based care costs ignored living arrangements of the elders, which was perhaps due to lack of reliable methods for projecting households and living arrangements. In this chapter, we apply the ProFamy extended cohort-component model for projecting family household structures and living arrangements; we then use the projections in combination with age-gender-race-living arrangement-specific disability rates and home-based care costs per disabled elder to project future trends and patterns in home-based care costs for different groups of the disabled elderly population (Zeng et al. 2014).[1] Extensive discussions, empirical tests, and justifications of the methodology, as well as a review of publications developing and applying the ProFamy extended cohort-component approach by us and other international scholars are presented in Chaps. 2, 3 and 4. In the next section of this chapter, we describe the data sources, estimates, and assumptions of the parameters for the projections. We describe the results of the projections with discussions in the third section. The last section summarizes the main findings and presents concluding remarks including contributions and limitations of this study as well as perspectives for further research.

10.2 Data Sources and Parameter Estimates and Assumptions

10.2.1 Demographic Age-Sex-Specific Standard Schedules and Summary Parameters

To make household and living arrangement projections using the ProFamy method and software, we need to input standard age-specific schedules of demographic rates, including age-sex-specific marriage/union formation and dissolution, children leaving parental home, domestic out-migration occurrence/exposure (o/e) rates, domestic in-migration and international net migration frequencies, and age-parity-specific marital and non-marital fertility o/e rates. If race categories are distinguished in the projection, as is the case in our present study, these standard schedules need to be race-specific as well. We also need the summary measures of total fertility rate, life expectancy at birth, number of migrants, mean age at first marriage, mean age at birth, and general rates of marriage/union formation and dissolution (see Appendix 1). Data sources for household and living arrangement projections in the United States were indicated in the last column of Table 3.1 of Chap. 3; these sources also are used here in our present study.

[1] Older adults with disabilities are more likely than those without any long-term care needs to live with their children if possible. The choice of living arrangements of elderly people also largely depends on demographic factors such as age, gender, race, marriage, divorce, cohabitation, and availability of children. Therefore, it is meaningful and practical to project elderly living arrangements first based on available demographic data and then project elderly disability status.

The estimates and assumptions for the race-sex-specific life expectancies at birth and the race-parity-specific Total Fertility Rates (TFR) in the baseline and future years are based on available data from the medium assumptions of the Census Bureau population projections (U.S. Census Bureau 2008). The numbers of international net migrants are estimated based on combined data from the American Community Survey (ACS) from 2000 to 2006; the migration parameters are assumed to be constant after 2006.

Four race groups (White non-Hispanic, Black non-Hispanic, Hispanic, Asian and other non-Hispanic) defined by the Census Bureau are distinguished in the projections. The estimated and assumed demographic summary measures in the baseline and future years for the United States are presented in Appendix 1.

10.2.2 Estimates of Disability and Home-Based Care Cost Parameters

We estimated age-sex-race-living arrangement-specific prevalence rates of disability and home-based care costs based on the 1999 wave of the National Long Term Care Survey (NLTCS) data.[2] Disability in activities of daily living (ADL) in the NLTCS is measured by six separate items including bathing, dressing, eating, indoor transferring, toileting, and continence. According to the NLTCS definition, a person is classified as ADL disabled if he/she needs help with any one of the six items for 3 months or more.

We use the NLTCS and a two-step procedure described in Appendix 2 to estimate the U.S. age-sex-race-living arrangement-specific rates of disability and care costs, which are listed in Appendix 3. The NLTCS provides an adequate database for projecting home-based care costs of ADL disabled Americans. Indeed, many crucial indicators of LTC released by governmental agencies are from the NLTCS (e.g., FIFARS 2000, 2004, 2008, 2010, 2012). The NLTCS gathered data on (1) number of hours of home-based care per week received by ADL disabled elderly respondents and (2) monthly payments to each helper who provided care for the ADL disabled elderly respondent living in the community. Note that, in this chapter, hours of care per week and its aggregation to yearly workdays of care for the disabled elderly include both paid and non-paid home-based care.[3] According to

---

[2] The 1999 wave of NLTCS had a response rate of 88.6 % in the screening stage. The response rate for detailed interview was 93.2 % (http://www.nltcs.aas.duke.edu/pdf/99_SourceAndAccuracy.pdf, accessed on August 21, 2012).

[3] There are two primary reasons why we did not separate paid and unpaid care hours. First, a single question for overall care hours (without distinguishing paid and unpaid) in the 1999 NLTCS questionnaire was asked first in addition to two separate questions about paid or unpaid care hours. We believe that data from the total care hours are more reliable, because the latter two paid and unpaid questions had much higher percentage of refusal or “do not know” answers. Second, the estimates would be unstable if we further divided the age-sex-race-disability status-living arrangement-specific hours of care received per disabled elder by paid and unpaid categories, due to small sample size for subpopulations of minority groups.

the estimates presented in Appendix 3, there is a clear-cut difference in prevalence rates of disability and home-based care hours received across living arrangements. Those living alone had the lowest disability rate and home-based care hours received, while those not living with a spouse/partner but with children and/or others had the highest. However, those living alone are likely to pay more for their home-based care. On average, rates of disability prevalence, home-based care hours per week, and the home-based care payments per month all increase with age (see Appendix 3).

Prior studies of trends in age-specific disability rates among the U.S. elderly have generally found a declining trend, but a few show modest increases (e.g., Freedman et al. 2004, 2002; Manton et al. 2006). Accordingly, we assume three different scenarios of future disability trends in our projections. The low disability scenario assumes an annual decline of 1 % in age-specific disability rates from 2010 to 2050. The medium disability scenario assumes an annual decline of 1 % in age-specific disability rates from 2010 to 2020, and constant rates after 2020. Note that both of our medium and low disability scenarios assume an annual decline of 1 % in age-specific disability rates from 2010 to 2020, with differences emerging after 2020. By comparison, our high disability scenario assumes an annual increase of 0.5 % in age-specific disability rates after 2010. We combine the parameters of the low, medium, and high disability scenarios with the demographic parameters of the medium family household and living arrangement projections to project future home-based care needs and costs, as described in the last subsection.[4]

10.3 Results

Our projections indicate that the aging of American households is remarkable,[5] in addition to reconfirming the rapid population aging trends presented in the Census Bureau (Vincent and Velkoff 2010) and Social Security Administration publications on old-age and survivors insurance (Board of Trustees 2012). For example, the proportion of elderly households (with householder aged 65+) among the total number of households would increase from 17.6 % in 2010 to 23.5 % and 25.3 % in 2030 and 2050. The proportion of the elderly aged 65+ living

[4] Although Zeng et al. (2013a) conducted projection scenarios of small, medium, and large family households, we do not include similar scenarios here, because the combinations of the low, medium, and high disability scenarios with small, medium and large family scenarios would result in nine (=3 × 3) composite scenarios, which would not permit a clear and meaningful presentation in one chapter.

[5] While we present in this chapter relatively detailed tables and graphics for projection results of numbers of disabled elders and their home-based care costs classified by age, race, and living arrangement, we only present and discuss a few summary indices of aging of households/living arrangements here due to space limitations; detailed tables are presented in Appendix 2 in Chap. 9.

alone among the total population in 2030 and 2050 will increase by 27.5 % and 40.5 %, compared to 2010. The projection results also demonstrate that the oldest-old aged 80+ living alone will increase even more dramatically after 2020. The percentage of the oldest-old living alone among the total population in 2030 and 2050 would be 1.77 and 2.41, respectively; representing a 18 % increase in 2030 and a 61 % increase in 2050, as compared to 2010. The rapid aging of the population, households, and living arrangements will certainly affect the number of disabled elderly and their needs and home-based care costs, as presented and discussed below. Relatively detailed numerical projection outcomes for number of disabled elders, yearly workdays of home-based care, and home-based care payments (in million dollars) classified by age, gender, and living arrangements in 2010, 2020, 2030, 2040 and 2050 in the U.S. are presented in Tables 10.1, 10.2 and 10.3 and Figs. 10.1, 10.2 and 10.3. Here we summarize the insights from the main indices of these decennial projections.

10.3.1 A Substantial Increase in Number of Disabled Elders and Remarkable Acceleration After 2020, Especially for the Oldest-Old

Table 10.1 presents the number of community-dwelling disabled elders, their annual increase rates in 2010–2050, and ratios of the numbers of the disabled elders in 2020 vs. 2010, in 2030 vs. 2020, and in 2050 vs. 2010, by living arrangement and under the medium disability trend scenario. The number of all disabled elders aged 65+ in 2050 would be 2.2 times as large as that in 2010, with annual growth rates of 2.0 %, during the period 2010–2050. The increase in the number of the disabled oldest-old aged 80+ would be even more dramatic than that of the disabled young-old aged 65–79: the disabled oldest-old and disabled young-old population in 2050 would be 2.7 and 1.7 times as large as that in 2010, with annual growth rates of 2.4 % and 1.4 % during the period 2010–2050, respectively.

The projection results show that the relative increase in disabled oldest-old aged 80+ would accelerate dramatically after 2020: a 1.7 % increase in 2020 compared to 2010, but an increase of 47.6 % in 2030 compared to 2020, and an increase of 165.5 % in 2050 compared to 2010 (see panel (C) Table 10.1). However, the pattern of change in the number of disabled young-old aged 65–79 is totally different from that of the oldest-old. More specifically, the number of disabled young-old would increase by 31.4 % in 2020 compared to 2010 and increase by 27.6 % in 2030 compared to 2020, but remain more or less stable after 2030 (almost no change in 2040 compared to 2030 and a 4.1 % increase in 2050 compared to 2040). The acceleration of the increase of disabled oldest-old in the later period 2020–2050 will be dramatic compared to the earlier period 2010–2020, but the quick increase in the disabled young-old will cease after 2030. Such trends and differential patterns are mainly due to the timing of post-world-war baby boomers entering young-old ages 65–79 in 2020–2030 and subsequently becoming oldest-old aged 80+ after 2020–2030.

Table 10.1 Number and relative increase of disabled elders by age and living arrangement in the U.S., 2010–2050

	2010	2020	2030	2040	2050	% inc in 2020 comp. to 2010	% inc in 2030 comp. to 2020	% inc in 2050 comp. to 2010	Ann. Inc. rate 2010–2050
(A) Aged 65+									
Two-genders	2,750,660	3,211,179	4,374,511	5,344,072	6,031,669	16.7	36.2	119.3	2.0 %
Living alone	687,626	728,005	958,124	1,190,786	1,328,057	5.9	31.6	93.1	1.7 %
With spouse	1,379,520	1,707,927	2,343,879	2,718,704	2,983,384	23.8	37.2	116.3	1.9 %
No spouse wh child	683,514	775,246	1,072,508	1,434,582	1,720,228	13.4	38.3	151.7	2.3 %
Males	1,011,264	1,229,174	1,702,760	2,085,544	2,388,477	21.5	38.5	136.2	2.2 %
Living alone	161,357	194,352	272,562	348,518	399,536	20.4	40.2	147.6	2.3 %
With spouse	694,668	834,788	1,127,620	1,304,928	1,433,829	20.2	35.1	106.4	1.8 %
No spouse wh child	155,239	200,035	302,579	432,099	555,112	28.9	51.3	257.6	3.2 %
Females	1,739,395	1,982,005	2,671,751	3,258,528	3,643,192	13.9	34.8	109.5	1.9 %
Living alone	526,268	533,653	685,562	842,269	928,521	1.4	28.5	76.4	1.4 %
With spouse	684,851	873,140	1,216,259	1,413,776	1,549,555	27.5	39.3	126.3	2.0 %
No spouse wh child	528,276	575,211	769,929	1,002,483	1,165,116	8.9	33.9	120.6	2.0 %
(B) Aged 65–79									
Two-genders	1,392,425	1,830,229	2,335,624	2,330,539	2,425,013	31.4	27.6	74.2	1.4 %
Living alone	228,081	283,523	355,743	341,579	338,857	24.3	25.5	48.6	1.0 %
With spouse	924,135	1,231,772	1,558,229	1,550,078	1,611,159	33.3	26.5	74.3	1.4 %
No spouse wh child	240,209	314,934	421,652	438,882	474,997	31.1	33.9	97.7	1.7 %
Males	568,494	756,142	968,849	973,717	1,034,090	33.0	28.1	81.9	1.5 %
Living alone	65,272	90,019	118,898	120,458	128,016	37.9	32.1	96.1	1.7 %
With spouse	443,115	578,062	722,799	709,628	735,041	30.5	25.0	65.9	1.3 %
No spouse wh child	60,107	88,061	127,152	143,632	171,032	46.5	44.4	184.5	2.6 %
Females	823,930	1,074,088	1,366,776	1,356,822	1,390,924	30.4	27.2	68.8	1.3 %
Living alone	162,809	193,505	236,845	221,121	210,841	18.9	22.4	29.5	0.7 %
With spouse	481,019	653,711	835,430	840,450	876,118	35.9	27.8	82.1	1.5 %
No spouse wh child	180,102	226,872	294,501	295,250	303,965	26.0	29.8	68.8	1.3 %
(C) Aged 80+									
Two-genders	1,358,235	1,380,949	2,038,886	3,013,533	3,606,656	1.7	47.6	165.5	2.4 %
Living alone	459,544	444,482	602,381	849,207	989,200	−3.3	35.5	115.3	1.9 %
With spouse	455,385	476,155	785,650	1,168,626	1,372,225	4.6	65.0	201.3	2.8 %
No spouse wh child	443,306	460,312	650,856	995,700	1,245,231	3.8	41.4	180.9	2.6 %
Males	442,770	473,032	733,911	1,111,827	1,354,388	6.8	55.2	205.9	2.8 %
Living alone	96,085	104,333	153,664	228,060	271,520	8.6	47.3	182.6	2.6 %
With spouse	251,553	256,726	404,820	595,300	698,788	2.1	57.7	177.8	2.6 %
No spouse wh child	95,132	111,973	175,427	288,467	384,080	17.7	56.7	303.7	3.5 %
Females	915,465	907,917	1,304,975	1,901,706	2,252,268	−0.8	43.7	146.0	2.3 %
Living alone	363,459	340,149	448,717	621,147	717,680	−6.4	31.9	97.5	1.7 %
With spouse	203,832	219,429	380,829	573,326	673,438	7.7	73.6	230.4	3.0 %
No spouse wh child	348,173	348,339	475,429	707,233	861,151	0.0	36.5	147.3	2.3 %

Note: (1) With spouse – Living with spouse but not living with children; No spouse wh child – Not living with spouse but with child(ren); (2) The numbers presented in this table are based on the medium disability scenario

Table 10.2 Yearly home-based care workdays (unit: weeks) and its relative increase for the disabled elderly, by age and living arrangement in the U.S., 2010–2050

	2010	2020	2030	2040	2050	% inc in 2020 comp. to 2010	% inc in 2030 comp. to 2020	% inc in 2050 comp. to 2010	Ann. Inc. rate 2010–2050
(A) All elderly aged 65+									
Two-genders combined-total	1,843,037	2,150,801	3,005,338	3,837,118	4,446,642	16.7	39.7	141.3	2.2 %
Living alone	262,455	274,394	366,986	477,713	548,026	4.5	33.7	108.8	1.8 %
With spouse (may with child)	982,754	1,204,076	1,696,530	2,051,028	2,298,594	22.5	40.9	133.9	2.1 %
Not with spouse, but with child	597,828	672,331	941,823	1,308,377	1,600,022	12.5	40.1	167.6	2.5 %
Males-total	736,745	891,540	1,260,875	1,601,152	1,876,735	21.0	41.4	154.7	2.3 %
Living alone	62,165	74,186	105,832	140,906	165,592	19.3	42.7	166.4	2.5 %
With spouse (may with child)	533,866	637,026	878,958	1,053,748	1,180,194	19.3	38.0	121.1	2.0 %
Not with spouse, but with child	140,714	180,328	276,085	406,498	530,949	28.2	53.1	277.3	3.3 %
Females-total	1,106,292	1,259,262	1,744,463	2,235,966	2,569,908	13.8	38.5	132.3	2.1 %
Living alone	200,290	200,208	261,153	336,807	382,434	0.0	30.4	90.9	1.6 %
With spouse (may with child)	448,888	567,050	817,572	997,279	1,118,400	26.3	44.2	149.1	2.3 %
Not with spouse, but with child	457,114	492,004	665,738	901,880	1,069,073	7.6	35.3	133.9	2.1 %
(B) Young-old aged 65–79									
Two-genders combined-total	815,694	1,085,718	1,405,159	1,430,487	1,515,316	33.1	29.4	85.8	1.6 %
Living alone	63,044	79,290	100,922	98,749	99,947	25.8	27.3	58.5	1.2 %
With spouse (may with child)	587,795	788,027	1,007,774	1,018,058	1,070,704	34.1	27.9	82.2	1.5 %
Not with spouse, but with child	164,855	218,401	296,463	313,680	344,665	32.5	35.7	109.1	1.8 %
Males-total	376,735	503,573	652,052	666,159	717,260	33.7	29.5	90.4	1.6 %
Living alone	20,719	28,714	38,425	39,657	42,892	38.6	33.8	107.0	1.8 %
With spouse (may with child)	309,375	406,202	513,391	511,713	536,166	31.3	26.4	73.3	1.4 %
Not with spouse, but with child	46,641	68,656	100,236	114,790	138,202	47.2	46.0	196.3	2.7 %
Females-total	438,959	582,145	753,107	764,327	798,057	32.6	29.4	81.8	1.5 %
Living alone	42,325	50,576	62,497	59,092	57,055	19.5	23.6	34.8	0.8 %
With spouse (may with child)	278,420	381,825	494,383	506,346	534,538	37.1	29.5	92.0	1.6 %
Not with spouse, but with child	118,213	149,744	196,227	198,890	206,464	26.7	31.0	74.7	1.4 %
(C) Oldest-old aged 80+									
Two-genders combined-total	1,027,344	1,065,084	1,600,179	2,406,632	2,931,326	3.7	50.2	185.3	2.6 %
Living alone	199,411	195,104	266,064	378,965	448,080	−2.2	36.4	124.7	2.0 %
With spouse (may with child)	394,959	416,049	688,755	1,032,969	1,227,890	5.3	65.5	210.9	2.8 %
Not with spouse, but with child	432,974	453,931	645,360	994,698	1,255,357	4.8	42.2	189.9	2.7 %
Males-total	360,010	387,967	608,823	934,993	1,159,475	7.8	56.9	222.1	2.9 %
Living alone	41,446	45,472	67,408	101,250	122,701	9.7	48.2	196.1	2.7 %
With spouse (may with child)	224,492	230,824	365,566	542,036	644,027	2.8	58.4	186.9	2.6 %
Not with spouse, but with child	94,073	111,672	175,848	291,707	392,747	18.7	57.5	317.5	3.6 %
Females-total	667,334	677,116	991,356	1,471,638	1,771,851	1.5	46.4	165.5	2.4 %
Living alone	157,965	149,632	198,656	277,715	325,379	−5.3	32.8	106.0	1.8 %
With spouse (may with child)	170,467	185,225	323,189	490,933	583,862	8.7	74.5	242.5	3.1 %
Not with spouse, but with child	338,901	342,259	469,511	702,990	862,610	1.0	37.2	154.5	2.3 %

Note: The numbers presented in this table are based on the medium disability scenario

Table 10.3 Yearly home-based care payments (in millions of dollars) and its relative increase for the disabled elderly by age and living arrangement in the U.S., 2010–2050

	2010	2020	2030	2040	2050	% inc in 2020 comp. to 2010	% inc in 2030 comp. to 2020	% inc in 2050 comp. to 2010	Ann. Inc. rate 2010–2050
(A) All elderly aged 65+									
Two-genders combined-total	6,548	7,147	9,785	12,550	14,199	9.1	36.9	116.8	1.9 %
Living alone	2,375	2,418	3,189	4,091	4,568	1.8	31.9	92.3	1.6 %
With spouse (may with child)	2,298	2,731	3,868	4,685	5,129	18.8	41.6	123.2	2.0 %
Not with spouse, but with child	1,876	1,997	2,728	3,774	4,502	6.4	36.6	140.0	2.2 %
Males-total	1,710	2,000	2,827	3,668	4,236	17.0	41.4	147.7	2.3 %
Living alone	567	659	928	1,220	1,391	16.2	40.8	145.3	2.2 %
With spouse (may with child)	814	936	1,288	1,550	1,702	15.0	37.6	109.1	1.8 %
Not with spouse, but with child	328	405	611	897	1,143	23.5	50.9	248.5	3.1 %
Females-total	4,839	5,147	6,958	8,882	9,963	6.4	35.2	105.9	1.8 %
Living alone	1,808	1,759	2,261	2,871	3,177	−2.7	28.5	75.7	1.4 %
With spouse (may with child)	1,484	1,796	2,580	3,135	3,427	21.0	43.7	130.9	2.1 %
Not with spouse, but with child	1,547	1,592	2,117	2,876	3,359	2.9	33.0	117.1	1.9 %
(B) Young-old aged 65–79									
Two-genders combined-total	2,114	2,742	3,446	3,362	3,425	29.7	25.7	62.0	1.2 %
Living alone	516	640	795	753	737	24.0	24.2	42.8	0.9 %
With spouse (may with child)	1,220	1,614	2,013	1,962	2,005	32.3	24.7	64.3	1.2 %
Not with spouse, but with child	378	488	639	648	684	29.1	30.9	81.0	1.5 %
Males-total	661	881	1,124	1,115	1,171	33.3	27.6	77.2	1.4 %
Living alone	168	230	299	297	310	36.9	30.0	84.5	1.5 %
With spouse (may with child)	409	528	651	627	640	29.1	23.3	56.5	1.1 %
Not with spouse, but with child	85	123	174	191	222	44.7	41.5	161.2	2.4 %
Females-total	1,453	1,861	2,323	2,247	2,254	28.1	24.8	55.1	1.1 %
Living alone	348	410	496	456	427	17.8	21.0	22.7	0.5 %
With spouse (may with child)	811	1,085	1,362	1,335	1,365	33.8	25.5	68.3	1.3 %
Not with spouse, but with child	294	365	465	457	462	24.1	27.4	57.1	1.1 %
(C) Oldest-old aged 80+									
Two-genders combined-total	4,434	4,405	6,339	9,187	10,774	−0.7	43.9	143.0	2.2 %
Living alone	1,859	1,778	2,394	3,338	3,831	−4.4	34.6	106.1	1.8 %
With spouse (may with child)	1,078	1,118	1,855	2,723	3,124	3.7	65.9	189.8	2.7 %
Not with spouse, but with child	1,497	1,509	2,090	3,126	3,819	0.8	38.5	155.1	2.3 %
Males-total	1,048	1,119	1,703	2,552	3,065	6.8	52.2	192.5	2.7 %
Living alone	399	429	629	923	1,082	7.5	46.6	171.2	2.5 %
With spouse (may with child)	405	408	637	923	1,062	0.7	56.1	162.2	2.4 %
Not with spouse, but with child	244	282	438	706	921	15.6	55.3	277.5	3.3 %
Females-total	3,386	3,286	4,635	6,635	7,709	−3.0	41.1	127.7	2.1 %
Living alone	1,460	1,349	1,765	2,415	2,750	−7.6	30.8	88.4	1.6 %
With spouse (may with child)	673	710	1,218	1,800	2,063	5.5	71.5	206.5	2.8 %
Not with spouse, but with child	1,253	1,227	1,652	2,419	2,897	−2.1	34.6	131.2	2.1 %

Note: The numbers presented in this table are based on the medium disability scenario

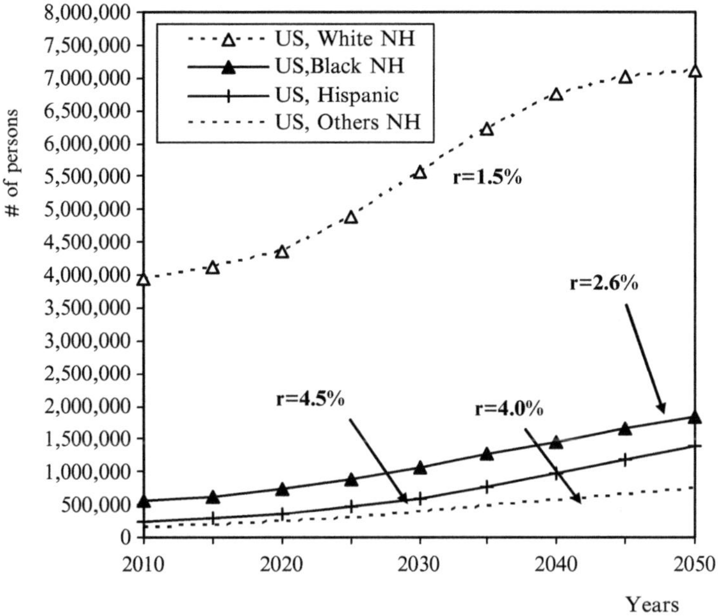

Fig. 10.1 Number of disabled elders aged 65+ and their annual increase rates (*r*) in 2010–2050, by race in the U.S. (medium disability scenario)

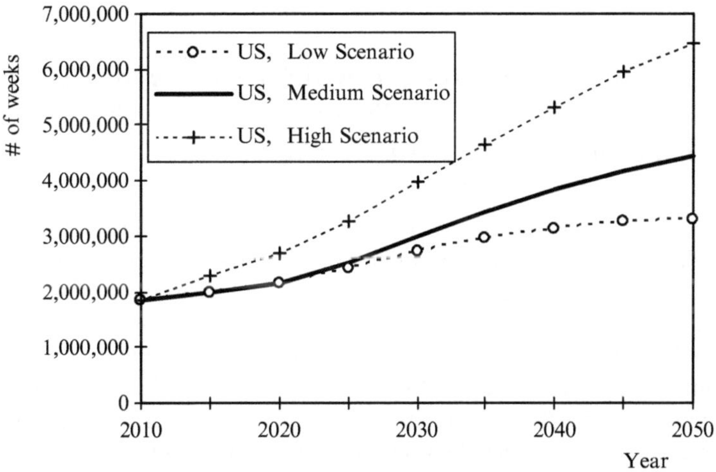

Fig. 10.2 Number of yearly home-based care workdays (unit: weeks) provided to disabled elders aged 65+ under the low, medium and high scenarios, 2010–2050

It is interesting to note that the number of disabled oldest-old aged 80+ was 2.5 % and 12.7 % smaller than the number of disabled young-old aged 65–79 in 2010 and 2030, but the disabled oldest-old will outnumber the disabled young-old

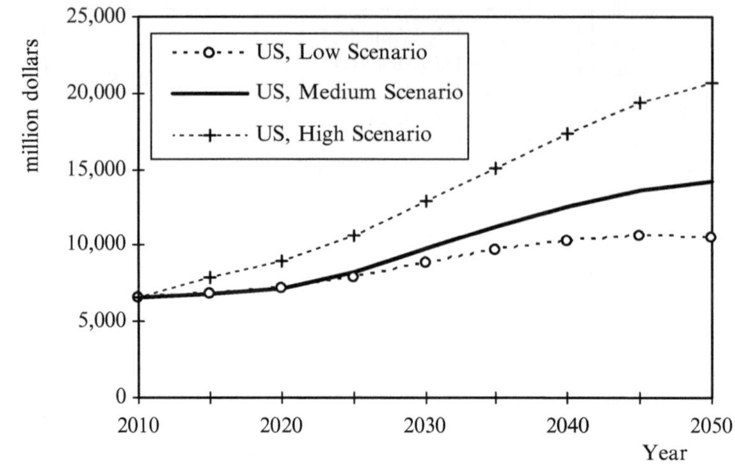

Fig. 10.3 Yearly payments (in million dollars) for home-based care provided to disabled elders aged 65+ under the low, medium and high scenarios, 2010–2050

by 29.3 % and 48.7 % in 2040 and 2050, respectively (estimated based on Table 10.1). Such a phenomenon certainly deserves serious governmental and societal attention because the disabled oldest-old need significantly more and quite different care services.

10.3.2 The Increase in Home-Based Care Costs for Disabled Elders Will Dramatically Accelerate After 2020, Especially for the Disabled Oldest-Old

In the present study, care hours per week were converted into yearly workdays (unit: weeks) and care costs per month were converted into yearly payment. Tables 10.2 and 10.3 present yearly workdays and yearly payments for home-based care for all disabled elders aged 65+, disabled young-old aged 65–79, and disabled oldest-old aged 80+ by gender and living arrangement. The number of yearly home-based care workdays for the disabled young-old aged 65–79 in 2050 would increase by 85.8 % compared to 2010, with an annual growth rate of 1.6 % in the period 2010–2050 (see panel (B) in Table 10.2). However, the number of yearly home-based care workdays for disabled oldest-old in 2050 would increase by 185.3 % compared to 2010, with an annual growth rate of 2.6 % in the period 2010–2050, which is significantly higher than among disabled young-old adults (see panel (C) in Table 10.2). Yearly home-based care payments will follow a similar trend and pattern of increase of more than doubling in the first half of this century, with an especially dramatic increase for the disabled oldest-old (see Table 10.3).

Our projection results also show a dramatic acceleration of yearly home-based care workdays and payments for the disabled oldest-old after 2020, in contrast to the roughly stabilized increase for the young-old. The yearly home-based care workdays and payments for the disabled oldest-old in 2020 will increase by −0.7–3.7 % in 2020 compared to 2010, but these figures would increase by 43.9–50.2 % in 2030 compared to 2020 (see panel (C) in Tables 10.2 and 10.3). However, for disabled young-old, the increase in yearly home-based care workdays and payments in both the 2010–2020 and 2030–3020 periods would remain more or less the same, in the range of 29.4–33.1 % or 25.7–29.7 % (see panel (B) in Tables 10.2 and 10.3). Clearly, the acceleration of the relative increase in yearly home-based care workdays and payments for the disabled oldest-old after 2020 as compared to the earlier period 2010–2020 will be much faster than that in the disabled young-old. This is again mainly due to the fact that the baby boomers will enter older ages and many of them will become oldest-old in the next couple of decades, which will dramatically affect home-based care costs for disabled elders, especially for those disabled oldest-old.

The share of yearly home-based care workdays for disabled oldest-old among the total yearly home-based care workdays for all disabled elders would increase from around half in 2010–2030 to around 63–66 % in 2040–2050 (see panels (C) and (A) in Table 10.2). The share of yearly home-based care payments for the disabled oldest-old among the total yearly home-based care payments for all disabled elders would increase from around 65 % in 2010–2030 to around three-fourths in 2040–2050 (see panels (C) and (A) in Table 10.3). These projection results indicate that the structure of the home-based care industry for the disabled elderly will substantially change after 2030 towards a much higher percentage of disabled oldest-old clients. Such trends certainly deserve strategic planning attention from governmental agencies and businesses.

10.3.3 Gender Differentials

The annual growth rate in the number of male disabled elders living with children/others (but not spouse/partner) will be about 3.2 %, much higher than the growth rate for male disabled elders living with a spouse/partner (1.8 %) in the first half of this century (see the last column of Table 10.1); very similar patterns appear among young-old and oldest-old male disabled persons. There are, however, no such differentials among the female disabled elders. Such trends for male disabled elders may be understood as one of the outcomes of the second demographic transition that occurred in the U.S. and many other developed countries in recent decades, a transition characterized by higher divorce rates and lower marriage rates (van de Kaa 2008). In the process of rapid population aging in the first half of this century, older cohorts, which experienced lower divorce rates and higher marriage rates, will be replaced by younger cohorts, which experienced higher divorce rates and lower marriage rates. This cohort-replacement process may result in more male elders, who usually have enough income to support themselves with no spouse, living with child(ren) or others when they are disabled. Consequently, the annual growth rates for the number of male disabled elders not with a spouse/partner will be higher than that for male disabled elders living with a spouse. On the other hand, divorced or windowed female elders may be more likely to be economically dependent and remarry or cohabit with another man; thus, the pattern of living arrangement among the disabled females is different from that among males.

As shown in the last columns of Tables 10.2 and 10.3, the annual growth rate in yearly home-based care workdays and payments for disabled male elders not living with a spouse/partner but with child/others will be substantially higher than that of disabled male elders living with a spouse. This may be due to the effects of the second demographic transition and cohort replacement, as discussed earlier, and also to the higher per-person home-based care costs for male disabled elders not-living with a spouse compared to their counterparts who live with a spouse.

Table 10.4 presents the male–female ratios for number of disabled elders, yearly home-based care workdays, and yearly home-based care payments (abbreviated as "the gender ratios" hereafter). The gender ratios for the total number of disabled elders, disabled elders living alone, or not-living with spouse but living with child/other are much smaller than one, and tend to increase over the next four decades. This trend is consistent with previous research findings that in general, there are more females, especially disabled females, at older ages, and a faster relative increase of male elders in the future (e.g., Lakdawalla and Philipson 2002).

It is interesting to note that, unlike all of the other ratios shown in Table 10.4 which are all substantially less than one and tend to increase, the gender ratios for number of disabled elders living with spouse and their yearly home-based care workdays received are close to or somewhat higher than one, and the ratios tend to moderately decrease from 2010 to 2030 and remain rather stable after 2030. These figures reconfirm that, in general, the proportion of male elders living with a spouse is higher than their female counterparts (e.g., Zeng et al. 2013a) but this gender differential would moderately change in the future due to the effects of the second demographic transition and cohort replacement, as discussed above.

10.3.4 Racial Differentials

Our estimates and projections have shown that the racial difference in disability and home-based care costs are noticeable. As indicated by the estimates presented in the Appendix 3, Black non-Hispanic elders have much higher disability rates as compared to the other three racial groups; White non-Hispanic elders are likely to have fewer care hours, but their care payment per disabled person per month is the highest; Asian and Other non-Hispanic elders have the lowest disability rate and their care payment per disabled person per month is the lowest.

Figure 10.1 shows that although there are some variations in the growth of absolute numbers of the community-dwelling disabled elderly by race across

Table 10.4 Male–female ratios for number disabled elders, yearly home-based care workdays, and yearly home-based care payments, by age and living arrangement

	All disabled elders aged 65+					Disabled young-old aged 65–79					Disabled oldest-old aged 80+				
	2010	2020	2030	2040	2050	2010	2020	2030	2040	2050	2010	2020	2030	2040	2050
Number of disabled elders															
Total	0.58	0.62	0.64	0.64	0.66	0.69	0.70	0.71	0.72	0.74	0.48	0.52	0.56	0.58	0.60
Living alone	0.31	0.36	0.40	0.41	0.43	0.40	0.47	0.50	0.54	0.61	0.26	0.31	0.34	0.37	0.38
With spouse (may with child)	1.01	0.96	0.93	0.92	0.93	0.92	0.88	0.87	0.84	0.84	1.23	1.17	1.06	1.04	1.04
Not with spouse, but with child	0.29	0.35	0.39	0.43	0.48	0.33	0.39	0.43	0.49	0.56	0.27	0.32	0.37	0.41	0.45
Yearly workdays															
Total	0.67	0.71	0.72	0.72	0.73	0.86	0.87	0.87	0.87	0.90	0.54	0.57	0.61	0.64	0.65
Living alone	0.31	0.37	0.41	0.42	0.43	0.49	0.57	0.61	0.67	0.75	0.26	0.30	0.34	0.36	0.38
With spouse (may with child)	1.19	1.12	1.08	1.06	1.06	1.11	1.06	1.04	1.01	1.00	1.32	1.25	1.13	1.10	1.10
Not with spouse, but with child	0.31	0.37	0.41	0.45	0.50	0.39	0.46	0.51	0.58	0.67	0.28	0.33	0.37	0.41	0.46
Yearly payments															
Total	0.35	0.39	0.41	0.41	0.43	0.45	0.47	0.48	0.50	0.52	0.31	0.34	0.37	0.38	0.40
Living alone	0.31	0.37	0.41	0.42	0.44	0.48	0.56	0.60	0.65	0.73	0.27	0.32	0.36	0.38	0.39
With spouse (may with child)	0.55	0.52	0.50	0.49	0.50	0.50	0.49	0.48	0.47	0.47	0.60	0.57	0.52	0.51	0.51
Not with spouse, but with child	0.21	0.25	0.29	0.31	0.34	0.29	0.34	0.37	0.42	0.48	0.19	0.23	0.27	0.29	0.32

years, there is a consistent general pattern of racial differential: the annual growth rate for White non-Hispanic is the lowest, the second lowest growth rate is for Black non-Hispanic, and Hispanics have the highest growth rate.

Additional projection results (not shown in the tables or figures) indicate that the racial differentials in the growth of yearly home-based care payments for the disabled elders are similar to those for number of disabled elders. Because disabled elders in Black, Hispanic, and Other non-Hispanic racial groups receive substantially more hours of home-based care services per person per week than White non-Hispanic elders (see Appendix 3), their needs for yearly workdays of care will grow much faster than that for White non-Hispanic elders.

10.3.5 High and Low Bounds of Home-Based Care Costs

So far, the projection results presented and discussed above are based on the medium disability scenario. Figures 10.2 and 10.3 provide the low and high bounds of home-based care costs for disabled elders (days and dollars), based on the low and the high disability scenarios, as described earlier (Sect. 10.2.2). As compared to the medium scenario of yearly home-based care workdays for disabled elderly in 2050, the low disability scenario estimate would be 25.1 % lower and the high disability scenario estimate would be 47.4 % higher (see Fig. 10.2). As compared to the medium scenario, yearly payments for home-based care for the disabled elderly in 2050, the low disability scenario would be 25.3 % lower and the high disability scenario would be 47.0 % higher (see Fig. 10.3).

10.4 Concluding Remarks

Based on data derived from census micro datasets and the National Long Term Care Survey and the ProFamy extended cohort-component method, we presented projections of numbers of ADL disabled elders and yearly home-based care workdays and payments for disabled elders by age, gender, race, and living arrangement from 2010 to 2050 for the United States. The high, medium, and low scenarios of disability trends were based on different assumptions about future disability patterns and combined with the medium scenario of demographic household and elderly living arrangement projections. The results show that population and household aging will lead to a remarkable acceleration in the increase in number of disabled elderly after 2020, with a much faster increase in the disabled oldest-old aged 80+ compared to the disabled young-old aged 65–79; the disabled oldest-old will outnumber the disabled young-old after 2030. Increases in yearly home-based care workdays and payments for disabled elders will dramatically accelerate after 2020, especially for disabled oldest-old. The home-based care costs for the disabled elderly population will be increasingly dominated by the disabled oldest-old after 2030. While these general trends are similar across the racial groups, our projections also show that some racial differentials are notable.

Clearly, given the fact that the elderly population (especially oldest-old) with disability and their home-based care costs will follow a remarkable increasing trajectory in the coming decades, new initiatives to develop more home-based care programs are necessary. These new programs ought to pay special attention to the home-based care needs and costs of disabled oldest-old aged 80+, as they will grow much faster and their service needs and costs are substantially larger compared to the disabled young-old aged 65–79.

Our low, medium, and high disability trend scenarios indicate that postponement of the onset of disability could substantially reduce home-based care needs and costs for elders. This saving is crucial to offset the long-term care budget constraints for federal and state governments and save LTC expenditures for millions of families with disabled elders. Given that the demographic trends of increasing human lifespan and rapidly growing numbers of the elderly (especially oldest-old) are inevitable, is it possible to realize the compression of morbidity (Fries 1980) demonstrated in our low disability scenario, or at least the dynamic equilibrium (Manton 1982) demonstrated in our medium disability scenario, and to avoid the expansion of disability (Gruenberg 1977) demonstrated in our high disability scenario? Why do some people survive to advanced ages with good health while others suffer severe disability and morbidity? So far, there are few answers to these critical questions, which determine federal and state governmental budgetary sustainability and quality of life not only for the elderly themselves but also for all members of society. Thus, it is crucially important and urgent for funding agencies and research communities to pay more attention to the effects of social, behavioral, and genetic factors and their interactions on healthy aging and develop more effective intervention programs to promote elderly health through interdisciplinary research.

By integrating living arrangements of the elderly, which is a crucially important determinant of home-based care costs, into projections of disability and yearly home-based care workdays and payments for disabled elders, our present study provides relatively more realistic and detailed information on future trends than previous studies which excluded living arrangements. Our more realistic and detailed projection outputs by living arrangement are useful for governmental policy analysis, strategic plans for future public services, and private sector market potential research.

We conclude by noting the following important limitations and unaddressed issues that need to be further investigated in future research. First, due to space limitations, we have presented here the general trends and patterns by age (oldest-old vs. young-old), gender, race, and major types of living arrangement. However, we have produced more detailed projection outputs of the numbers of disabled elders, the yearly workdays and payments of home-based care by 5-year age groups, marital/union status, and more detailed living arrangements, not only for the U.S. as a whole but also for California, Florida, North Carolina, and Minnesota.[6] A full presentation and discussion of all of these results would require preparation of five different reports for the U.S. as a whole and for each of the four states, which could be useful in more geographically localized and detailed academic research, governmental panning, and business market analysis in future studies. We are open to collaborations with interested colleagues.

Second, there are many sources of influences (demographic, economic, social, and political factors) on long-term care spending. This chapter uses comprehensive demographic projections and analysis to investigate how changes in households and living arrangements may affect future home-based care costs for disabled elders in the United States; it does not take into account other important determinants of future long-term care spending such as changes in income and service prices. One reason is that, as outlined in the literature review, including income and service prices in the projection model would require forecasting these microeconomic variables first, which is very difficult and may not be feasible, and may lead to forecast instability (Lee and Miller 2002). Another reason is that exclusion of these microeconomic variables allows us to adequately analyze the effects of changes in demographic and disability rates on home-based care costs, which is the main objective of this chapter, rather than effects of a mixture of demographic, disability, and uncertain economic trends. Although the analyses presented in this chapter serves our current research purpose well, future analyses need to also take into account the impacts of income and service prices.

Third, we did not include institutional care costs in this article, mainly because the NLTCS did not have large enough sub-sample size to estimate the age-gender-race-disability status-specific transition rates of entering into and discharging out of institutional care units. Given the paucity of existing literature on home-based care costs, our analysis is a relevant first step. Institutional care costs are certainly important for long-term care spending studies and should be investigated in further research. Further research may also estimate the cost-benefit effects of home-based care versus institutional care. For example, does home-based care save money and satisfy the disabled elders' needs compared to institutional care? If so, how much and to what extent do living arrangement trends result in savings along this dimension? Such further analyses would be particularly relevant and useful for policy makers.

Fourth, further in-depth research may also investigate how much of the future increase in home-based care costs for the disabled elderly is due to increases in longevity. How much due to gender differentials in longevity and service use? How much is due to divorce, remarriage, fertility, and other sources? The relative

[6] The ProFamy extended cohort-component model and its associated software produces a large amount of output for household status and living arrangement projections cross-classified by race, sex, age, marital/union status, number of co-residing children, and living with no, one, or two parents, for each of the projection years (see Table 2 in Zeng et al. 2006).

contribution of different forces to home-based care costs will help to identify specific areas for intervention.

Finally, as we discussed in the other chapters, we emphasize again that projections with less than 20 years of time horizon may be used as forecasting with reasonable accuracy for business and governmental planning, but any results beyond that should be considered to be simulations only, due to large uncertainties after 20 years. Thus, the projection results after 2030 presented in this chapter should be mainly regarded as simulations. Such simulations are useful for academic and policy analysis to answer the "what, if" questions about effects of changes in demographics and disability prevalence rates on the future general trends and patterns of home-based care needs and costs for disabled elders in the U.S., but they cannot be considered to be accurate forecasts.

Appendix 1: The Estimated and Assumed Demographic Summary Measures in the Baseline and Future Years for the United States

	White non-Hispanic			Black non-Hispanic			Hispanic			Asian and other non-Hispanic		
	2010	2025	2050	2010	2025	2050	2010	2025	2050	2010	2025	2050
Male life exp. e_0	75.3	76.3	77.5	68.8	70.1	73.6	77.4	78.4	79.3	77.2	77.4	78.3
Female life exp. e_0	80.4	81.1	82.1	75.8	77.1	80.0	82.9	83.7	84.4	80.5	81.4	83.3
TFR-all births	1.86	1.90	1.89	2.02	1.91	1.88	2.65	2.53	2.29	1.86	1.90	1.89
TFR(1)-1st birth	0.82	0.86	0.86	0.84	0.82	0.84	0.95	0.95	0.95	0.88	0.95	0.95
TFR(2)-2nd birth	0.61	0.61	0.60	0.68	0.63	0.60	0.88	0.81	0.69	0.57	0.55	0.55
TFR(3)-3rd birth	0.29	0.29	0.28	0.28	0.26	0.25	0.47	0.43	0.37	0.27	0.26	0.26
TFR(4)-4th birth	0.10	0.10	0.10	0.13	0.12	0.11	0.21	0.20	0.17	0.10	0.09	0.09
TFR(5)-5+ birth	0.05	0.05	0.05	0.09	0.08	0.08	0.14	0.13	0.11	0.05	0.04	0.04
General marriage rate	0.05	0.05	0.05	0.02	0.02	0.02	0.05	0.05	0.05	0.05	0.05	0.05
General divorce rate	0.02	0.03	0.03	0.02	0.02	0.02	0.02	0.02	0.02	0.02	0.02	0.02
General cohabiting rate	0.09	0.10	0.10	0.07	0.08	0.08	0.09	0.09	0.10	0.10	0.10	0.10
General union break rate	0.26	0.26	0.26	0.32	0.32	0.32	0.19	0.19	0.29	0.29	0.29	0.29
Male mean age 1st mar.	28.0	29.00	29.00	31.0	32.15	32.15	27.9	28.50	28.50	31.6	32.80	32.80
Female mean age 1st mar.	26.5	27.8	27.8	30.4	32.1	32.1	27.0	28.1	28.1	29.8	31.2	31.2
Mean age at births	28.7	29.8	29.8	25.5	25.9	25.9	26.0	26.2	26.2	29.1	29.9	29.9
Mean age at births	29.2	30.3	30.3	25.9	26.3	26.3	26.4	26.7	26.7	29.6	30.4	30.4

Appendix 2: A Two-Step Procedure to Estimate Age-Sex-Race-Living Arrangement-Disability Status-Specific Care Hours and Care Costs Per Elder

Because of too small sub-sample size for minority sub-groups, especially for the oldest-old aged 80+, we did not use the unreliable directly-computed age-sex-race-living arrangement-disability status-specific care hours and care costs per elder. Instead, we applied a "two-step" approach to obtain our estimates after careful investigations and tests:

Step one: We first calculated baseline sex-age-specific values with all of the other attributes combined (including racial groups, disability status, and living arrangement types) for home-based care costs per elder from the data directly;

Step two: We then estimated the sex-age-attribute-specific care costs per elder by multiplying the baseline sex-age-specific care costs per elder by the multivariate regression estimates of the corresponding odds ratios of care needs/costs among persons with different attributes, as compared to the baseline.

Why did we adopt the two-step approach rather than the "one-step" approach of multivariate regression models to directly estimate the age- race- sex- living arrangement- disability status-specific care need/costs per elder? In general, multivariate regression models are powerful in explanatory analysis of associations with socio-economic and demographic covariates. When the primary purpose is, however, to estimate the age-specific schedules (or trajectories) and propensities of the occurrence of the events rather than explanatory analysis, the classic regression approach may not be ideal. This is because the estimate of the age covariate coefficients in the regression model may not accurately represent the age trajectory, unless the age trajectory follows precisely linear or log-linear or another kind of analytical distribution, which is unlikely (Land et al. 1994: 304), especially in the case of age-race- sex-living arrangement- disability status-specific care costs per elder. Furthermore, regression models presume that all sources of individual-level variations are explained by the covariates that enter the regressions. That is, the regression models assume that no "hidden heterogeneity" is present in the age and other covariate specific rates estimated based on the regression coefficients. This specification is almost surely not true empirically, especially for extended periods of more than 1 year (Land et al. 1994: 304).

We have empirically tested the "one-step" approach of the multivariate regression model to directly estimate the race- sex- age- living arrangement- disability status-specific care costs per elder, without the estimates of the sex-age-specific baseline care needs/costs per elder. The results are out of an empirically plausible range for some age groups. Even after correcting the logic errors by introducing some constraints to the regression, the estimates are still unreasonable. In short, our empirical tests and theoretical considerations lead us to believe that the "two-step" approach is much more robust than the "one-step" approach in estimating the race-sex- age- living arrangement- disability status-specific care costs per elder.

Appendix 3: Age-Sex-Race-Living Arrangement-Specific Disability Rates, Home-Based Care Hours, and Care Costs ($), Based on Data from NLTCS 1999 Wave, the United States

	Males					Females				
Age	65–69	70–74	75–79	80–84	85+	65–69	70–74	75–79	80–84	85+
Disability (%)										
Living alone										
White non-Hispanic	2.31	2.54	3.84	6.12	11.13	2.65	3.32	4.87	7.90	15.26
Black non-Hispanic	3.93	4.33	6.48	10.16	17.92	4.51	5.62	8.15	12.91	23.88
Hispanic	2.19	2.41	3.64	5.81	10.59	2.51	3.14	4.62	7.50	14.55
Asia and other non-Hispanic	1.94	2.13	3.23	5.17	9.47	2.23	2.79	4.10	6.69	13.07
Living with spouse/partner, may (or may not) live with children/others										
White non-Hispanic	4.32	4.76	7.11	11.10	19.44	4.95	6.17	8.92	14.06	25.74
Black non-Hispanic	7.28	8.00	11.75	17.84	29.73	8.29	10.25	14.53	22.03	37.79
Hispanic	4.10	4.51	6.74	10.56	18.57	4.70	5.85	8.48	13.40	24.68
Asia and other non-Hispanic	3.64	4.00	6.00	9.44	16.74	4.17	5.20	7.56	12.03	22.42
Not living with spouse/partner but living with children/others										
White non-Hispanic	6.14	6.75	9.98	15.31	25.99	7.01	8.69	12.41	19.09	33.52
Black non-Hispanic	10.21	11.20	16.21	23.96	38.26	11.57	14.22	19.76	28.95	47.06
Hispanic	5.82	6.40	9.48	14.60	24.91	6.65	8.25	11.81	18.25	32.27
Asia and other non-Hispanic	5.18	5.69	8.46	13.11	22.61	5.92	7.36	10.58	16.48	29.57
Home-based care hours per week received by disabled elders										
Living alone										
White non-Hispanic	9.95	11.36	12.43	14.57	17.27	7.99	9.20	10.71	12.01	21.21
Black non-Hispanic	13.31	15.48	16.83	19.62	23.10	10.46	12.21	14.49	16.21	27.68
Hispanic	13.63	15.79	17.21	19.98	23.56	10.75	12.51	14.78	16.48	27.76
Asia and other non-Hispanic	12.15	14.11	15.36	17.95	21.19	9.57	11.15	13.21	14.81	25.74
Living with spouse/partner, may (or may not) live with children/others										
White non-Hispanic	23.43	26.62	28.57	32.10	36.31	18.84	21.51	25.18	27.32	40.17
Black non-Hispanic	29.32	32.57	34.69	38.34	42.39	24.22	27.04	31.17	33.28	45.72
Hispanic	29.74	33.09	35.30	38.64	42.77	24.62	27.41	31.39	33.33	45.30
Asia and other non-Hispanic	27.46	30.68	32.73	36.47	40.55	22.49	25.30	29.37	31.57	44.38
Not living with spouse/partner but living with children/others										
White non-Hispanic	24.85	28.08	30.07	33.72	37.91	20.08	22.83	26.69	28.87	41.84
Black non-Hispanic	30.78	33.98	36.12	39.84	43.81	25.61	28.44	32.69	34.80	47.18
Hispanic	31.24	34.55	36.77	40.18	44.23	26.04	28.84	32.93	34.87	46.78
Asia and other non-Hispanic	28.91	32.10	34.16	37.99	42.01	23.84	26.67	30.89	33.10	45.89
Care payment ($) per month of home-based care for disabled elders										
Living alone										
White non-Hispanic	192.16	240.46	272.40	348.29	389.17	117.20	216.24	251.62	297.58	429.53
Black non-Hispanic	117.27	146.27	167.61	223.70	259.53	71.66	131.54	154.62	190.83	285.12
Hispanic	169.40	209.93	238.58	304.40	342.06	104.23	190.73	222.54	263.17	381.74

(continued)

	Males					Females				
Age	65–69	70–74	75–79	80–84	85+	65–69	70–74	75–79	80–84	85+
Asia and other non-Hispanic	116.79	142.22	162.78	212.27	242.87	72.26	130.84	153.97	185.60	275.05
Living with spouse/partner, may (or may not) live with children/others										
White non-Hispanic	67.56	87.98	101.73	133.42	160.46	95.98	178.16	207.62	253.40	372.39
Black non-Hispanic	38.46	50.85	58.94	77.83	94.97	57.27	106.49	124.78	158.10	238.66
Hispanic	58.91	76.45	88.47	114.78	137.95	84.22	155.05	180.98	220.14	324.40
Asia and other non-Hispanic	38.87	49.43	57.28	75.26	90.62	56.76	103.51	121.67	149.94	224.73
Not living with spouse/partner but living with children/others										
White non-Hispanic	114.03	147.11	169.18	222.55	262.30	102.75	190.82	222.75	270.83	397.03
Black non-Hispanic	66.07	86.06	99.51	133.86	161.77	61.53	113.73	133.89	168.99	255.59
Hispanic	98.80	126.52	145.73	190.02	224.40	90.50	166.50	194.82	236.16	347.56
Asia and other non-Hispanic	65.38	82.04	94.78	125.31	149.00	61.51	111.73	131.75	161.72	242.45

Chapter 11
Projections of Household Vehicle Consumption in the United States

11.1 Introduction

American automakers have been experiencing a continuous downturn in production. For example, the output of passenger cars in the United States has decreased steadily from 5.6 million in 1999 to 2.2 million in 2009, a drop of approximately 60 % (OICA website). Moreover, in spite of a fast increase in household vehicle ownership in the 1980s, the trend has significantly slowed down in most recent decades (Hu and Reuscher 2004; Davis and Diegel 2009): the annual growth rate of the average number of vehicles per household was 2.53 % in the period 1970–1980, but decreased to 0.48 % in the period 1990–2000. Knowledge of vehicle consumption dynamics in the current market environment and demographic context is important to today's manufacturers, dealers, and other stakeholders in this sector. Projections of household-based vehicle consumption are of particular significance due to its sheer size in the vehicle market; indeed, it has already received extensive attention in market research (e.g., Bhat and Sen 2005).

Vehicle consumption at the household level is inherently associated with household demographic and socio-economic status. Household surveys have consistently shown that household purchase of vehicles depends largely on the type, size, and income of the household, and on the age, sex, and race of the householder (Hu and Reuscher 2004; Vance and Buchheim 2004). These factors often affect the household purchases of specific types of vehicle (Bhat and Sen 2005; Golob and Brownstone 2005). For example, households with more children are more likely to own and use vans rather than cars and trucks, and ownership of vans is over-represented in middle-income households (Bhat and Sen 2005).

Changes in population and household structures can have major influences on automobile consumption. O'Neill and Chen (2002) found that household changes in the United States had a substantial impact on aggregate demand for vehicles over the past several decades. It is also interesting to note that American female baby-boomers have a far higher proportion of licensed drivers compared to the current generation of elderly women, portending an increase in vehicle consumption

Y. Zeng et al., *Household and Living Arrangement Projections*, The Springer Series on Demographic Methods and Population Analysis 36, DOI 10.1007/978-90-481-8906-9_11, among elderly groups in the near future (Spain 1997). A similar trend has also been reported for vehicle consumption in Germany (Buettner and Grubler 1995).

The existing literature clearly suggests that a more reasonable projection of household-based vehicle consumption should be based on household forecasts by type/size/income of household and age/sex/race of householder; however, few studies have attempted to forecast vehicle consumption by these household and householder characteristics. Instead, most researchers use aggregated population (rather than household) and economic data to forecast future vehicle demand, following either regression or other specific modeling approaches (e.g., California Energy Commission 2003; Cao and Mokhtarian 2003; Natural Resources of Canada 2004). Some vehicle consumption forecasts have used household information, but are based on the classic headship rate method; the headship rate approach is very limited and has been widely criticized for more than two decades (Mason and Racelis 1992; Murphy 1991; Spicer et al. 1992; see Sect. 4.2 of Chap. 4 of this book).

This chapter applies the ProFamy extended cohort-component method (see Chaps. 2, 3 and 4 for details) to project household vehicle consumption in four regions of the United States from 2000 to 2025 (Feng et al. 2011). In the following section, we demonstrate the main components and procedure of the projection, including the major categorical definitions, estimates of vehicle ownership rates, and the household vehicle consumption forecasts for the four regions of the United States from 2000 to 2025. We also validate the ProFamy method by comparing the projected results with officially released household vehicle consumption data in the period 2000–2009. The major findings of this study are summarized and discussed in the conclusion section.

11.2 Data Sources and Model Specification to Account for Regional, Income, and Racial Differentials

The four regions used in this projection are the Northeast, Midwest, South, and West, as defined by the U.S. Census Bureau in the 2000 census (see Appendix 1 for specific states in each region). To take into account the regional differentials in demographics, income distributions, and consumption patterns, we must forecast households and vehicle consumption for each of the four regions; however, this does not mean simply applying the ProFamy method in an isolated way to each region. The four regions are related to one another through domestic migration, and a series of consistency examinations need to be ensured (see Appendix 2).

Four household income categories (high income, middle income I, middle income II and low income) are defined based on income quartiles from the micro dataset of the 2000 census; the ranges for the high, middle I, middle II, and low income categories are set as $> \$71{,}158$, \$40,712–\$71,158, \$20,245–\$40,712 and $< \$20{,}245$, respectively. We used income quartiles rather than the absolute number in dollars to create household income categories because: (1) the absolute number in dollars always changes over time, even after standardizing; (2) technically, it is very hard to predict household income distributions by using a measure of absolute numbers in dollars, but relatively easy by using percentile-based income categories; and (3) it is also feasible to forecast household vehicle consumption for different percentile-based household income categories through time series analysis or expert opinions on economic growth, income diversity, and consumer behavior changes. A procedure was developed to ensure consistency for the distribution of income categories in the projection (see Appendix 3).

Four racial/ethnic groups are distinguished for householders in the projection: White non-Hispanic, Black non-Hispanic, Hispanic, and Asian and other non-Hispanic. Household vehicles in this study are grouped into three basic types, including car (passenger car, station wagon, SUV, and other cars), van (minivan, cargo-van, and passenger van), and truck (pickup and other trucks).

In order to estimate the household vehicle ownership rate of cars, vans, trucks, and all vehicles, this study used various sources, including the micro dataset of the 2000 census, American Community Survey (ACS) 2000–2002, American Housing Survey (AHS) 2001 and 2003, and the National Household Travel Survey (NHTS) 2001. The data sources and estimation procedures of the household forecasts for the United States were presented in Sect. 8.1 of Chap. 8 and elsewhere (Zeng et al. 2013a) and thus will not be repeated here.

11.3 Estimation of Household Vehicle Ownership Rates by Household Characteristics, Race, and Region

Due to space limitations, we present only a summarized description of the estimates of vehicle ownership rates here, with the details given in Table 11.1. This table shows that one-couple households (households with one couple only or one couple with child(ren)/others) have the highest vehicle ownership rate. In general, the vehicle ownership rates increase with household size, except for trucks. The van ownership rate of one-couple households is much higher than that of single-person households (households of one person only or a single person with child(ren)/others), and the gap enlarges sharply with the increase in household size. The truck ownership rate of single-man households is higher than that of single-woman households, while there is almost no difference observed for van ownership rates between these two household types (see Table 11.1).

For all types of vehicles except for trucks, the higher the household income, the more vehicles the household owns. With regard to trucks, middle income I households have the highest ownership rate and there is almost no difference in vehicle ownership rates between high income and middle income II households (see Table 11.1).

Table 11.1 Estimates of household vehicle ownership rates by household characteristics, age group, income group, race, region and vehicle type, 2000

	Whole country				Northwest				Midwest				South				West			
	All	Car	Van	Truck	All	Car	Van	Truck	All	Car	Van	Truck	All	Car	Van	Truck	All	Car	Van	Truck
Household type/size																				
Single man only, hh. size 1	1.13	0.78	0.08	0.26	0.95	0.74	0.07	0.14	1.21	0.82	0.09	0.29	1.15	0.75	0.09	0.30	1.16	0.82	0.06	0.27
Single man, hh. size 2	1.61	1.16	0.10	0.36	1.39	1.17	0.06	0.16	1.71	1.19	0.12	0.40	1.61	1.18	0.11	0.33	1.68	1.08	0.10	0.50
Single man, hh. size 3	1.83	1.28	0.15	0.40	1.57	1.26	0.14	0.17	1.95	1.39	0.20	0.35	1.81	1.20	0.14	0.48	1.94	1.33	0.12	0.49
Single man, hh. size 4	1.99	1.46	0.18	0.35	1.66	1.33	0.16	0.17	2.09	1.50	0.21	0.38	1.99	1.44	0.16	0.39	2.11	1.54	0.18	0.39
Single man, hh. size 5+	2.09	1.57	0.19	0.33	1.67	1.38	0.15	0.14	2.15	1.58	0.22	0.34	2.08	1.53	0.17	0.38	2.36	1.76	0.20	0.40
Single woman only, hh. size 1	0.86	0.73	0.04	0.09	0.72	0.64	0.03	0.05	0.88	0.75	0.05	0.08	0.90	0.74	0.04	0.12	0.92	0.79	0.04	0.10
Single woman, hh. size 2	1.27	1.04	0.07	0.17	1.11	0.95	0.05	0.11	1.32	1.06	0.08	0.18	1.28	1.02	0.07	0.18	1.37	1.12	0.06	0.19
Single woman, hh. size 3	1.37	1.10	0.10	0.17	1.20	1.00	0.08	0.11	1.42	1.10	0.12	0.20	1.36	1.10	0.09	0.16	1.52	1.20	0.09	0.23
Single woman, hh. size 4	1.54	1.15	0.21	0.17	1.28	0.99	0.18	0.11	1.64	1.18	0.22	0.23	1.53	1.19	0.16	0.17	1.70	1.20	0.33	0.17
Single woman, hh. size 5+	1.69	1.26	0.26	0.17	1.33	1.03	0.21	0.09	1.72	1.26	0.23	0.23	1.71	1.27	0.25	0.18	1.94	1.44	0.33	0.16
One couple only, hh. size 2	1.92	1.37	0.13	0.42	1.73	1.37	0.12	0.24	1.99	1.41	0.15	0.43	1.94	1.32	0.13	0.49	1.96	1.42	0.11	0.44
Couple size, hh. 3	2.20	1.58	0.17	0.45	2.03	1.58	0.17	0.29	2.32	1.62	0.22	0.48	2.20	1.54	0.15	0.51	2.20	1.58	0.17	0.44
Couple size, hh. 4	2.25	1.51	0.30	0.44	2.10	1.52	0.32	0.26	2.38	1.55	0.36	0.47	2.25	1.47	0.27	0.50	2.26	1.53	0.28	0.46
Couple size, hh. 5	2.24	1.36	0.47	0.41	2.07	1.33	0.52	0.23	2.39	1.36	0.56	0.47	2.23	1.35	0.41	0.47	2.26	1.40	0.43	0.42
Couple size, hh. 6+	2.25	1.33	0.53	0.38	1.97	1.25	0.55	0.17	2.37	1.17	0.73	0.47	2.26	1.31	0.53	0.42	2.32	1.50	0.41	0.41

Age groups																				
<25	1.48	1.13	0.10	0.25	1.20	1.01	0.08	0.10	1.56	1.17	0.12	0.27	1.51	1.13	0.09	0.29	1.55	1.18	0.10	0.27
25–34	1.69	1.21	0.16	0.32	1.49	1.18	0.15	0.16	1.77	1.22	0.21	0.35	1.71	1.20	0.14	0.37	1.73	1.25	0.15	0.33
35–44	1.93	1.27	0.27	0.40	1.73	1.22	0.27	0.24	2.04	1.29	0.32	0.43	1.94	1.25	0.24	0.45	1.97	1.31	0.25	0.42
45–54	2.04	1.44	0.20	0.40	1.89	1.44	0.20	0.25	2.14	1.48	0.23	0.43	2.02	1.40	0.18	0.45	2.09	1.49	0.19	0.42
55–64	1.81	1.27	0.16	0.39	1.67	1.25	0.15	0.27	1.88	1.30	0.18	0.41	1.81	1.22	0.16	0.43	1.89	1.33	0.15	0.40
65–79	1.39	1.06	0.11	0.22	1.23	1.02	0.11	0.10	1.44	1.08	0.13	0.23	1.40	1.02	0.11	0.27	1.48	1.13	0.10	0.25
80+	0.91	0.79	0.04	0.08	0.75	0.71	0.03	0.01	0.93	0.80	0.05	0.08	0.93	0.77	0.04	0.12	1.02	0.89	0.04	0.10
Income groups																				
High income	2.31	1.70	0.25	0.36	2.13	1.68	0.25	0.20	2.40	1.72	0.29	0.39	2.33	1.68	0.22	0.42	2.35	1.74	0.23	0.39
Middle income I	1.99	1.35	0.21	0.44	1.76	1.28	0.20	0.28	2.10	1.38	0.25	0.47	2.04	1.35	0.20	0.49	2.01	1.36	0.20	0.44
Middle income II	1.60	1.10	0.16	0.34	1.36	1.02	0.14	0.19	1.67	1.12	0.18	0.36	1.65	1.11	0.15	0.40	1.62	1.11	0.15	0.35
Low income	1.07	0.78	0.09	0.20	0.83	0.67	0.07	0.08	1.11	0.80	0.11	0.20	1.12	0.80	0.09	0.23	1.13	0.82	0.09	0.22
Race																				
White non-Hispanic	1.82	1.27	0.18	0.37	1.69	1.28	0.18	0.23	1.89	1.29	0.21	0.39	1.83	1.23	0.16	0.43	1.86	1.31	0.16	0.39
Black non-Hispanic	1.33	1.07	0.11	0.15	0.98	0.86	0.10	0.02	1.27	1.01	0.13	0.13	1.45	1.14	0.11	0.19	1.41	1.16	0.10	0.15
Hispanic	1.59	1.09	0.22	0.27	1.01	0.78	0.16	0.07	1.63	1.08	0.28	0.27	1.67	1.11	0.23	0.33	1.73	1.20	0.23	0.30
Asian and other non-Hispanic	1.69	1.23	0.20	0.26	1.34	1.07	0.18	0.09	1.69	1.19	0.23	0.27	1.78	1.25	0.20	0.34	1.80	1.31	0.19	0.29
Total	1.74	1.23	0.18	0.33	1.56	1.19	0.17	0.19	1.82	1.25	0.21	0.36	1.75	1.21	0.16	0.38	1.81	1.29	0.17	0.35

Note: "hh." means "household"

Black non-Hispanic households have the lowest vehicle ownership rates for all types of vehicle, while White non-Hispanic households have the highest rates except for vans. Hispanic households own more vans in comparison with other racial groups. The truck ownership rate of white non-Hispanic households is much higher than that of other races (see Table 11.1).

The age pattern of vehicle ownership rate is similar across vehicle types. The ownership rate initially increases with age, reaches its peak value in the middle ages such as 35–44 or 45–54, and then decreases substantially after the mid-50s (see Table 11.1).

The Northeast has the lowest ownership rates across all kinds of vehicles, especially trucks. The South has the highest rate of truck ownership; the Midwest has the highest ownership rates of vans; and the West has the highest ownership rates of cars (see Table 11.1).

11.4 Results of Household Vehicle Consumption Forecasts

11.4.1 Household Projection Outcome

The results of the household forecasts are summarized in Table 11.2. As can been seen in that table, from 2000 to 2025, the population size of the United States will increase from 281 million to 344 million and the number of households will increase from 105 million to 130 million. In 2000, the South had the largest population size and number of households, whereas the Northeast had the smallest population size and number of households. The population size and number of households in the Midwest and West were very close to each other. From 2000 to 2025, this pattern will mostly remain except that the West is projected to exceed the Midwest in population size and household number. The average household size in each region will generally witness a steady and slow decline from 2000 to 2025.

As shown in Table 11.2, the married-couple household category is dominant among all household types, accounting for 50.2–53.0 % in 2000 for each region. This household type will continue to be dominant in 2025, although its percentage will be reduced to 48.1–51.0 %. The South will have the largest decline in the percentage of married-couple households, while the West will have the smallest decline. The one-person-only households were 24.2–27.2 % of the total number of households in 2000, and will steadily increase to 25.9–29.3 % in 2025. By comparison, the percentage of single-parent households will decrease from 12.1–14.8 % in 2000 to 10.3–13.2 % in 2025. The percentage of cohabiting couple households will increase from 4.4–5.7 % in 2000 to 5.5–6.3 % in 2025.

Based on the household projection results summarized above and the estimates of vehicle ownership rates discussed in Sect. 11.2, we generated household vehicle consumption forecasts. In forecasting the future household vehicle consumption, we held constant the age-sex-race-household type/size-income-region-specific

Table 11.2 Summary of household forecasts from 2000 to 2025

Region	Population size	Number of households	Average household size	Married- couple households (%)	One-person only households (%)	Single-parent households (%)	Cohabiting-couple households (%)
Northeast							
2000	53,594,380	20,285,610	2.55	50.18	27.21	13.59	5.09
2010	54,741,484	21,163,530	2.50	48.69	28.09	12.39	5.77
2015	54,745,540	21,268,024	2.49	48.48	28.26	12.28	5.81
2020	54,751,312	21,241,202	2.49	48.33	28.29	12.35	5.79
2025	54,774,228	21,161,876	2.50	48.09	28.37	12.46	5.77
Midwest							
2000	64,392,772	24,734,530	2.52	53.02	26.86	12.09	4.80
2010	67,293,184	26,515,424	2.47	51.66	28.46	10.39	5.51
2015	68,470,448	27,280,866	2.44	51.40	28.78	10.19	5.56
2020	69,544,488	27,890,148	2.42	51.13	29.00	10.17	5.56
2025	70,519,144	28,403,604	2.41	50.71	29.26	10.25	5.54
South							
2000	100,236,816	38,015,212	2.56	51.65	25.33	14.79	4.40
2010	113,848,880	43,663,700	2.53	49.51	26.75	12.67	5.60
2015	120,802,040	46,537,776	2.52	49.02	27.06	12.56	5.72
2020	127,728,984	48,775,108	2.55	48.78	27.05	12.76	5.75
2025	134,841,200	50,318,712	2.60	48.41	27.03	13.22	5.71
West							
2000	63,197,932	22,444,732	2.73	51.80	24.22	13.23	5.72
2010	72,110,904	26,184,768	2.69	51.48	25.51	11.15	6.37
2015	76,046,448	27,817,182	2.67	51.39	25.71	10.94	6.36
2020	79,913,616	29,257,780	2.67	51.23	25.81	10.98	6.31
2025	83,746,896	30,558,684	2.68	51.02	25.85	11.14	6.25
Total							
2000	281,421,900	105,480,084	2.59	51.72	25.82	13.59	4.91
2010	307,994,452	117,527,422	2.55	50.29	27.10	11.76	5.78
2015	320,064,476	122,903,848	2.53	49.99	27.34	11.62	5.85
2020	331,938,400	127,164,238	2.54	49.78	27.40	11.72	5.85
2025	343,881,468	130,442,876	2.56	49.47	27.46	11.96	5.81

Table 11.3 A comparison between our projected number of home-use passenger cars and the official statistics of the US Department of Transportation, 2000–2009

Year	ProFamy projections	Official statistics	Forecast % error
2000	128,043,495	132,247,286	−4.4
2001	133,836,606	136,340,945	−2.8
2002	132,001,744	134,604,524	−3.0
2003	133,732,685	134,336,851	−1.4
2004	135,398,896	135,007,031	−0.8
2005	136,997,767	135,192,288	0.3
2006	138,523,397	134,012,369	2.3
2007	139,995,950	134,510,252	2.9
2008	141,409,761	135,637,845	3.1
2009	142,928,590	133,437,105	7.1
Mean algebraic % error			0.3
Mean absolute % error			2.8

vehicle ownership rate. It is a common approach to hold some attribute-specific rates constant throughout the projection period (e.g., Day 1996). Smith et al. (2001) provided some scientific and practical grounds for holding some rates and proportions constant (see Sect. 8.2 of Chap. 8 of this book).

11.4.2 Validation Tests: A Comparison Between Our Projected Numbers and Official Statistics on Number of Home-Use Passenger Cars, 2000–2009

In order to validate this application of the ProFamy method, we next compare the ProFamy-projected number of household passenger cars with actual numbers of non-public-owned passenger cars reported from the US Department of Transportation from 2000 to 2009 (US Department of Transportation 2000–2009). To our knowledge, this is the best available data source for the validation test.

As shown in Table 11.3, the differences between the ProFamy projected number of cars and the official statistics of the US Department of Transportation are within a reasonable range, with a mean absolute per cent error of 2.8 %. This validates the ProFamy method in household vehicle consumption forecasting. We also notice that our projection tends to moderately overestimate household vehicle consumption in very recent years, especially 2009. This is probably due to the fact that American households reduced their vehicle consumption because of the Great Recession of 2008–2009, an event that was not accounted for in our projections based on household ownership rates observed in 2000.

Table 11.4 Home vehicle consumption forecasts, 2000–2025 (unit: 1,000)

	All vehicles		Cars		Vans		Trucks	
Year	Number	% Cum. inc.	Number	% Cum. inc.	Number	% Cum. inc.	Number	% Cum inc.
2000	179,590	–	128,043	–	17,695	–	33,852	–
2002	186,299	0.04	132,002	0.03	18,977	0.07	35,320	0.04
2004	191,440	0.07	135,264	0.06	19,712	0.11	36,464	0.08
2006	196,276	0.09	138,247	0.08	20,443	0.16	37,586	0.11
2008	200,784	0.12	140,986	0.10	21,156	0.20	38,642	0.14
2010	205,315	0.14	143,748	0.12	21,894	0.24	39,673	0.17
2012	210,713	0.17	146,514	0.14	23,103	0.31	41,096	0.21
2014	215,003	0.20	149,009	0.16	23,912	0.35	42,082	0.24
2016	220,198	0.23	151,421	0.18	25,267	0.43	43,510	0.29
2018	224,141	0.25	153,582	0.20	26,163	0.48	44,396	0.31
2020	227,700	0.27	155,426	0.21	27,066	0.53	45,208	0.34
2022	231,056	0.29	157,114	0.23	27,968	0.58	45,974	0.36
2025	235,087	0.31	159,248	0.24	29,064	0.64	46,775	0.38

% Cum. inc. is percentage of cumulative increase, and cumulative increase is defined as the difference between the number of household vehicles in the projected year and the number of household vehicles observed in the baseline year 2000

11.4.3 A General Description of the Forecasts

Table 11.4 shows the size and composition of household vehicle consumption in the United States from 2000 to 2025. The forecast results show that the total number of household vehicles in 2025 will reach 235 million, 55 million more than in 2000, which represents a 31 % increase over 25 years. Of these 55 million increase of the household vehicles, 56.2 % is due to the increase in the consumption of cars, 23.3 % due to trucks, and 20.5 % due to vans. The number of household cars is forecasted to increase by 24.4 % in 2025, as compared to 2000. In contrast, household vans and trucks will increase by 64.2 % and 38.2 %, respectively. In 2025, the structure of household vehicle consumption will be 67.7 % for cars, 12.4 % for vans and 19.9 % for trucks, respectively. In comparison with the structure in 2000 (71.3 % for cars versus 9.9 % for vans and 18.8 % for trucks), the rise of vans in household vehicle consumption is evident.

11.4.4 Forecast by Age and Race of Householders

In the following projections, we demonstrate the cumulative increase in household vehicle consumption by household characteristics and the four regions. The 'cumulative increase' is defined as the difference between the number of household vehicles in the projected year and the number of household vehicles observed in the baseline year 2000.

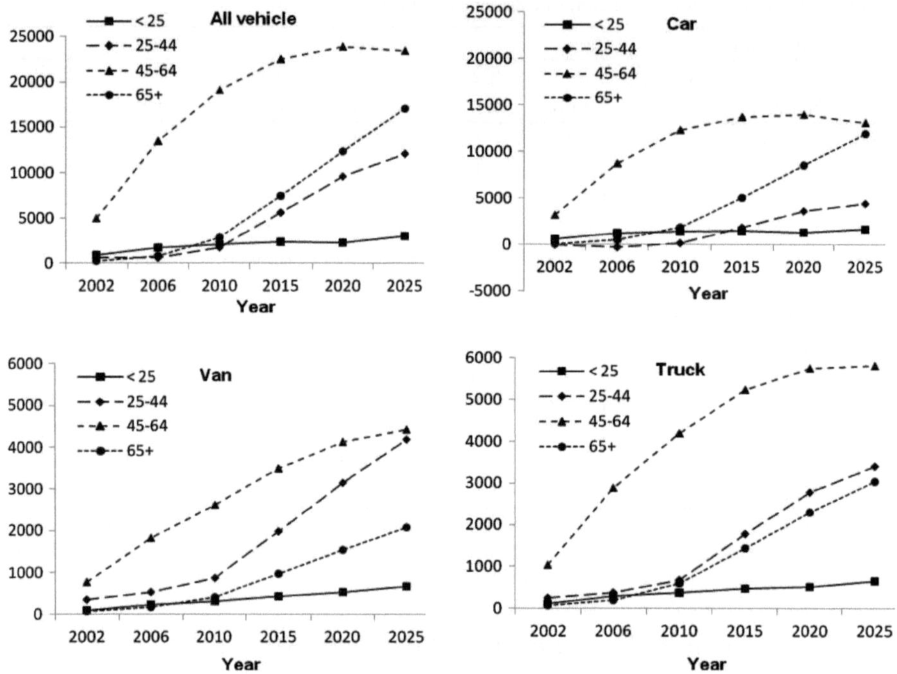

Fig. 11.1 Forecasts of cumulative increase in household vehicle consumption by age of householders (unit: 1000)

Figure 11.1 shows that, among different age groups, the future cumulative increases in household vehicle consumption will mainly come from householders aged 45–64. However, the number of older (aged 65+) and younger (aged 25–44) owners of cars, vans and trucks will increase rapidly after 2010.

Figure 11.2 shows the racial differentials in the cumulative increase of household vehicle consumption. The cumulative increases in cars, vans and trucks are largest among White non-Hispanic, followed by Hispanic, while the cumulative increases in Black non-Hispanic and Asian and other non-Hispanic racial groups are relatively smaller and are nearly identical.

The interaction between race and age factors and its impact on household vehicle consumption in future years have important business implications. As shown in Fig. 11.3, there is a significant increase in vehicle consumption for householders aged 65+ among White non-Hispanic households. In contrast, for Hispanic households, the largest vehicle consumption increase comes from the 25–44 age group. Among Black non-Hispanic and Asian and other non-Hispanic households, the age patterns are similar except that the difference between the 65+ age group and the 25–44 age group is larger in Black non-Hispanic than Asian and other non-Hispanic.

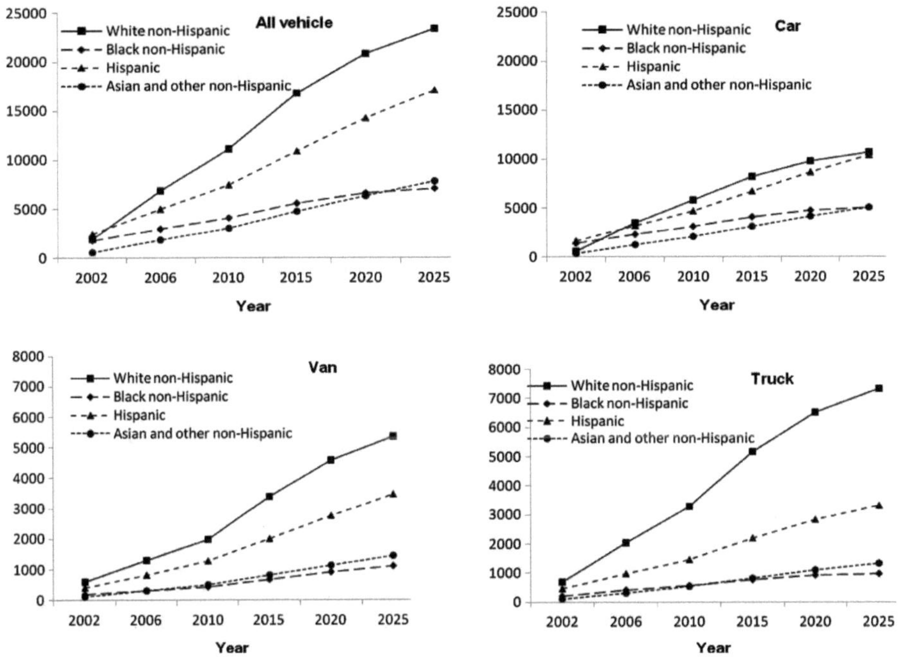

Fig. 11.2 Forecasts of cumulative increase in household vehicle consumption by race of householders (unit: 1000)

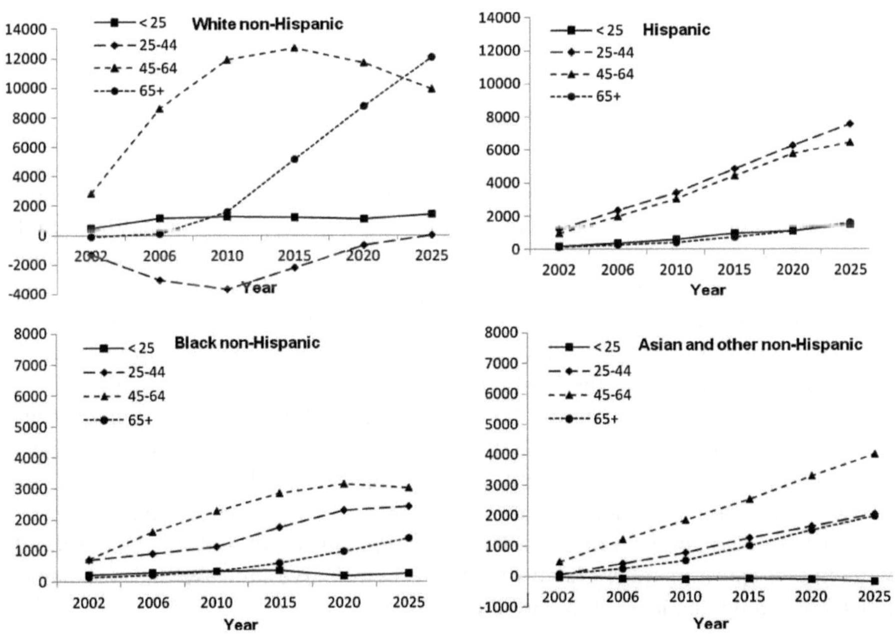

Fig. 11.3 Forecasts of cumulative increase in household vehicle consumption by race and age of householders (unit: 1000)

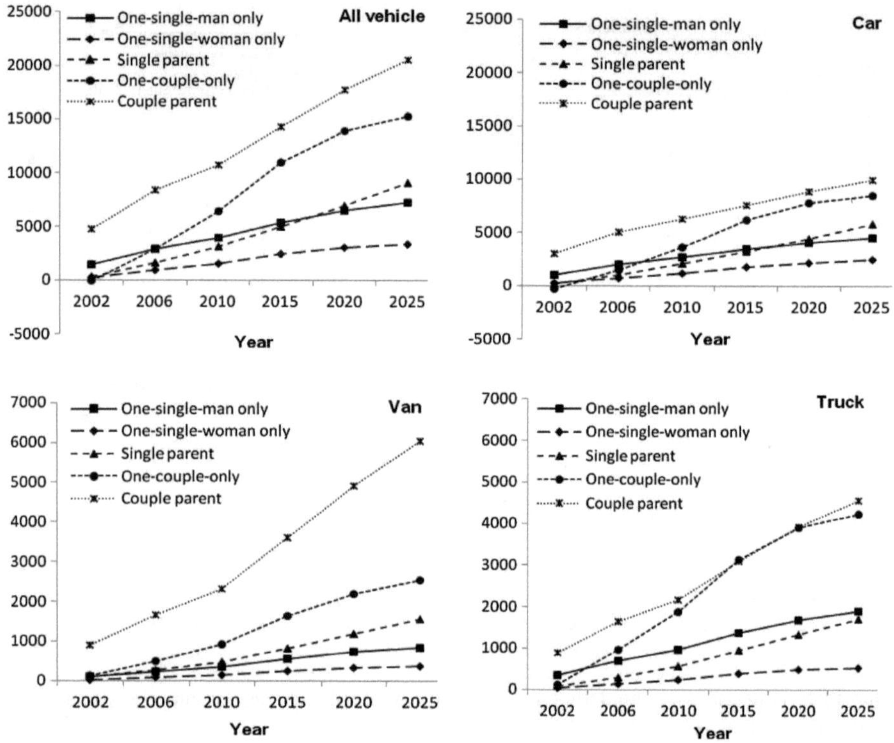

Fig. 11.4 Forecasts of cumulative increase in household vehicle consumption by household type (unit: 1000)

11.4.5 Forecast by Type, Size and Income of Households

With regard to vehicle consumption by household type, the largest and the second largest cumulative increase happens for the couple-parent households (household with a couple and child(ren) and/or others) and the one-couple-only households, respectively (Fig. 11.4). In comparison, the cumulative increase in vehicle consumption by single-man-only households and single-parent households (households with one-single-person and child(ren) and/or others) are smaller. Single-woman-only households will have the least cumulative increase from 2000 to 2025.

According to Fig. 11.5, across households with different sizes, the largest and second largest cumulative increase of car consumption from 2000 to 2025 comes from two-person households and one-person households, respectively; and households with five or more members have one of the least cumulative increases. This pattern also applies to trucks. However, in consumption of vans, households with five or more members have the largest cumulative increase and one-person households have the least cumulative increase.

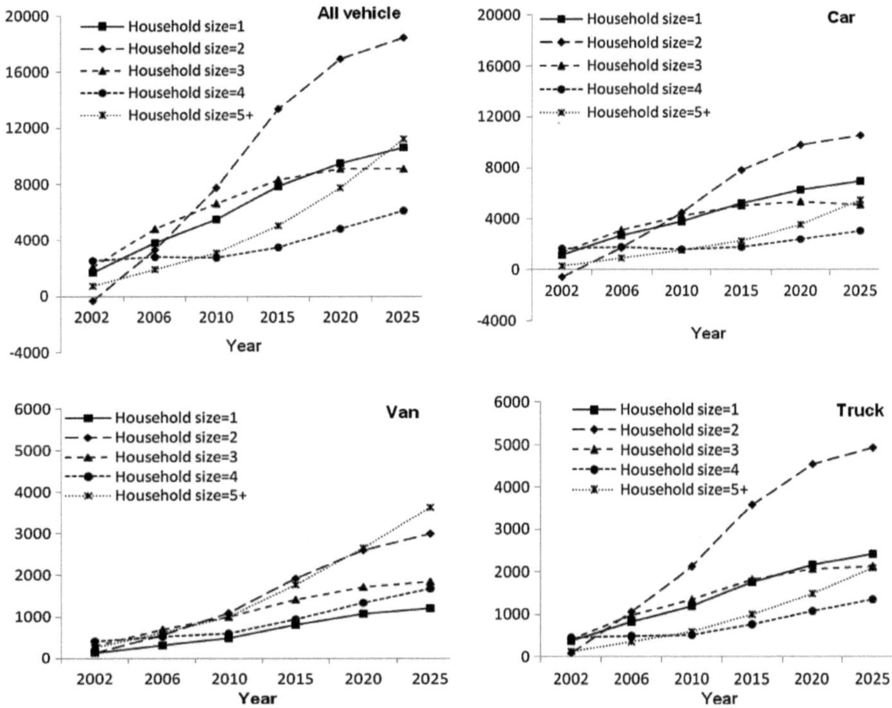

Fig. 11.5 Forecasts of cumulative increase in household vehicle consumption by household size (unit: 1000)

Figure 11.6 presents the consumption forecast results by household income category. The patterns for cars and vans are similar: the higher the household income level, the greater the cumulative increase in vehicle consumption will be from 2000 to 2025. For trucks, however, the pattern is slightly different. That is, the category of High income ranks second to Middle I income, and is followed by Middle II income and Low income. This may be because, instead of purchasing a truck directly, high-income households may choose to buy services for jobs where a truck is needed.

11.4.6 Forecast by Region

In Fig. 11.7, the cumulative increase of household vehicle consumption from 2000 to 2025 is shown for each of the four regions of the United States (the Northeast, Midwest, South, and West). The pattern is consistent for all three kinds of vehicles: the South and the West have the largest and second largest cumulative increases, the Midwest ranks third, and the Northeast has the lowest cumulative increase, from 2000 to 2025.

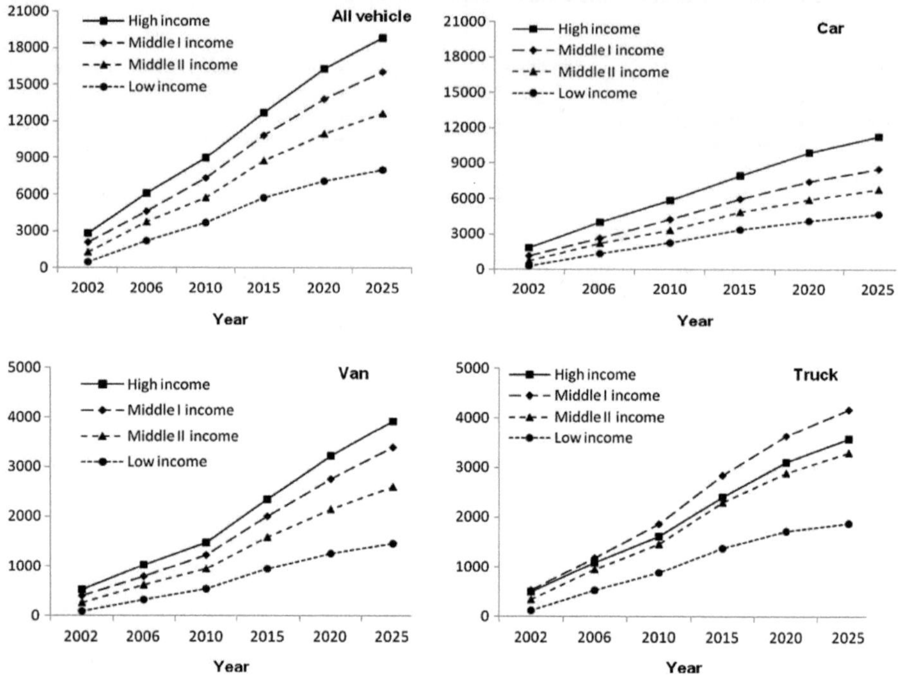

Fig. 11.6 Forecasts of cumulative increase in household vehicle consumption by household income (unit: 1000)

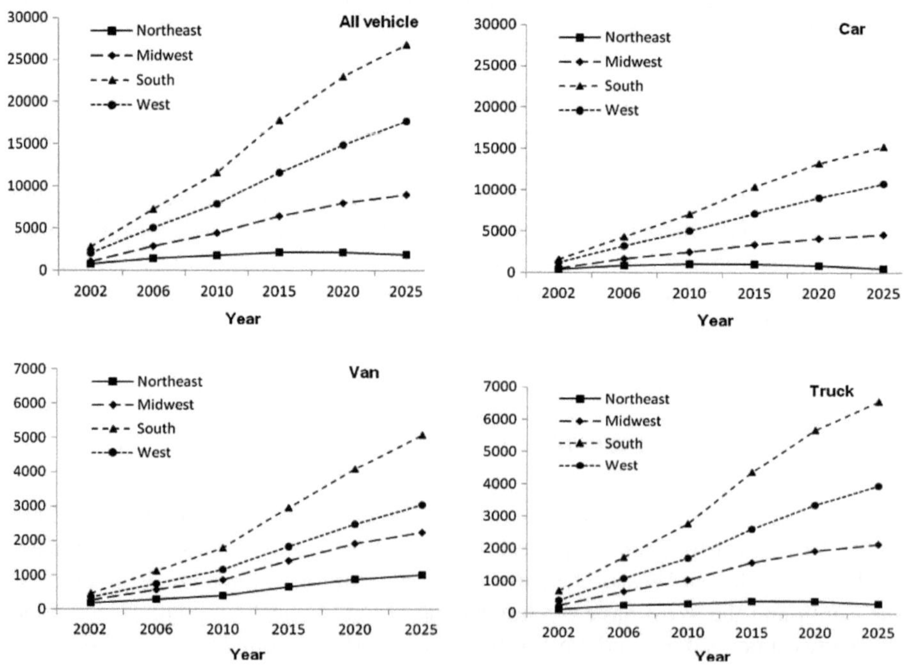

Fig. 11.7 Forecasts of cumulative increase in household vehicle consumption by region (unit: 1000)

11.5 Conclusion and Discussion

This chapter applied the ProFamy method to forecast household vehicle consumption by type/size/income of households and by age/sex/race of householder in four regions of the United States from 2000 to 2025. Many previous studies (Bhat and Sen 2005; Hu and Reuscher 2004; Vance and Buchheim 2004) have recommended the incorporation of detailed household characteristics into household vehicle forecasts. The ProFamy method and its detailed forecasts of household vehicle consumption presented in this chapter have delivered useful messages for researchers and automobile producers and distributors.

The forecasts have shown that more than half of the cumulative increase in household vehicle consumption is due to the increased consumption of cars, suggesting that future vehicle production and marketing may still need to focus on cars. Interestingly, it has also been projected that the consumption of vans will increase relatively and substantially quicker than the other kinds of vehicles from 2000 to 2025, and thus deserves special attention for business considerations. Regarding the characteristics of future van consumers, the study also provides valuable clues: the rise of van consumption mainly comes from householders aged 25–64, from white non-Hispanic households and Hispanic households, and from households with more than five members.

This study demonstrated that the 45–54 age group will make the largest contribution to the increase in vehicle consumption. Another important finding, that vehicle owners aged 65+ will increase rapidly after 2010, indicates that household vehicle owners are aging quickly in the United States. This is expected as American baby-boomers born after the Second World War are entering these older age groups during the projection period of 2000–2025. Such a finding matches with the facts that the American population is aging quickly and older Americans use personal vehicles as much as younger people (Collia et al. 2003). The forecasts in this paper also suggest Hispanic households as a critical target group for future automobile marketing. The projections have shown that Hispanic households will play a significant role in the vehicle consumption increase of the next decades. In particular, Hispanic householders aged 25–44 will make up the major proportion of vehicle consumption.

Note that, for example, Prskawetz et al. (2004) found that the classic headship rate method yielded misleading over-forecasts of the vehicle consumption increase in Austria by using the forecasted number of households without household size information. The likelihood of the upward bias of the headship rate method can be intuitively understood, because future Austrian and American households will comprise many more one- and two-person households, which mostly need only one vehicle, than do today's average households; however the classic headship rate approach does not forecast households by household size. Therefore, it is to be expected that the ProFamy extended cohort-component method will produce relatively more accurate forecasts of future vehicle demands than those produced by the traditional headship rate method.

Our forecasts should not be understood as vehicle sales forecasts. We used the cumulative increase in vehicle consumption to measure the changes in level and composition of vehicle consumption between 2000 and future years, but they are not forecasts of changes in sales of vehicles for three reasons. First, we only forecast vehicles owned by private households, thus excluding vehicles owned by enterprises, governmental agencies, public institutions and the military, all of which are beyond the scope of this project. Second, our forecast changes are based on changes in household numbers and composition, but do not include sales for vehicle replacement. Finally, these cumulative increases are based on cross-sectional comparisons and represent changes in overall market scale potentials, but they do not count for the cohort and period effects of changes in life styles, preferences, public transportation policy and traffic conditions, etc. In the short, the term ‘consumption’ as used in this chapter refers only to “market potentials” for vehicles and vehicle-related services and materials rather than actual “sales”.

Forecasts of household vehicles consumption potentials could be also affected by various factors beyond what we have examined in the current projection. Obviously, legislative, policy, socio-economic, and technological changes could all influence household vehicle consumption. However, many events that shape vehicle consumption are random, predictions for policy trends are difficult to make, and thus all vehicle consumption projections are subject to a lot of uncertainties. For instance, increasing oil prices and the global financial difficulties of recent years have profoundly influenced household vehicle consumption in the United States: vehicle sales continue to decline due to the economic downturn (Vlasic and Bunkley 2008), and consumers tend to choose vehicles with better fuel economy such as hybrid models, so fewer SUVs and pickup trucks are sold (MSNBC 2008). These factors are important but hard to integrate into vehicle projection models because of the unpredictability of such events. However, well-constructed forecasting models are still desirable because they provide valuable insights into the magnitude of future trends in the market potential of vehicle consumption, as has been presented in this chapter.

Appendix 1: The Four Regions Defined By the US Census Bureau

Northeast (9 states): Maine, New Hampshire, Vermont, Massachusetts, Rhode Island, Connecticut, New York, New Jersey, Pennsylvania;

Midwest (12 states): Ohio, Indiana, Illinois, Michigan, Wisconsin, Minnesota, Iowa, Missouri, North Dakota, South Dakota, Nebraska, Kansas;

South (16 states and DC): Delaware, Maryland, District of Columbia, Virginia, West Virginia, North Carolina, South Carolina, Georgia, Florida, Kentucky, Tennessee, Alabama, Mississippi, Arkansas, Louisiana, Oklahoma, Texas;

West (13 states): Montana, Idaho, Wyoming, Colorado, New Mexico, Arizona, Utah, Nevada, Washington, Oregon, California, Alaska, Hawaii.

Appendix 2: Consistency Examination Across the Four Regions

1. The sum of all domestic immigrants and domestic out-migrants in the four regions in each of the projection years is equal to zero;
2. The sum of the net international migration of the four regions in each of the projection years is equal to the whole country's net international migration in the corresponding year forecasted by the Census Bureau;
3. In the years of the forecasting, the estimated region-race-specific summary measures must pass the consistency check. That is, the weighted race-specific average summary measures for the whole country, which are derived from the estimated region-race-specific summary measures and the regional proportions of each race group, are equal to the directly estimated summary measures of the whole country.

Appendix 3: Consistency Examination for Percentile Distribution of Income Categories

The age-race-specific proportions of the high, middle I, middle II, and low incomes for each household type/size category from 2001 to 2025 are assumed to be the same as those obtained from the 2000 census 5 % sample dataset, as has been justified in the existing literature and discussed in the text. But the aggregated race-specific proportions of the four income categories for each household type of all ages combined are not constant over time because they are a weighted average of the proportions across ages, and the age structure of the householder (i.e., the weights of the aggregate proportions) change over time. Similarly, the age-specific proportions of each income category of all races combined for each household category are not constant over time because they are a weighted average of the proportions across races, and the race compositions of households change over time. In sum, the overall proportions of each income category for each race and all ages combined, the age-specific proportions of each income category for all races combined, and the overall all-age-race-combined proportion of income category are dynamic from 2000 to 2015, due to changes in the household distributions and age structure of the householder. At the same time, the census-based or American Community Survey (ACS)-based age-race-household category-specific proportions of each income category, which measure the race-age-sex-region differentials of income distributions, are basically kept constant.

The procedure for the consistency check is as below:

$I_k(t)$, the percent of income category k;

$P_k(i, x, t, r, j)$, proportion of households of kth income category among households of type/size i with householder of age group x, race group r, and region j in year t; one may assume that $P_k(i, x, t, r, j)$ in the projection year is the same as that observed in the most recent year or assume some systematic changes. In any case, $\sum_k P_k(i, x, t, r, j) = 1.0$

$H(i, x, t, r, j)$, number of projected households of type/size i with householders of age group x, race group r, and region j in year t;

$H(i, x, t, r, j)\, P_k(i, x, t, r, j)$, the first estimate of the number of households with income category k, household type/size i, and householder of age x, race group r, and region j in year t.

Because of the changes in compositions of households of different types/sizes, and age structure of householders in projection year t, $\sum_i \sum_x \sum_r \sum_j H(i, x, t, r, j) P_k(i, x, t, r, j) / \sum_i \sum_x \sum_r \sum_j H(i, x, t, r, j)$ may not be exactly equal to $I_k(t)$ although the discrepancy is usually not large. Thus some adjustments are needed as below.

$$\frac{C_k(t)\sum_i \sum_x \sum_r \sum_j H(i,x,t,r,j) P_k(i,x,t,r,j)}{\sum_i \sum_x \sum_r \sum_j H(i,x,t,r,j)} = 0.25$$

$$C_k(t) = \frac{0.25 \sum_i \sum_x \sum_r \sum_j H(i,x,t,r,j)}{\sum_i \sum_x \sum_r \sum_j H(i,x,t,r,j) P_k(i,x,t,r,j)} \tag{11.1}$$

$$P'_k(i,x,t,r,j) = C_k(t) P_k(i,x,t,r,j) \tag{11.2}$$

$$P''_k(i,x,t,r,j) = P'_k(i,x,t,r,j) \frac{1.0}{\sum_k P'_k(i,x,t,r,j)} \tag{11.3}$$

We compute the quartiles of high, middle I, middle II, and low income again if their relative differences from 0.25 are all less than 0.01, say, or another criterion, we accept $P''_k(i, x, t, r, j)$. More specifically, if

$$\left\{\left[\frac{\sum_i \sum_x \sum_r \sum_j H(i,x,t,r,j) P''_k(i,x,t,r,j)}{\sum_i \sum_x \sum_r \sum_j H(i,x,t,r,j)} - 0.25\right] / 0.25\right\} < 0.01 \quad \text{for all}$$

income categories k (for example, k = 1, 2, 3, 4), we accept $P''_k(i, x, t, r, j)$. Otherwise, we repeat the adjustment procedure expressed in formulas (11.1), (11.2), and (11.3) until the criterion is met.

Part III
Applications in China

Chapter 12
Household and Living Arrangement Projections in China at the National Level

12.1 Introduction

A number of studies based on 1982, 1990, and 2000 census micro data reveal that household composition in China has been changing substantially in the past couple of decades (Zeng and Wang 2003; Guo 2003). These changes are reflected, for example, in the decrease in average household size and substantial increase in one-person and one-couple-only households. The proportion of the elderly not living with children and the proportion of one-couple-only households among the elderly population have also increased considerably. These changes in household composition will continue to reduce the elderly care capacity of Chinese families, which will increasingly affect social services and economic development. Clearly, family support for the elderly is facing grave challenges in the process of rapid population aging and substantive changes in household structures.

Applying the ProFamy method (described in Chaps. 2, 3 and 4), its associated computer software (described in Chaps. 16 and 17), and census and survey data, this chapter projects future changes in family household size and structure, elderly living arrangements, and population aging in the period of 2000–2050 under medium scenarios of fertility, mortality, rural-urban migration, marriage, and divorce for rural and urban areas in China (Zeng et al. 2008). Compared to previous demographic projection studies concerning the future population in China, this study has two unique features. First, we simultaneously project age-sex-specific population distributions, family household types, and elderly living arrangements, which are useful for socioeconomic planning and policy analysis. Second, we classify the projections by rural and urban sectors and take into account the large rural-urban differentials in fertility, mortality, and marriage formation and dissolution, as well as the massive migration from rural to urban areas in the process of economic and social development. Dynamic and integrated projection of rural and urban family households, elderly living arrangements, and population aging is important to understand the future population structure of Chinese society and investigate appropriate strategies for sustainable development.

Y. Zeng et al., *Household and Living Arrangement Projections*,

The Springer Series on Demographic Methods and Population Analysis 36,
DOI 10.1007/978-90-481-8906-9_12,

12.2 Data and Estimates

The data needed to project family households using the ProFamy method are listed in Table 3.1 of Chap. 3. When rural-urban classifications are requested, as for the Chinese projections presented in this chapter, the standard schedules and the summary measures described in Table 3.1 should be rural-urban-specific. The Chinese 2000 census and the 2005 mini-census (1 % survey covering all of China) collected the date (year and month) of first marriage for all persons aged 15 and older. Using these data, we have estimated the sex-age-specific standard schedules of the occurrence/exposure (o/e) rates of first marriage in rural and urban areas. The In-Depth Fertility Surveys conducted in Shanghai, Hebei, and Shannxi in 1985 and the Chinese Longitudinal Healthy Longevity Survey (CLHLS), conducted in 2002 and 2005 collected event history data of divorce and remarriage. Using the In-Depth Fertility Survey and the CLHLS data sources, we estimated the standard schedules of the sex-age-specific o/e rates of divorce and remarriage in rural and urban areas. Using the sex-age-specific standard schedules of first marriage, divorce, and remarriage, the 2000 census micro data, and the published total number of marriages (including 1st and remarriages) and divorces, we estimated the standardized general rates of marriage and divorce for the rural and urban areas of China, employing the simple procedure presented in Appendix 4 of Chap. 3.

Age-parity-specific o/e rates of marital fertility in rural and urban areas are estimated based on the 1997 and 2001 national sampling surveys on reproductive health. We assume that births outside of marriage are negligible. The sex-age-specific frequency distributions of rural—urban net migration are estimated based on the 2000 census data. The sum of the age-specific frequencies of rural—urban net migration is equal to one. We derive the sex-specific total number of rural-urban net migrants in future years using the projected proportion of the urban population and the projected sex ratio of migrants. Multiplying the total number of these projected sex-specific rural-urban migrants by the standard schedules of the sex-age-specific frequencies, we derive the sex-age-specific numbers of rural-urban net migrants in future years. Based on the 1990 and 2000 census micro data files, we used the ProFamy program to estimate standard schedules of the age-sex-specific net rates of leaving home, using a method initially proposed by Coale (1984, 1985) and Coale et al. (1985), and generalized by Stupp (1988).

The 2000 Chinese census reported an extremely low TFR of 1.22. This rate is too low due to under-reported births, but there is no consensus on the true fertility level in China today. Based on our review of existing related studies, we assume a 25 % under-reporting of birth rates, which implies that the TFR for rural and urban areas combined in 2000 was about 1.63; 1.9 in rural areas and 1.15 in urban areas. We expect that China will gradually relax its one-child policy in response to the current very low fertility level and the government's concern about future population aging problems.[1] Because young people in China are delaying their marriages and births (Fig. 12.1),

[1] Various provinces in China have already started to slightly relax the one-child policy, such as allowing couples with both parties who are an only-child (i.e., no siblings) to have two children.

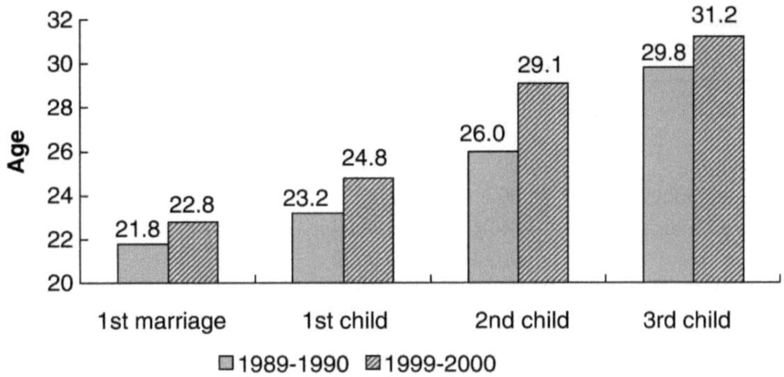

Fig. 12.1 Average age at 1st marriage and 1st, 2nd, and 3rd birth: a comparison between 1989–1990 and 1999–2000 (Data source: Chinese censuses conducted in 1990 and 2000)

most likely due to rapid socioeconomic development, we assume that the age at the first, second or higher-order births will increase by 0.9 and 1.8 years during the years from 2015 to 2030, which constitutes an annual growth rate of 0.05 and 0.1 years, respectively. According to the Bongaarts-Feeney method (Bongaarts and Feeney 1998; Zeng and Land 2001, 2002), the projected period TFR of the first, second or higher order births in the years 2012 and 2030 will be 5 % and 10 % lower than the parity-specific cohort TFR, due to the fertility timing effects; the overall TFR (all parities combined and rural/urban combined) are assumed to be 1.89 in 2015 and 1.81 in 2030. We assume that the fertility restrict will be released by and after the year 2035, and the period TFR will be slightly increased and will remain constant thereafter. Considering the process of urbanization, the weighted average period TFR for the rural and urban areas combined will be 1.81 in 2030 and 1.82 in 2050 (see Table 12.1).

The general rates of marriage and divorce are assumed to be constant in the next a few decades. One common approach in forecasting is to hold some of the current rates constant throughout the forecasting horizon (e.g., Day 1996; Treadway 1997). Smith et al. (2001: 83–84) argue that holding some of the rates and proportions constant when forecasting can be justified on either of two grounds. The first is when future rates and proportions are unlikely to differ much from the current level. The second is when neither the direction nor the magnitude of future changes can be predicted accurately. The argument here is not so much that the current rates will remain constant, but rather that scientific theories and past history do not provide reliable bases for predicting how those rates will change. If upward or downward movements are equally likely, the current rates provide a reasonable forecast of future rates.

Our educated projections of the values of the demographic parameters over the next four decades are listed in Table 12.1. A sensitivity analysis on the impact of high or low fertility, mortality, marriage, and divorce scenarios would be out of the scope of this chapter, given that its main objective is to provide a general profile of our projections of the future trends of Chinese family household and elderly living arrangements in the context of rapid population aging.

Table 12.1 Main demographic parameters used in the projections

Year	2000	2015	2030	2035	2050
Rural					
Total fertility rate	1.90	2.15	2.15	2.21	2.21
Male life expectancy	68.00	70.10	72.20	72.90	75.00
Female life expectancy	72.00	74.10	76.20	76.80	78.90
General marriage rate	0.0674	0.0674	0.0674	0.0674	0.0674
General divorce rate	0.0022	0.0022	0.0022	0.0022	0.0022
Mean age at birth	25.20	25.20	26.50	26.50	26.50
Urban					
Total fertility rate	1.14	1.67	1.67	1.72	1.72
Male life expectancy	72.00	74.00	75.90	76.60	78.60
Female life expectancy	76.00	78.00	79.90	80.60	82.50
General marriage rate	0.0601	0.0601	0.0601	0.0601	0.0601
General divorce rate	0.0056	0.0056	0.0056	0.0056	0.0056
Mean age at birth	26.00	26.00	27.30	27.30	27.30
% of urban population	37	52	60	64	75
Rural and urban combined					
Total fertility rate	1.63	1.89	1.81	1.84	1.77
Male life expectancy	69.5	72.1	74.4	75.3	77.7
Female life expectancy	73.5	76.1	78.4	79.2	81.6
General marriage rate	0.0647	0.0636	0.0630	0.0627	0.0619
General divorce rate	0.0035	0.0040	0.0042	0.0044	0.0048
Mean age at birth	25.5	25.6	27.0	27.0	27.1

12.3 Profile of Future Trends

12.3.1 Rapid Population Aging

The very large size of the elderly population is a unique characteristic of population aging in China. In 2000, there were 88 million elderly persons aged 65 and older. By the years 2030 and 2050, there will be 238 million and 338 million elderly people in China, respectively, under the medium mortality assumption and based on baseline data derived from the 2000 census (see Fig. 12.2).

Although the proportion of elderly aged 65 and older is not very high yet in China, at 6.9 % in 2000 and 8.9 % in 2010, the speed of population aging will be extremely fast in the first half of the twenty-first century. Under the medium fertility and conservative medium mortality assumptions, the Chinese elderly aged 65 and older will account for 16.4 % and 23.9 % of the total population by 2030 and 2050, respectively (see Fig. 12.3). The projections under the medium fertility and medium mortality assumptions by the United Nations and other scholars inside and outside China show similar general trends in the future size and proportion of the Chinese

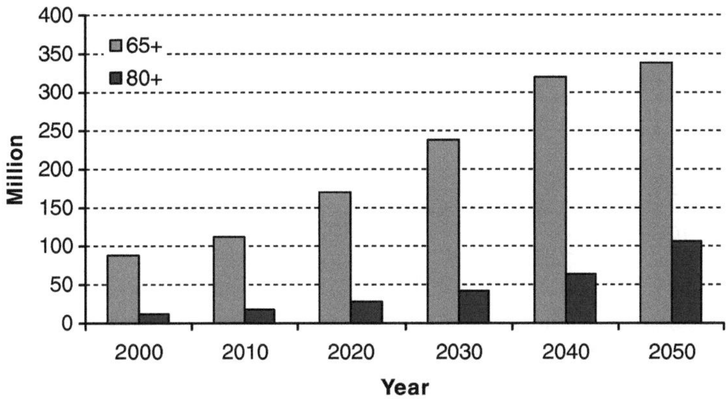

Fig. 12.2 Projected size of elderly population aged 65+ and the oldest-old aged 80+, rural-urban combined, 2000–2050

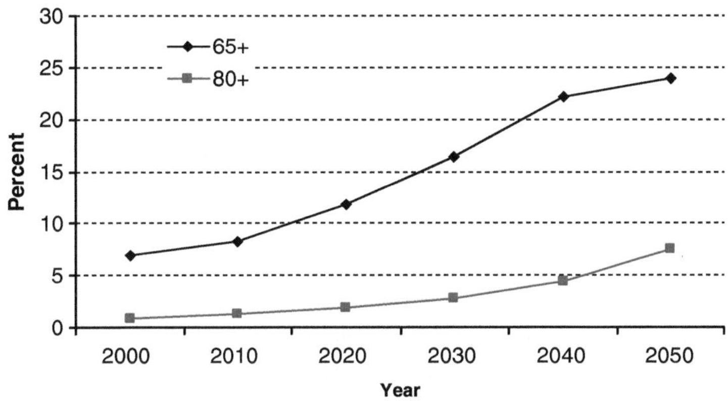

Fig. 12.3 Projected percentage of elderly population aged 65+ and the oldest-old aged 80+ among total population under the medium fertility and mortality scenario: rural and urban combined, 2000–2050

elderly population (e.g. U.N. 2013), confirming the likelihood of extremely rapid population aging in China in the first half of the twenty-first century.

There were about 12 million and 20 million oldest-old aged 80 and older in 2000 and 2010, respectively, but the number of oldest-old will climb rapidly to about 65 and 107 million in the years 2040 and 2050, respectively, under the medium mortality assumption (see Fig. 12.2). The average annual increase in the rate of the oldest-old between 2000 and 2050 will be 4.5 %. The percent share of the oldest-old among the elderly population in 2030 and 2050 will be 1.26 and 2.33 times as large as that in 2000. From 2000 to 2040, this share will increase by approximately 1.7

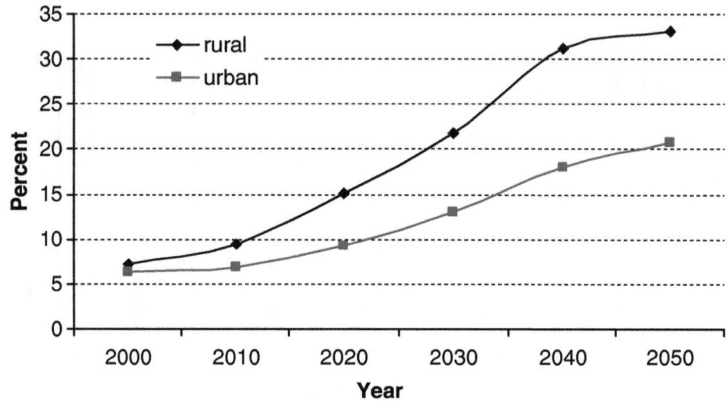

Fig. 12.4 Projected percentage of elderly population aged 65+ among the total population under the assumption of medium fertility and medium mortality: a comparison between rural and urban areas, 2000–2050

percentage points per 10 years. But in the 10 years from 2040 to 2050, that share will increase by 11.5 percentage points, mainly because China's baby boomers born in the 1950's and 1960's will fall into the oldest-old age category at that time. Despite the uncertainties in accurately forecasting the oldest-old population, it is certain that the oldest-old, who most likely need assistance and care in daily life, will increase tremendously in the next century in China (also see Gu and Vlosky 2008; Mayer et al. 1992: 81–82; Zeng and George 2010), and that the middle of the next century will be a hard time for the country due to serious problems associated with population aging.

In European societies, the aging transition has been spread over one century or more. In China, however, this change will take place within a few decades to reach more or less the same level of population aging as most developed countries by the middle of this century. The proportion of elderly population in China will increase much faster than in almost all other countries in the world. According to the latest U.N. population projections (U.N. 2013), it will take about 20 years for the elderly population to increase from 10 % to 20 % of the population in China (2017–2037), compared to 23 years in Japan (1984–2007), 57 years in Germany (1951–2008), 69 years in Sweden (1947–2016), and 57 years in the United States (1972–2029).

Although fertility in rural areas in China is much higher than in urban areas, aging problems will be more serious in rural areas because of the continuing massive rural-urban migration of young people. Under the medium fertility and medium mortality assumptions, the proportion of the elderly will be 33.1 % in rural areas and 20.8 % in urban areas by the middle of this century (see Fig. 12.4). In 2050, the percent of oldest-old in rural areas may be twice as high as that in urban areas (see Fig. 12.5). It is important to note that these projected figures are under the assumption that the age distribution of the rural-urban migrants in the next few decades will be the same as that observed in the 2000 census. This suggests that if

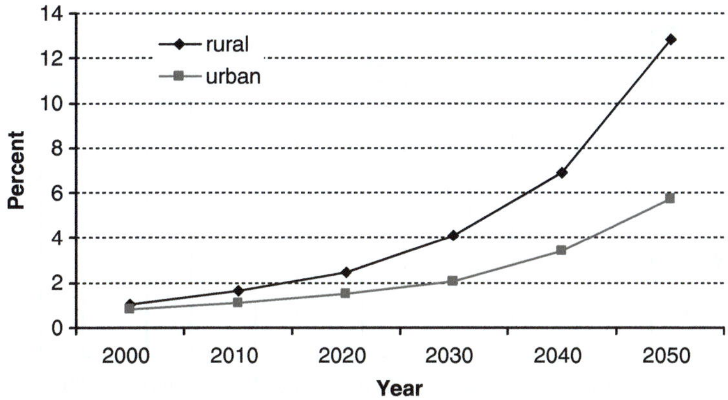

Fig. 12.5 Projected percentage of the oldest-old aged 80+ among the total population under the assumptions of medium fertility and medium mortality: a comparison between rural and urban areas, 2000–2050

Chinese rural-urban migration continues to include young people only, with elderly parents remaining in rural areas, as was indicated in the 2000 and 2010 censuses, in just a few decades the percentage of elderly population in general and the oldest-old in particular will be too high for rural society to manage. Thus, China needs to investigate and adopt policies to encourage rural-to-urban family migration or family reunions after the young migrants settle in urban locales, to avoid the "elderly village" phenomenon spreading throughout rural areas, and to prevent resulting serious social problems (Qiao 2001; Zeng et al. 2008).

12.3.2 Projection of Family Household Structure and Size

According to our projections, the average household size of China will continuously decrease from 3.46 persons per household in 2000 to 2.96 in 2020 and 2.67 in 2050 (see Fig. 12.6). Our annual projections of household size distributions show that there will be proportionally more small households (one or two persons) and fewer large households (six or more persons) in future years as compared to the year 2000. Note that the Chinese family household has dramatically transferred from a larger unit before the late 1970s to a smaller one in the early twenty-first century, and will continue to evolve to an even smaller size in the next few decades. We believe that this phenomenon was caused by a tremendous fertility decline plus substantial changes in social attitudes and economic mobility related to co-residence between elder parents and adult children. Clearly, the government's policy on birth control is one of the preeminent causes of the family revolution in China, characterized mainly by the trend toward much lower fertility, later

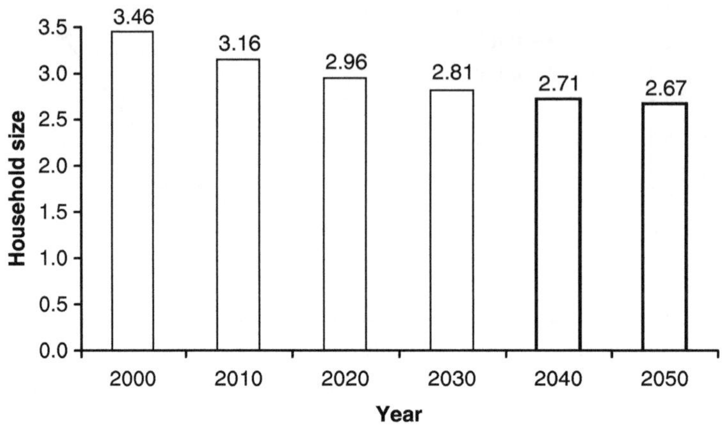

Fig. 12.6 Changes in average family household size (rural and urban combined) under the medium fertility and medium mortality assumptions

marriage, and smaller household size. This is, in general, in agreement with the arguments of Wolf (1986).

The projected proportions of various household types among the total number of households under the medium fertility and mortality assumptions are given in Table 12.2. Our projections show that the proportion of the households with at least one person aged 65+ (we may call such households "elderly households") will increase dramatically in China in the next few decades. In 2030 and 2050, the percentage of households with one elder living alone will be 2.17 and 4.08 times as high as that in 2000, reaching 10.4 % in 2050; the percentage of households with one-elderly-couple without other family members living together will be 2.02 and 2.98 times as high as that in 2000, reaching 8.2 % in 2050. The overall proportion of elderly households (including elder(s)-only households and two-generation and three-generation households with at least one elder) in 2030 and 2050 will be 30.0 % and 84.6 % higher, respectively, than that in the year 2000.

Households with one non-elderly person only will increase from 6.8 % in 2000 to 12.4 % in 2050. The percentage of households with one couple and child(ren), without elderly parents, will decrease from 51.8 % in 2000 to 34.6 % in 2030 and 27.7 % in 2050. The share of the non-elderly one couple only households and non-elderly single-parent with child(ren) households will somewhat increase in the period 2000–2030 and then decline considerably thereafter. The overall percentage of households without elders aged 65+ will decrease from 76.0 % in 2000 to 68.8 % in 2030 and 55.7 % in 2050 (see Table 12.2). The substantial decrease in the percentage of young or relatively young households (i.e., those households without an elder member(s) aged 65+) and the dramatic increase in elderly households clearly indicate that Chinese family households will be aging rapidly and substantially in the next few decades.

Table 12.2 Percentages of households by type among the total number of households under the medium fertility and medium mortality assumption

Household types	2000	2030	2040	2050
Households without elderly persons				
One person	6.78	13.16	12.93	12.36
One couple (no elderly parents, or children)	10.75	13.38	11.72	10.19
Couple with at least one child, but no elderly parents	51.81	34.55	28.83	27.65
Single parent with at least one child, but no elderly parents	6.70	7.71	5.86	5.52
Sub total	76.04	68.80	59.34	55.72
Households with at least one elderly person				
One elderly person only	2.54	5.51	8.06	10.37
One elderly couple only	2.74	6.06	8.08	8.16
One elderly couple with at least one child	0.03	3.69	5.62	5.59
Single elderly parent with at least one child	0.06	4.86	8.90	10.68
Three or more generations with one or two elderly grandparents	18.61	11.06	10.02	9.48
Sub total	23.98	31.18	40.68	44.28
Total	100.00	100.00	100.00	100.00

12.3.3 Proportion of Elderly Who Live in Empty-Nest Households

The rural-urban combined proportion of elderly aged 65 and older living in empty-nest households (i.e., without children) among the total population will be 2.9 and 4.6 times of that in 2000 by the year 2030 and 2050, respectively, under the medium fertility and medium mortality assumptions. The increases in percentages of the oldest-old aged 80+ living in empty-nest households will be even more dramatic: 3.4 and 11.1 times in 2030 and 2050 as high as that in the year 2000. Figures 12.7 and 12.8 depict the rural-urban differentials in percentages of the elderly aged 65+ and the oldest-old living in empty-nest households under medium fertility and medium mortality assumptions. The rural and urban curves are very close to each other in 2000, but the gap becomes larger after 2030 for both age groups.

Note that in our household and population projection, we have assumed that the preference of co-residence between old parents and adult children declines rather slowly over time. Thus, the large increase in the percentage of elderly living in empty-nest households is mainly due to the effects of fertility decline. The fertility decline not only substantially increases the overall proportion of the elderly among the total population, but also results in a much smaller resource of offspring, which is one of the main determinants of co-residence between older parents and children. In other words, in the future in China, considerable numbers of elderly persons may not be able to co-reside with children, even if they wish to do so, due to the shortage of available children.

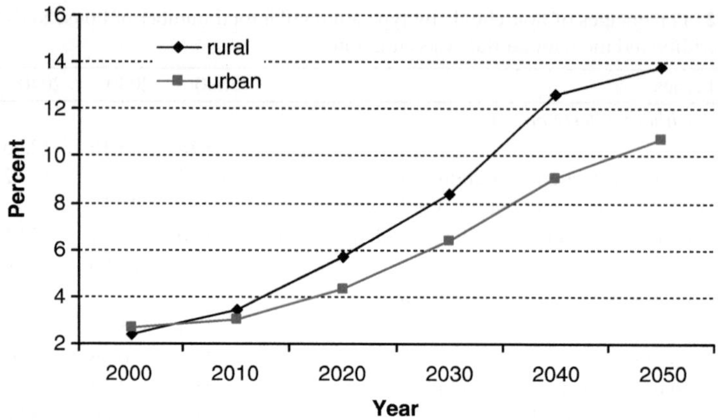

Fig. 12.7 Projected percentage of elderly population aged 65+ living in empty-nest households among total population: a comparison between rural and urban areas, 2000–2050

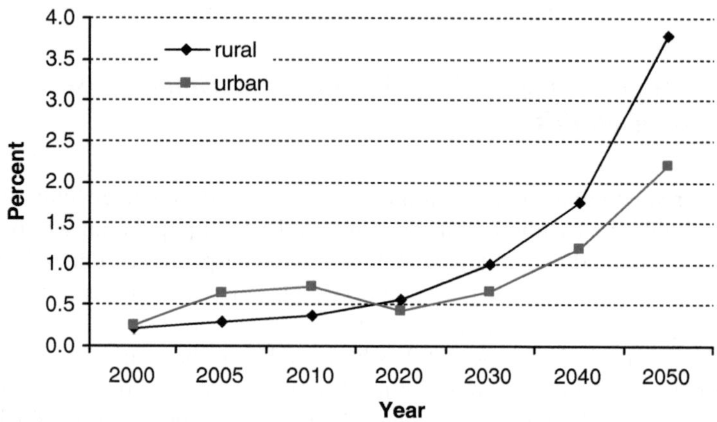

Fig. 12.8 Projected percentage of the oldest-old aged 80+ living in empty-nest households among the total population: a comparison between rural and urban areas, 2000–2050

12.3.4 Dependency Ratios

One widely used demographic index to estimate the relative productive and dependency potential of a society is the dependency ratio. The child dependency ratio is defined as the ratio of children under age 15 to the population aged 15–64; the old-age dependency ratio is defined as the ratio of elderly aged 65 and older to the population aged 15–64. The sum of the child and the old-age dependency ratios is called the total dependency ratio (Smith 1992: 14). This conventionally defined dependency ratio assumes that an average child and an average elderly person depend equally on the working-age population. This assumption may not be accurate in reality, especially in societies where the remaining lifespan of elders will be continuously and substantially prolonged. Clark and Spengler (1978) reported that a broad survey of the costs of public support in the U.S. revealed that the average ratio of government expenditures for elderly to government expenditures for children was 1:0.33. The average ratio of total expenditures, including governmental and family costs, for the elderly to those for children was found to be 1:0.58 in Germany (U.N. 1973), and 1:0.31 in France (Rix and Fisher 1982). Three different surveys conducted in China gave estimates of the average ratio of total expenditures for the elderly to those for children as 1:0.40 (Liu 1984), 1:0.55 (World Bank 1985) and 1:0.53 (Yu 1992). We take the average of these three estimates, namely 1:0.5, as the elderly/children expenditure ratio in China.[2] Keeping the sum of the weights for the old-age dependency ratio and child dependency ratio equal to 2, which is the same as that of the un-weighted total dependency ratio, and using the average ratio of the total expenditures for elderly to those for children (1:0.5) as weight, we can compute:

Weighted total dependency ratio = (0.6667 × child dependency ratio) + (1.333 × elderly dependency ratio).

Note that not all of the persons aged 15–64 work, not all elderly persons aged 65+ do not work, and the average estimate of 1:0.5 as the elderly/children expenditure ratio, based on the three Chinese survey studies, may not be accurate for future years. Therefore, the estimates of the dependency ratios to be discussed below can only be regarded as proxies for the productive and dependency potential in the first half of the twenty-first century.

With the medium fertility and medium mortality assumptions, the old-age dependency ratio in China in 2020 and 2030 will increase by 7 percentage points and 15 percentage points, as compared to the year 2000. However, the child dependency ratio will decrease by 12 percentage points and 14 percentage points in the same period. Offsetting the increase in the old-age dependency ratio by a decrease in child dependency ratio, the un-weighted total dependency ratio in China in 2020 will be smaller than that in 2000 by 5 percentage points, and in 2030 the un-weighted total dependency ratio will be the same as that in 2000. When we take into account that the elderly need more support than children, the Chinese weighted total dependency ratio will increase by 2 percentage points and 11 percentage points in 2020 and 2030, respectively, as compared to 2000. Both the weighted and un-weighted total dependency ratios will increase dramatically after 2030 due to a large increase in the old-age dependency ratio and a stabilization of the child dependency ratio (see Table 12.3).

Clearly, the demographic dividend is open up to 2030, due to the increasing large labor force population, decreasing number of children, and not yet very high proportion of elderly population during this period. However, the Chinese demographic dividend will be gone after 2030 due to rapid population aging and a large decrease in labor force, family support capacities, and resources.

---

[2] Although we did not find more recent survey data about the average ratio of total expenditures for the elderly to those for children, we believe that the average of these three survey estimates may be used as a reasonable approximation, as the recent increase in costs of child-rearing may be offset by the substantial increase of the elderly life span up to oldest-old ages which need more care.

Table 12.3 Un-weighted and weighted total dependency ratios

	2000	2010	2020	2030	2040	2050
Child dependency ratio	0.46	0.34	0.34	0.32	0.32	0.34
Elderly dependency ratio	0.11	0.12	0.18	0.26	0.38	0.42
Un-weighted total dependency ratio	0.57	0.46	0.52	0.57	0.69	0.75
Weighted total dependency ratio	0.45	0.39	0.47	0.56	0.72	0.79

12.4 Summary and Concluding Remarks

Fertility in China has declined dramatically from more than six children per woman in the 1950s and 1960s to about 1.6 children per woman today, significantly lower than that in the U.S. Average life expectancy at birth for both sexes combined in China has increased from about 43.3 years in 1950 to 69.5 years in 1990, 72.1 years in 2000, and 74.1 years in 2005, and will continue to increase in the future (U.N. 2013). Large cohorts of baby boomers born in the 1950s and 1960s will become elders in a couple of decades. Demographic regimes have determined that China, the most populous country in the world with more than 1.3 billion people in 2010, is aging at a rapid speed and to a large scale, especially the oldest-old population aged 80+.

The Chinese family household has transformed from a larger unit before the late 1970s to a much smaller one in the early twenty-first century, and will continue to evolve to an even smaller size in the next few decades. We believe that this phenomenon was caused by, and will continue to be influenced by, a tremendous fertility decline plus substantial changes in social attitudes and economic mobility related to co-residence between elderly parents and adult children. Under our medium fertility and medium mortality scenarios, we show that the proportion of elderly households with at least one person aged 65+ will increase dramatically in China. For example, the overall proportion of elderly households in 2030 and 2050 will be 30 % and 85 % higher than that in the year 2000. At the same time, the percentage of young or relatively young households (i.e., those households without old member(s) aged 65+) will substantially decrease. This indicates that Chinese family households will be aging rapidly and substantially in the first half of the twenty-first century.

By the year 2030 and 2050, the proportion of the elderly aged 65 or older living in empty-nest households without co-residing children among the total population will be 2.9 and 4.6 times of that in 2000. The increase in the percentage of the oldest-old aged 80+ living in empty-nest households will be even more dramatic: 3.4 and 11.1 times in 2030 and 2050 as in the year 2000.

Although fertility in rural areas in China is much higher than that in urban areas, aging problems with respect to proportions of elderly and elderly households, as well as proportions of elderly living in empty-nest households, will be much more serious in rural areas because of the continuing massive rural-urban migration. Such anticipated trends are based on the assumption that the age distribution of rural-urban migrants in the future will be the same as that observed in the recent census, namely, that almost all rural-urban migrants will be young people. Our study strongly suggests that China needs to adopt policies to encourage rural-to-urban family migration or family reunion after the young migrants are settled in urban areas, to avoid the "elderly village" phenomenon in rural areas, which portends serious social problems in the future.

Our projection also indicates the Chinese demographic dividend may continue to exist up to 2025–2030, due to a large labor force, decreasing numbers of children, and not yet burdensome proportion of elderly population. During this "golden-age" period, it is possible for China to mobilize a large amount of individual savings and state capital to build a solid financial and institutional base for social security programs in both rural and urban areas. However, we have to act now because this demographic widow of opportunity will be closed around 2030, at which time it will be too late to start to resolve China's looming serious social and economic problems.

Chapter 13
Dynamics of Households and Living Arrangements in the Eastern, Middle, and Western Regions of China

13.1 Introduction

Most prior publications concerning future trends of population and household aging in China conducted projections and analyses either for China as whole or for a single province or region with rural and urban populations combined. Very few previous publications accounted for the huge regional and urban-rural heterogeneities in demographics, fertility policy, and socioeconomic conditions across the Eastern, Middle, and Western regions in China (Zeng et al. 2013b). This chapter addresses the research question: If the current migration pattern continues, what future differentials in population and household aging will be evident across the Eastern, Middle, and Western regions and between urban and rural areas?

13.2 Method, Data Sources, and Parameter Assumptions

The method used in this chapter is the ProFamy extended cohort-component approach, which was presented and justified in Part I of this book. Note that this chapter applies the ProFamy model to household and population projections in multiple regions for rural and urban areas using an integrated framework of demographic analysis. To do so, we need to pay a close attention to two points that are critical to the consistency of the projections. First, we should take regional variations in demographic rates into account and keep the weighted average of regional demographic rates equal to the projected national average rates. Second, the total number of female and male domestic in-migrants across all regions must be equal to the total number of female and male domestic out-migrants, respectively.

Note that the conventional multiregional model classifies the population by both initial and current residence regions, emphasizes cross-regional migration flow directions, and distinguishes migration from region i to region j and vice versa (Rogers 1975). If N regions are distinguished in the ProFamy multistate population

Y. Zeng et al., *Household and Living Arrangement Projections*,
The Springer Series on Demographic Methods and Population Analysis 36,
DOI 10.1007/978-90-481-8906-9_13, © Springer Science+Business Media Dordrecht 2014

and household projection following the conventional multiregional model, one needs to classify $N \times N$ residence statuses (i.e., N current residence regions cross-classified by N future residence regions), also classified by single year of age, sex, marital status, number of co-residing children and parents, and whether living in private households or institutions. One would also estimate $N \times N$ sets of region-to-region age-sex-specific-migration rates and $N \times N$ projected summary parameters for the number of migrants from region i to region j and vice versa in future years. This would substantially increase the data requirements and the complexity of the model, as well as increase the difficulty and uncertainty of projections of the flow-direction-specific numbers of migrants from region i to region j and vice versa in future years. Moreover, such extra data requirements and complications would not add any strength to our focus on population and household aging projections, which is not related to the initial residence and mobility flow directions of the migrants. Therefore, we adopted a procedure that projects out–migration to any other regions and immigration from any other regions for each of the regions in our present study. This procedure serves our research objectives well, largely simplifies the data requirements, and avoids the unnecessary complexity of the $N \times N$ cross-dimensional classifications of the population and the input demographic rates.

We used the micro-data sample from the 2000 census to extract the baseline population classified by rural/urban residence, single year of age, sex, marital status, number of co-residing children and parents, whether living in private or institutional household, and region.[1] In order to take into account the regional differentials, we divided the 31 provinces into three regions (Eastern, Middle, and Western regions), mainly based on socioeconomic development levels and geography, classified by the National Bureau of Statistics of China. The Eastern region includes Beijing, Tianjin, Hebei, Liaoning, Shanghai, Jiangsu, Zhejiang, Fujian, Shandong, Guangdong, and Hainan provinces; the Middle region includes Shanxi, Jilin, Heilongjiang, Anhui, Jiangxi, Henan, Hubei, and Hunan provinces; and the Western region includes Chongqing, Sichuan, Guizhou, Yunnan, Tibet, Shaanxi, Gansu, Qinghai, Ningxia, Xinjiang, Inner Monglia, and Guangxi provinces. In historical and contemporary China, the Eastern region contains the most developed coastal provinces of China; the Western region contains the least developed provinces including ethnic-minorities-concentrated autonomous districts; and the remaining provinces are grouped into the Middle region, with a modest socioeconomic development level. For example, the percentage shares of the urban population among the total population in 2010 in Eastern, Middle and Western regions were 59.15 %, 44.02 %, 39.72 %, respectively.

Based on the micro-data files of the 2000 census and the 2005 1 % sample survey, we estimated the region-rural/urban-single age-sex-specific occurrence/exposure (o/e) rates of first marriage and fertility by parity, and age-sex-specific

[1] The micro-data sample of the 6th census of China conducted in 2010 is not yet available for scholars and the public to use.

net migration rate between rural and urban areas. The model standard schedules of age-sex-specific o/e rates of divorce and remarriage were estimated from the Chinese In-depth Fertility Survey and the Chinese Longitudinal Healthy Longevity Survey. We also estimated the region-rural/urban-specific general marriage rate and divorce rate based on the age-sex-specific standard model schedules of marriage and divorce rates, the 2000 census micro-data sample, and provincial total numbers of marriages and divorces in 2000–2010 published by the Ministry of Civil Affairs of China. The rural/urban age-sex-specific probabilities of children leaving the parental home were estimated based on the 1990 and 2000 census micro-data samples and the iterated interpolating method within cohorts proposed by Coale (1985) and further extended by Stupp (1988).

Based on the 2000 census and the 2005 sample survey data, we estimated region-rural/urban-specific male and female life expectancy at birth, parity-specific total fertility rates, average ages at first marriage, average age at childbearing, and sex ratios at birth, as well as the region-specific proportions of urban population among the total population (see Appendix 1). Based on the most recent census and survey data and references to others' and our own research, we estimated that the current period TFRs in rural Eastern, Middle and Western areas are 1.69, 1.91, and 2.11, respectively, and that the period TFRs in urban Eastern, Middle and Western areas are 1.10, 1.15, and 1.21, respectively (Wang et al. 2004; Zeng 2012).

Instead of constant assumptions, time-dependent changes were specified for some of the parameters for the period 2000–2010, so that the projected values for 2010 were consistent with the 2010 census' corresponding main results. These parameters include: rural/urban-sex-specific life expectancy at birth, rural/urban-age-sex-specific proportion of persons who live in group quarters, rural/urban -sex-specific proportion of those aged 45–49 who do not live with parents, rural/urban-household size-specific average number of other relatives (other than spouse/partner, parents, or children) and non-relatives living in the same household, and proportion of urban residents among total population. This is similar to the practice adopted by other demographic projections when an earlier census year is the starting point of the projection and the most recent census year is within the projection period; in these cases, cross-tabulation results of the most recent census are published but detailed micro-data to derive the base population of the projections are not yet available (Bureau of Census 2008).

In the analysis presented in this chapter, we adopted the medium fertility assumption, which we believe is the most likely scenario. More specifically, we assume a regionally diversified two-child with adequate spacing policy transition period until 2015 when, on average, a couple in urban areas of Eastern, Middle, and Western regions would have 1.75, 1.83, and 1.92 children in their lifetime; on average, a couple in rural areas of Eastern, Middle and Western regions would have 2.0, 2.27, and 2.51 children in their lifetime, respectively. We assume that, due to delay of marriages and births under rapid socioeconomic development and the encouragement of governmental policies, the average age at first and second or higher order births will increase by 0.75 and 1.5 years by 2030 as compared to 2015, which constitutes an annual growth rate of about 0.05 and 0.1 years for ages at 1st and 2nd births in rural and urban areas of the three regions during the years 2015–2030. As a result, the period TFRs at the first- and second- or higher-order births would be 5 % and 10 % lower than the parity-specific lifetime cohort TFR in rural and urban areas of the three regions, based on estimates employing the widely recognized Bongaarts-Feeney formula (Bongaarts and Feeney 1998). This would lead to period TFRs in rural areas of the Eastern, Middle, and Western regions being 1.86, 2.11, and 2.33, respectively, and the period TFRs in urban areas of these three regions being 1.62, 1.69, and 1.78, respectively during the period 2015–2030 (see Appendix 1). Age at birth is assumed to be constant after 2030 and the period TFRs for rural and urban areas are assumed to be the same as the lifetime cohort TFR by 2035 and constant afterwards.

We conducted a validation test for projections of family households and population from 2000 to 2010 by comparing the results of the projections with the 2010 census observations in the Eastern, Middle, and Western regions in China (see Table 4.4 in Chap. 4). The differences are within a reasonable range, which validate the ProFamy method and the basic data prepared for our multi-regional projections.

13.3 Results of the Comparative Regional Projections

13.3.1 The Middle Region Will Face the Most Serious Challenges of Population and Household Aging

The Chinese census data in 2000 and 2010 demonstrated that population aging was most severe in the Eastern region among the three regions. However, the results of our projections show that the most serious population and household aging by and after the year 2020 will not occur in the Eastern region, but rather in the Middle region. Figures 13.1, 13.2, 13.3, 13.4, 13.5, and 13.6 show that, under the medium fertility assumption, the proportion of elderly population (aged 65+) and of the oldest-old (aged 80+), the proportion of elders and oldest-old living in empty-nest households, and the proportion of elders and oldest-old living alone will be substantially higher in the Middle region than in the Eastern and Western regions by and after 2020. This may seem counterintuitive because the Middle region has substantially higher fertility than the Eastern region, which should result in less severe aging at first glance. We believe that our results are mainly due to continuous Middle-to-Eastern cross-regional migration, mainly by young people, assuming that the age-specific-distributions of cross-regional migrants are the same in the projections as observed in the recent census. The Middle region's continuous outflow of young persons to the Eastern region will lead to an acceleration of population and household aging.

Fig. 13.1 Percentage of elders aged 65+ among total population by region (r = annual growth rate)

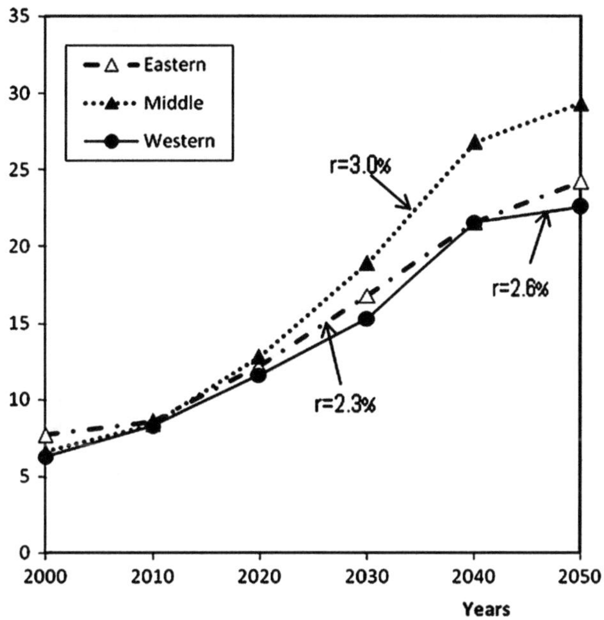

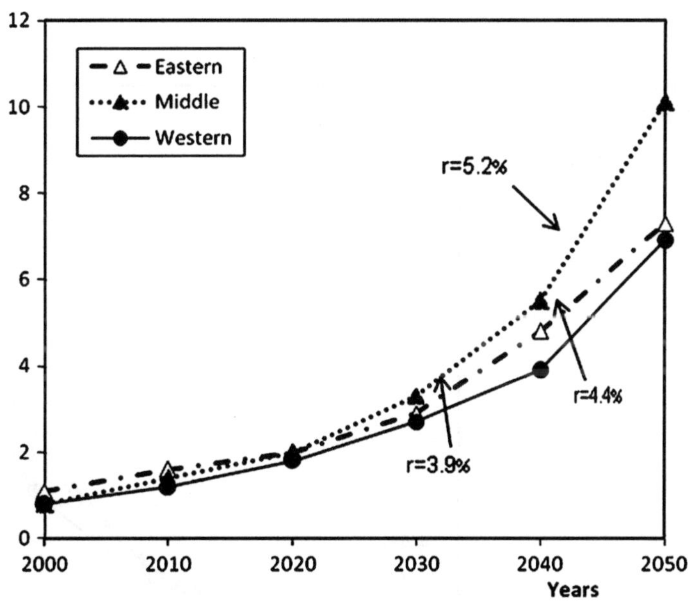

Fig. 13.2 Percentage of oldest-old aged 80+ among total population by region (r = annual growth rate)

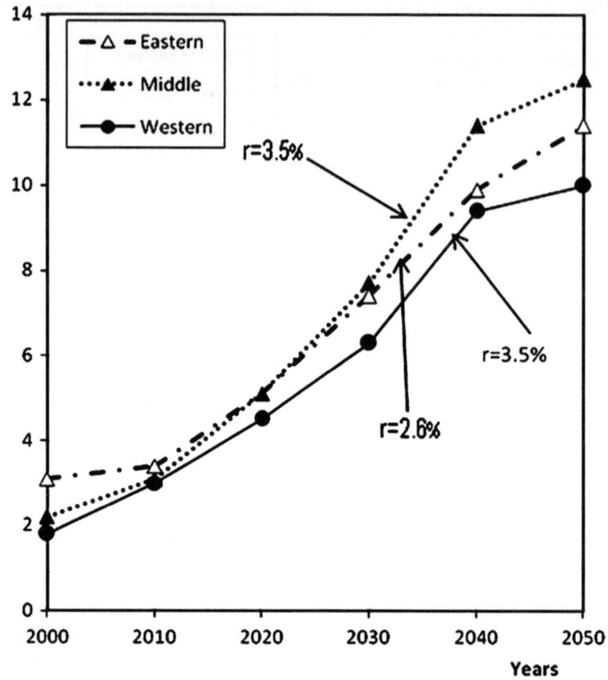

Fig. 13.3 Percentage of elders aged 65+ living in empty-nest households among the total population by region (r = annual growth rate)

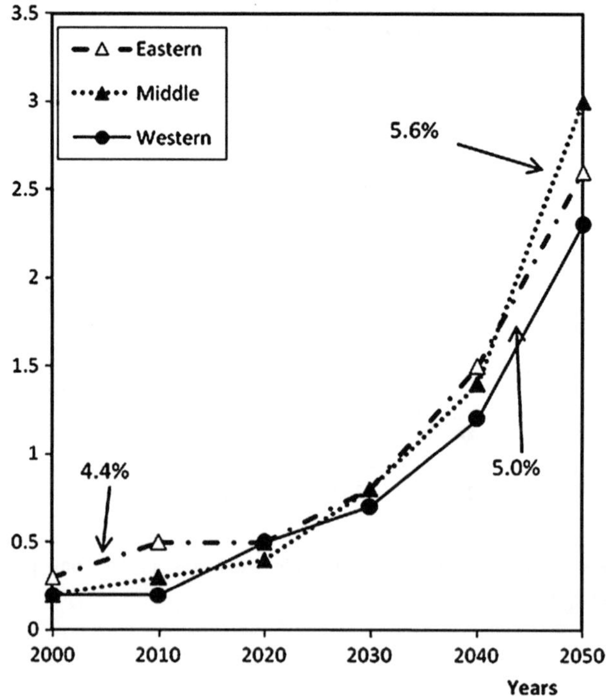

Fig. 13.4 Percentage of oldest-old aged 80+ living in empty-nest households among the total population by region (r = annual growth rate)

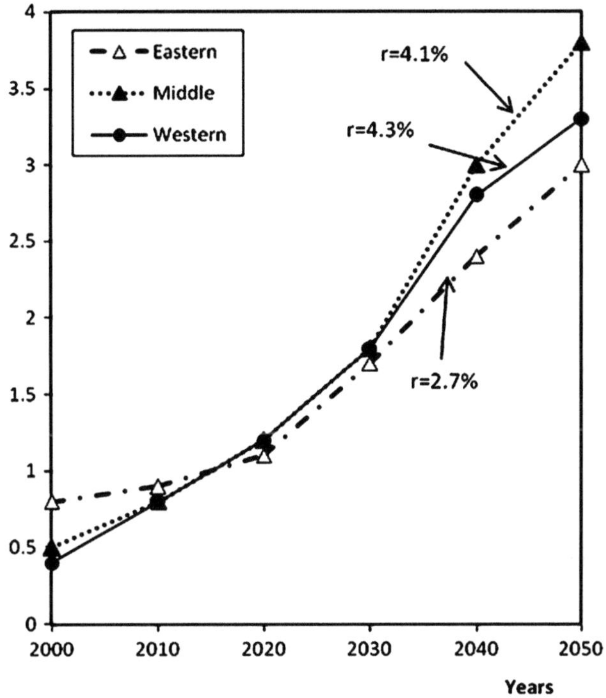

Fig. 13.5 Percentage of elders aged 65+ living alone among the total population by region (r = annual growth rate)

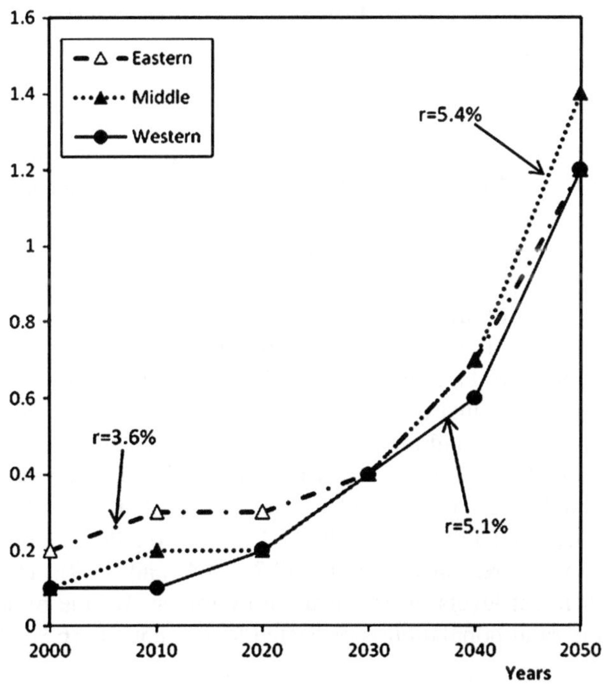

Fig. 13.6 Percentage of oldest-old aged 80+ living alone among the total population by region (r = annual growth rate)

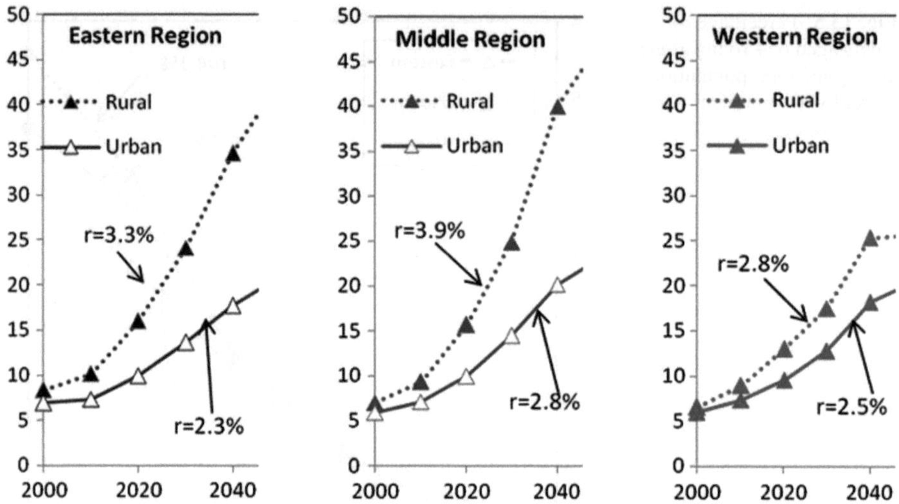

Fig. 13.7 Percentage of elders aged 65+ among total population by rural or urban area and by region (r = annual growth rate)

13.3.2 Population and Household Aging Will Be More than 1.5 Times or Twice as Severe in Rural Areas Compared to Urban Areas in the Middle and Eastern Regions

It has been demonstrated that Chinese rural areas will experience more serious population aging than urban areas because numerous young rural migrants move into the cities (Zeng et al. 2008), but very few of the previous studies have looked at rural–urban differentials across regions of China. In this study, we further project the urban-rural differentials in population and household aging by regions. We assume that the single-age-specific distributions of rural–urban migrants estimated in each of the three regions from the 2000 census micro-data sample remain unchanged, with a high concentration of young migrants. Figures 13.7, 13.8, and 13.9 show that, under the medium fertility assumption, the percentage of the total population that is elderly aged 65+, the percentage of elders living in empty-nest households, and the percentage of elders living alone in rural areas after 2030 are more than 1.5 times or nearly twice as high as that in the urban areas in the Middle and Eastern regions. Note that the rural–urban differentials in population and household aging in Middle and Eastern regions are much larger than those in the Western region (see Figs. 13.7, 13.8, and 13.9). The major reason lies in the different levels of urbanization by region. By the middle of twenty first century, the rural population is projected to account for 16 %, 26 % and 42 % of the total

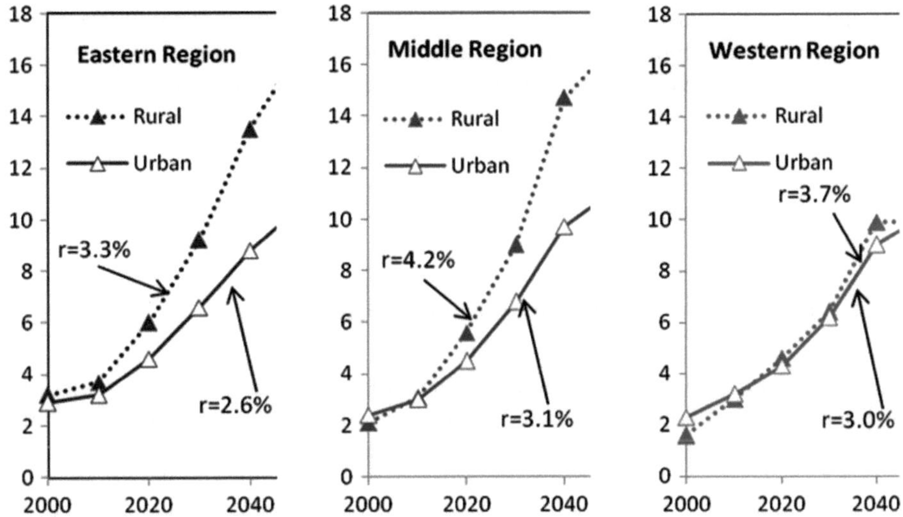

Fig. 13.8 Percentage of elders aged 65+ living in empty-nest households among total population by rural or urban area and by region (r = annual growth rate)

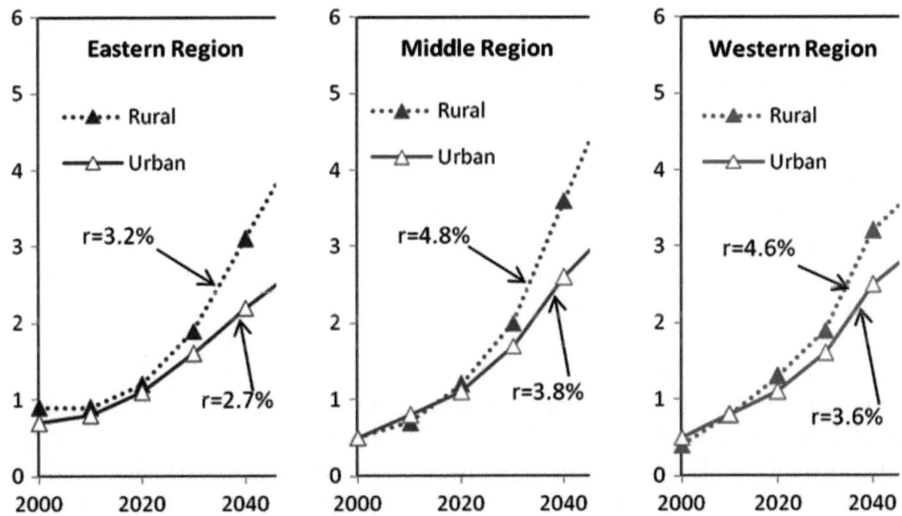

Fig. 13.9 Percentage of elders aged 65+ living alone among total population by rural or urban area and by region (r = annual growth rate)

population in the Eastern, Middle, and Western regions, respectively; about two fifths of the rural population in the Eastern and Middle regions represents the so-called left-behind elderly people, which is much higher than that in the Western region. The large outflows of rural young people during the urbanization process contribute to the accelerating population aging in rural areas in the Eastern and Middle regions of China.

13.4 Discussion and Policy Considerations

Our projections show that the Middle region will face the most serious challenges of population and household aging, followed by the Eastern region. The results show that the degree of population and household aging in rural areas will be more than 1.5 times or nearly twice as high as that in the urban areas in the Middle and Eastern regions after 2030. These projection results are based on the assumption that the current age distribution of migrants, which is dominated by young people, remains unchanged. However, the serious problems of population aging in the Middle region and in the rural areas would be substantially reduced if the future age pattern of migrants becomes more demographically balanced due to family migration including both young and old persons. Thus, to avoid the over-aging problems in the Middle region and the rural areas, we recommend that the Chinese government take two policy actions: (1) encourage young and middle-aged migrants who are already settled in urban areas to bring their old parents from villages to co-reside or live in close proximity in urban areas, and (2) encourage future family migration that includes members of different ages instead of only younger migrants.

The family migration policy actions encourage adult children to live together with or near to their old-age parents, allowing elders to receive home-based care from children whenever needed and also provide care for grandchildren. Shen (2011) discovered that older parents who live with their adult children are significantly more advantaged in cognitive function, self-rated health, and self-rated life satisfaction, based on the large nationwide sample data from the Chinese Longitudinal Healthy Longevity Survey (CLHLS) and instrumental variable (IV) analyses correcting for the endogeneity of living arrangements. Using a similar IV method and CLHLS data, Shen, Yan, and Zeng (2013) found that co-residence with elderly parents significantly increases the labor force participation of female adult children by 23 %. Co-residence with parents increases the hours per week that adult children work by 19.9 %, an effect that is more evident for females and for rural adult children. Subsequent empirical analyses reveal the mechanism: in intergenerational families, elderly parents help to relieve the housework burden of their adult children, especially daughters, allowing the children to devote more time to working. The study also found that co-residence with elderly parents is beneficial for the self-rated health of adult children, especially for daughters (Shen et al. 2013). Co-residence with or close proximity to adult children will also decrease home-based care expenditures for disabled older adults (Zeng et al. 2013b). Living in close proximity, as opposed to co-residence, could increase the happiness of elderly parents by avoiding intergenerational conflicts between elders and their children or grandchildren concerning eating, entertainment, etc.

Clearly, it is beneficial for both old parents and adult children to promote family migration which involves intergenerational co-residence or proximate living arrangements. Encouraging such a mutually advantageous living arrangement may be a sound policy response to the rapid population and household aging in the Middle region and many rural areas in China. We believe that such a program is highly feasible, given the thousands of years of the Chinese cultural tradition of filial piety to old parents, which is still deeply rooted in China (Laidlaw et al. 2010).

While we are satisfied with the unique and meaningful contributions we have made in this study, we are aware that there are some important limitations of the multi-regional projections presented in this study that need to be investigated in future research. Mainly due to space limitations, we have presented here the general trends and patterns of both-sexes-combined numbers of elders aged 65+ and the oldest-old aged 80+, by rural/urban residence and living arrangement in each of the three regions. In fact, our ProFamy multistate model has produced more detailed projections of the number of elders, middle-aged and young adults, and children by gender, single year of age, marital status, number of co-residing children, number of co-residing parents, living in one-, two- or three-generation household, and private or institutional household. Full presentations and discussions of these detailed results, including not only population and household aging but also labor force resource and child dependency, could be useful in more detailed future academic research, governmental panning, and business market analysis.

Finally, as stated in the other chapters about projection applications, we emphasize again that projections for a time horizon of less than 20 years may be used as forecasting for business and governmental planning, but any results beyond that should be considered to be simulations only, due to large uncertainties after more than 20 years. Thus, the unique multi-regional projection results for the first half of this century presented in this chapter should be mainly regarded as simulations. Such simulations are useful for academic and policy analysis to answer the "what if" questions about effects of changes in demographic parameters and fertility policies on future trends and patterns of population and household aging, but they cannot be considered to be accurate forecasts.

Appendix 1: Parameters of Population and Household Projection at the Regional Level

	Rural					Urban				
Year	2000	2015	2030	2035	2050	2000	2015	2030	2035	2050
Eastern region										
Total fertility rate	1.69	1.86	1.86	2.00	1.95	1.10	1.62	1.62	1.75	1.75
Male life expectancy	69.97	72.18	74.48	75.27	77.64	72.66	74.67	76.56	77.26	79.10
Female life expectancy	74.57	76.78	79.07	79.75	82.10	76.83	78.84	80.73	81.42	83.17
General marriage rate	0.0761	0.0759	0.0758	0.0755	0.0753	0.0684	0.0701	0.0698	0.0697	0.0691
General divorce rate	0.0022	0.0022	0.0021	0.0021	0.0021	0.0057	0.0057	0.0056	0.0056	0.0056
Mean age at birth	25.73	25.71	27.03	27.05	27.13	26.57	26.53	27.81	27.79	27.74
% of urban population						46	58	70	73	84
Middle region										
Total fertility rate	1.91	2.11	2.111	2.27	2.21	1.15	1.69	1.69	1.83	1.83
Male life expectancy	68.87	71.04	73.31	74.09	76.42	71.46	73.44	75.30	75.98	77.80
Female life expectancy	72.18	74.31	76.52	77.18	79.46	75.33	77.30	79.16	79.83	81.54
General marriage rate	0.0562	0.0561	0.0559	0.0558	0.0556	0.0445	0.0456	0.0454	0.0453	0.0449
General divorce rate	0.0020	0.0020	0.0020	0.0019	0.0019	0.0050	0.0050	0.0050	0.0050	0.0050
Mean age at birth	25.03	25.01	26.29	26.31	26.40	25.47	25.43	26.65	26.64	26.59
% of urban population						33	45	58	62	74
Western region										
Total fertility rate	2.11	2.33	2.33	2.51	2.45	1.21	1.78	1.78	1.92	1.93
Male life expectancy	65.07	67.13	69.26	70.00	72.21	71.26	73.23	75.09	75.77	77.58
Female life expectancy	69.08	71.12	73.24	73.87	76.05	75.03	77.00	78.84	79.52	81.22
General marriage rate	0.0703	0.0701	0.0700	0.0697	0.0695	0.0590	0.0605	0.0602	0.0602	0.0596
General divorce rate	0.0025	0.0025	0.0024	0.0024	0.0024	0.0062	0.0062	0.0062	0.0062	0.0062
Mean age at birth	24.83	24.81	26.08	26.10	26.18	25.47	25.43	26.65	26.64	26.59
% of urban population						29	38	47	50	58

Chapter 14
Application of Household and Living Arrangement Projections to Policy Analysis in China

14.1 Introduction

The tremendous reduction in fertility rates combined with baby boomers entering the labor force has resulted in a demographic "dividend" in China since the 1980s, which features a low child dependency ratio, still not-yet high elderly dependency ratio, and a rich labor force supply. Although high social and political costs were paid, the demographic dividend contributed significantly to China's economic boom in the past 20 years. However, this dividend will disappear in a couple of decades (Cai 2006; Wang and Mason 2006), producing many questions about China's future. What will the social and economic consequences be if China continues to implement its current strict fertility control policy? Is it necessary to change the current fertility policy? If so, what are the optimistic and feasible options? This chapter addresses these important questions with an application of the ProFamy extended cohort-component method. The unique features of this study include a comparative analysis of possible options for fertility policy transition based on demographic projections of population aging, households and elderly living arrangements, dependency ratios, pension deficits, labor force supply, the marriage squeeze, and economic costs under different fertility policy scenarios.

The chapter is organized as follows. A brief summary of related previous research and recent debates and discussions about China' options for fertility policy transitions are presented in the next section. In the third section, we briefly discuss the data resources for this study and design the four fertility policy scenarios, referring to the debating policy options in China. The fourth section presents a detailed comparative analysis between different fertility policy scenarios with respect to various aspects of demographics and their socioeconomic consequences. The challenges and opportunities concerning rural old age insurance programs and retirement age will be discussed in Sects. 5 and 6. In the final section, based on projection results and empirical data analysis, we will show why China needs to transfer to a universal two-child with adequate spacing policy as soon as possible. We will also discuss why it is crucially important to further develop the old-age

Y. Zeng et al., *Household and Living Arrangement Projections*, The Springer Series on Demographic Methods and Population Analysis 36,
DOI 10.1007/978-90-481-8906-9_14, © Springer Science+Business Media Dordrecht 2014

insurance program in rural areas and to gradually increase the age at retirement in order to face the serious challenges of population and household aging and create opportunities for sustained development.

14.2 A Brief Review of Related Policy Research and Debates on Fertility Policy Transition Options in China

The strict one-child policy was implemented without sophisticated demographic research support in 1980, a time when the Chinese government was eager to rid itself of rapid population growth given the very poor economic conditions at that time and very limited natural resources per capita. The implementation of the initial strict one-child policy met strong and wide resistance from peasants in rural areas. Thus, in 1984 the Chinese government relaxed the "one-child policy" for most rural areas and implemented a "1.5-child policy" instead. Six provinces and autonomous regions adopted a policy of allowing all rural couples to have a second child a few years after the birth of the first child. Minority nationalities were generally allowed to have two or more births per couple. This relaxed version of the one-child policy has remained stable to the present time; statistical analysis based on the local official fertility policy regulations has shown that the current overall Chinese fertility rate is on average about 1.47 children per couple (Guo et al. 2003).

The first domestically proposed alterative was the "two-child plus spacing" by Liang (1979) and Ma and Zhang (1984). The first internationally proposed two-child policy was published by Bongaarts and Greenhalgh (1985) and Greenhalgh and Bongaarts (1987). Zeng and Vaupel (1989), Zeng (1990), and Vaupel and Zeng (1991) published Chinese rural–urban dynamic models and policy analyses, and concluded that the "two-child and late childbearing" policy was the best alternative to avoid excess population size and severe aging problems in the future. Li (1997) reached a similar conclusion. Based on a theoretical and empirical analysis of microeconomic rational choice of fertility behavior, Johnson (1994) concluded that, even if the relaxed policy produces a relatively larger population, it would not have a negative impact on the food supply, average income per capita, and social welfare. The State Family Planning Commission (SFPC) of China project team (2000) published a research report and proposed that the current fertility policy should be stabilized with some minor adjustments to allow couples to have a second child if either or both partners are an only-child. In 2004, a research consortium (PI: Baochang Gu), which also included the first author of this book, presented a report which proposed gradually transferring to a two-child policy (The Research Group 2004). Wang (2005) presented an analysis on the social, economic, and political costs and negative consequences if China keeps its current fertility policy unchanged. After reviewing the effects of China's one-child policy on population growth, sex ratio at birth, and old age dependency ratio,

Hesketh et al. (2005) concluded that a relaxation of the one-child policy would be desirable. Zeng (2005a, 2006) proposed a "smooth transition to the two-child policy using late-childbearing as a lever", based on analysis of recent trends in fertility quantum/tempo and demographic projections.

Most demographers believe that the current Chinese fertility policy must be revised, although their opinions on how to do so are very diverse. Yet most policy-makers, administrators, and a considerable number of scholars (especially in the natural and engineering sciences) in China advocate for keeping the current fertility policy unchanged, with some mini-adjustment, on a long-term basis.[1] Their rationale is mainly two-fold: (1) The base population of China is too large and natural resources per capita are too small compared to other countries; "too many people" is the main cause of problems of environmental pollution, traffic jams, low GDP per capita, etc.; (2) The current low fertility in China is not sustainable if the policy is relaxed because if you allow people to have two births they could then have a third or more. Some people even advocate for a further tightening of the current fertility policy (e.g., Li 2004).

In addition to the option of retaining the current fertility policy unchanged, three main options for fertility policy transitions have been internally debated/discussed in China among scholars and some policy makers. One popular option is "the mini-adjustment (w*ei tiao*) and natural transition (*zhi ran guo du*)"; namely, couples with at least one party being an only-child are allowed to have two children, while the current 1.5-child policy still holds for rural couples (if the first child is a girl the couple is allowed to have a second child). This option (see, e.g., SFPC project team 2000) is abbreviated as the "Two-child solely for only-child couples" hereafter.

Another option is to relax or eliminate the birth spacing policy first and then gradually transition to a two-child policy later on. With the policy conditions for allowing for a second birth remain unchanged, three provinces eliminated and six other provinces relaxed birth spacing policies around 2005 (Zeng 2005a). This is mainly because many people in China, including policy makers and some scholars, do not fully understand the important impact of changes in fertility tempo on the period fertility quantum and population growth. On one hand, Guo (2000), Ding (2003), and Zeng (2005a) have academically analyzed the important impact of changes in fertility tempo on the period fertility quantum as quantified by the method proposed by Bongaarts and Feeney (1998). On the other hand, some influential unpublished internal policy research reports advised the government that eliminating the birth spacing policy would have very little impact on the period number of births and population growth under the current low fertility. Thus, policy makers in nine provinces have chosen to eliminate or relax the birth spacing policy,

[1] For example, in a national family planning policy/management meeting held in early 2005, the Vice Prime Minister asked the director generals of all 31 provincial committees of population and family planning whether they would agree if the current fertility policy were relaxed. All of the director generals except the one from Shanghai said "no".

which is politically much easier to do than relaxing the one-child policy, in order to reduce the difficulties in the fertility policy implementation (Shu 2005).

Yet another option is to smoothly transfer to a universal two-child policy in both rural and urban areas, while continuing the implementation of current birth spacing policy with educational and socioeconomic incentive programs (Zeng 2005a, b, 2006, 2007, 2011, 2012).

14.3 The Data Sources, Policy Scenario Designs, and Parameter Assumptions

We employ the ProFamy extended cohort-component method, which was presented in detail in Part I of this book, to perform the family household and population projections to be presented in this chapter. The data sources and general issues of the estimates are presented and discussed in Sect. 12.2 of Chap. 12 and will not be repeated here.

The Chinese period total fertility rate (TFR) for years since the end of the 1990s, estimated by various domestic and international demographers and statistical offices using all kinds of relevant data and direct or indirect methods, ranged from 1.5 to 1.8 (e.g., Guo 2004; Zhang and Zhao 2006). According to the 2010 population census data, the observed period TFR in China in 2010 was 1.2, which is too low and unbelievable. Based on the 2010 census data for ages 10–19, which is accurate because children born more than 10 years ago without "birth quota" were already registered for schooling, and the "backward forecasting" method, we estimated that the average under-reporting rate for ages 0–9 in the China 2000 census was about 9.5 %. We used the 2000 population census as a source of basic data for women of reproductive age and various possible values of TFR in the period 2000–2010 to "forward-forecast" the number of children aged 0–9 in 2010. We then compared the forecasted number with the actual observed population aged 0–9 in the 2010 census, taking into account the 2010 census under-reporting rate for ages 0–9, which was widely believed to decrease to some extent compared to that in the 2000 census because the governmental public propaganda emphasized no penalty for reporting children who were born "out of the birth quota" and fieldwork management was stronger in the 2010 census. This demographic evaluation procedure resulted in an estimated TFR in China in 2010, adjusted for the under-reporting of births, of 1.63,[2] which is near the middle of the TFR range 1.5–1.8. Using the

[2] Our estimate of the TFR as 1.63 in 2010 in China implies an under-reporting rate of new births of about 25 %. Note that the average under-reporting rate of about 9.5 % in 2000 mentioned above was an average for ages 0–9. However, the underreporting rate of new births is much higher than other childhood ages, especially ages 6–9, when most of the under-reported children were registered for school.

census' observed rural–urban TFR differential, we estimated that the period TFRs in 2010 were 2.01 for rural areas and 1.24 for urban areas.[3]

Referring to the earlier described options of fertility policy transitions in China, we designed the four demographic scenarios described below. In making the assumptions of future fertility levels in these scenarios, we also consider the effect of rapid economic development, where more young couples in cities choose to have one child only or no children (Double-Income with no Kids (DINK)). For example, based on a survey of 20,649 persons aged 18–30 in Shanghai in 2003, the average desired number of children was 1.1 (Xinhuanet 2003). It was reported that about 10 % of the young married couples in Beijing said that they did not intend to have any children (Xinhuanet 2004). We also consider the tremendous differences in socioeconomic development levels and fertility attitudes and levels between the rural and urban areas (nearly half of the Chinese are still rural residents). However, these are the educated assumptions only for qualitatively answering the questions of "what if" under different fertility policy options, and are not intended for any kind of accurate forecasting.

1. The "*two-child with encouragement for adequate spacing policy*" scenario assumes that the two-child policy will be started soon with a smooth transition period. Around 2015, all couples in China would be allowed to choose to have a second child and encouraged to have appropriate spacing between children. Considering the much lower level of socioeconomic development in rural areas and the presence of rural minority ethnic groups that are allowed to have three children, the cohort life-time TFRs by, and after, the year 2015 are assumed to be 2.27 in rural areas and 1.8 in urban areas. Programs that encourage late-childbearing and the effects of socioeconomic development will result in delays of marriages and births, so we assume that ages at first- and second- or higher-order births will increase by 0.75 and 1.5 years in 2030 as compared to 2015, which constitutes an annual growth rate of 0.05 and 0.1 years of age, respectively, during the years 2015–2030. According to the method proposed by Bongaarts and Feeney (1998), the projected period TFR of the first- and second- or higher-order births in the years between 2015 and 2030 will be 5 % and 10 % lower than the parity-specific cohort TFR, due to fertility tempo effects. Thus, the TFR of all parities combined in rural and urban areas in 2015–2030 will be 2.15 and 1.67 (in contrast to the cohort TFR 2.27 and 1.8), respectively (see Table 14.1). We expect that China will "soft-land" to allow its citizens to freely choose family size and fertility timing around 2030–2035, and the period TFR will slightly increase in 2035 and then gradually and slightly decrease after 2035.
2. The "*two-child with constant mean age at birth*" scenario assumes that the cohort TFR would be the same as that in scenario (1) in, and after, 2015, but the mean age at birth would remain constant due to relaxing or eliminating the

[3] Our rural and urban TFR estimates in 2010 were slightly higher than those estimated for 2000 shown in Table 12.1, because various provinces in China had started to allow couples with both parties who are an only-child (i.e., no siblings) to have two children.

Table 14.1 Total fertility rates under different fertility policy scenarios

Year	2010	2015	2030	2035	2050	2080
Rural						
Two-child policy with spacing	2.01	2.15	2.15	2.21	2.13	2.11
Two-child policy with constant mean age at birth	2.01	2.27	2.27	2.27	2.27	2.27
Two-children solely for only-child couples policy	2.01	2.12	2.18	2.20	2.20	2.20
Current fertility policy unchanged	2.01	2.01	2.01	2.01	2.01	2.01
Urban						
Two-child policy with spacing	1.24	1.67	1.67	1.72	1.72	1.71
Two-child policy with constant mean age at birth	1.24	1.80	1.80	1.80	1.80	1.80
Two-children solely for only-child couples policy	1.24	1.69	1.50	1.48	1.48	1.48
Current fertility policy unchanged	1.24	1.24	1.24	1.24	1.24	1.24
Rural and urban combined						
Two-child policy with spacing	1.63	1.89	1.81	1.84	1.77	1.74
Two-child policy with constant mean age at birth	1.63	2.01	1.94	1.92	1.86	1.84
Two-children solely for only-child couples policy	1.63	1.89	1.73	1.70	1.62	1.62
Current fertility policy unchanged	1.63	1.58	1.47	1.44	1.33	1.27

spacing policy which may offset the impact of socioeconomic development on delaying marriages and births and thus the period TFR would be the same as the cohort TFR in and after 2015 (see Table 14.1).

3. The "*two-child solely for only-child couples policy*" scenario implies that the current policy would be relaxed to some extent to allow all couples with at least one only-child partner and rural couples whose first child is a girl to have two children. Because this policy option requires more complicated conditions of having one or no sibling(s) for bearing the second child, no macro simulation models including ProFamy could perform a population projection that would accurately follow its policy requirements. Thus, we invited Wenzhao Shi, the Technical Director of the Chinese National Population Administration and Decision Information System (PADIS), and his group to conduct the population projection for China 2010–2080 under the two-child solely for only-child couples policy scenario, using a micro-simulation approach. The micro simulation after 2010 under the two-child solely for only-child couples policy reported that TFR would increase slightly in rural areas but increase more substantially in urban areas, due to the fact that the percentage of couples with at least one only-child partner in rural areas is substantially smaller than in urban areas (see Table 14.1).
4. The "*current fertility policy unchanged*" scenario assumes that the period rural and urban TFR in, and after, 2015 would remain the same as that in 2010 (see Table 14.1).

The medium mortality assumption adopted in all scenarios in this study assumes that there will be gradual improvement in mortality in China during the period 2010–2080 – from a life expectancy of 74 years old for both sexes combined in 2010, to 81.8 years old in 2050, and 84.05 years old in 2080; gender differentials in life expectancy are assumed to remain the same as that observed in 2010. Given that the main interest of this study is to explore the consequences of different fertility policy scenarios, we assume that the rural and urban general marriage and divorce rates in future years will remain the same as those observed in 2010 (Refer to Table 12.1 of Chap. 12 for the assumptions about the future years' life expectancy at birth, the general rates of marriage and divorce, and proportion of urban population).

14.4 Comparative Analysis Under Different Fertility Policy Scenarios

14.4.1 Population Growth

The "current policy unchanged" scenario would reach a smaller peak population of about 1.4 billion around 2023 and then rapidly decline, with a negative annual growth rate of −6 per thousand in 2040–2050 and −10.7 per thousand in 2050–2080. It is widely recognized among social scientists that rapid population decline will not only cause problems in aging, but also result in serious problems such as labor shortages and pension deficits, which will be discussed later.

While the average number of children per individual couple in the two two-child policy scenarios are exactly the same, the population size of the "two-child with for adequate spacing" soft-landing scenario would be smaller than the "two-child with constant mean age at birth" by about 58, 98, and 148 million in 2030, 2050, and 2080, respectively. The "two-child with adequate spacing" soft-landing option will enable China to never exceed a total population size of 1.45 billion, while individual couples' demands for two children will be fully meet. But the other two-child policy option, which eliminates the policy of encouraging adequate birth spacing, would result in a substantially larger population size and concomitant resource pressure, which is the decisive concern of Chinese policy makers and the public.

The labor force quantum under the "two-child with adequate spacing" scenario is relatively close to that of "the two-child with constant mean age at birth" scenario until 2050 or so, with somewhat larger differences afterwards. However, our scenarios demonstrate that relaxing or eliminating the birth spacing policy first and then gradually transferring to a "two-child" policy later on is a poor option. This conclusion is based on the substantial difference in population growth between scenario (1) and (2), purely due to assumed changes (or no changes) in mean ages at birth in the policy transition period up to 2030. This exercise also shows that "two-child with adequate spacing" policy scenario will enable China to achieve the dual goals of avoiding excessive population growth and allowing individual couples to fully meet their demand for two children. Again, this is a "what if" policy analysis, rather than any kind of forecasting.

Because the relative differences in projected demographic indicators of population aging and labor force between the two scenarios with two-child policies but

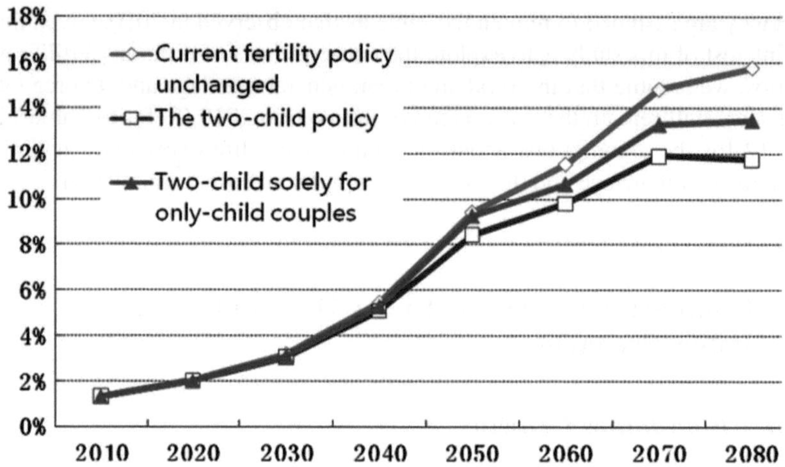

Fig. 14.1 Percentage of elderly aged 65+ among total population, under different fertility policy scenarios

different mean age at birth are rather small, we will focus the policy analysis in the rest of this chapter on three options: the "universal two-child with encouragement for adequate spacing policy" (abbreviated as "the two-child policy" hereafter), the "two-child solely for only-child couples policy", and the "current fertility policy unchanged".

14.4.2 Percentage of Elderly and Elderly Living in Empty-Nest Households

As shown in Figs. 14.1 and 14.2, the Chinese population ages quickly under all scenarios. However, under the current policy unchanged scenario, the percentage of elderly population aged 65+ in 2050 and 2080 will be 28.6 and 37.2, and the corresponding percentages for the oldest-old aged 80+ will be 9.4 and 15.8, respectively. The proportion of the elderly aged 65+ under the current policy unchanged scenario will be higher than that in the two-child scenario by 11.7 % and 31.2 % in 2050 and 2080, respectively (see Fig. 14.1). The proportion of the oldest-old in the current policy unchanged scenario will be higher than that in the two-child scenario by 11.8 % and 34.4 % in 2050 and 2080 (see Fig. 14.2).

Under the two-child solely for only-child couples policy scenario, the percentage of the elderly aged 65+ will reach 28.0 and 33.8 in 2050 and 2080, higher than that under the two-child policy by 9.4 % and 19.1 %, respectively. The percentage of the oldest-old aged 80+ will be 9.3 and 13.4 under this policy in 2050 and 2080, higher than that under the two-child policy by 9.8 % and 14.4 %, respectively.

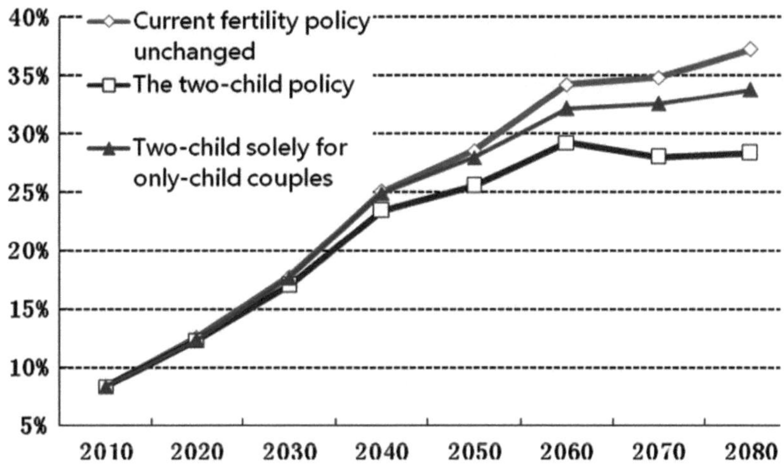

Fig. 14.2 Percentage of oldest-old aged 80+ among total population, under different fertility policy scenarios

The proportion of the elderly aged 65+ living in empty-nest households under the current policy unchanged scenario is 11.2 and 29.9 % higher than in the two-child scenario in 2050 and 2080, respectively (see Fig. 14.3). Under the current policy unchanged scenario, the percentage of oldest-old who live in empty-nest-households will be 11.4 and 33.0 % higher in 2050 and 2080, respectively, as compared to the two-child scenario (see Fig. 14.4).

14.4.3 Labor Force and the Elderly Dependency Ratio

As shown in Fig. 14.5, the labor force aged 18–64 under the current policy unchanged scenario will quickly shrink from 910 million in 2030 to 730 million in 2050 and 470 million in 2080. From 2030 to 2080, the labor force will be reduced by about 90 million every 10 years; As compared to 2030, the labor force in 2050 and 2080 will be reduced by 20 % and 48.4 %. Furthermore under the current policy unchanged, the proportion of "old workers" aged 55–64 among the total labor force will rise quickly from 16.4 % in 2010 to 25.2 and 29.7 % in 2030 and 2050. In contrast, the labor force in the two-child scenario will be substantially larger than in the current policy unchanged scenario by 30, 60, and 200 million persons in 2040, 2050, and 2080, respectively (see Fig. 14.5). The aging of the labor force will be much less serious in the two-child policy scenario than that under the current policy unchanged.

The elderly dependency ratio under the current policy unchanged scenario will be substantially higher than that under the two-child policy after 2035 (See Fig. 14.6). Under the current policy unchanged scenario, the number of

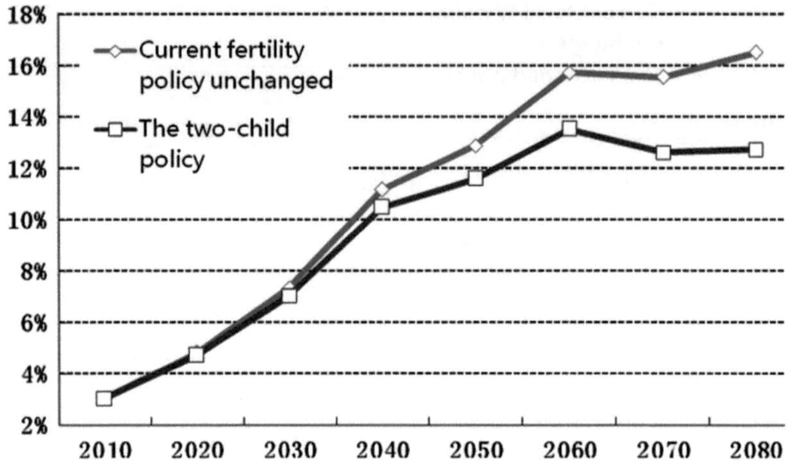

Fig. 14.3 Percentage of elderly aged 65+ living in empty-nest households among total population, under different fertility policy scenarios (Note: The micro simulation projection under the "two-child solely for only-child couples" scenario by Shi (2012) did not include households and living arrangement; thus, we present comparisons between the two-child policy and current fertility policy unchanged only in Figs. 14.3 and 14.4)

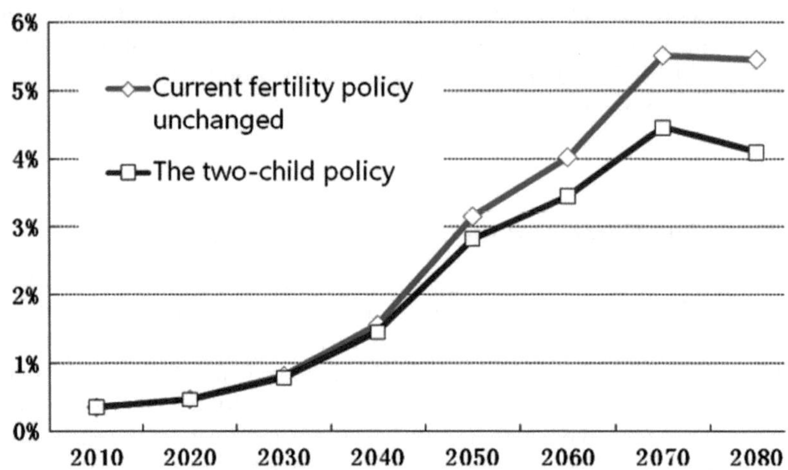

Fig. 14.4 Percentage of oldest-old aged 80+ living in empty-nest households among total population, under different fertility policy scenarios

working-age persons per elder (reciprocal of the elderly dependency ratios shown in Fig. 14.6) will dramatically decrease from 8.0 in 2010 to 3.5, 2.0 and 1.4 in 2030, 2050, and 2080, respectively; the elderly dependency ratio in 2030, 2050, and 2080 will be 2.3, 4.1 and 5.8 times as high as that in 2010.

Figures 14.5 and 14.6 demonstrate that, although somewhat better than that under the current policy unchanged, the shrinking of the labor force and increasing elderly

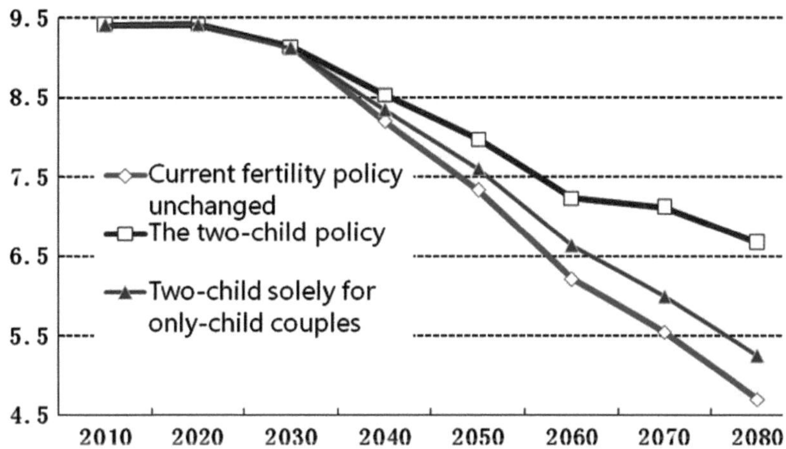

Fig. 14.5 Number of labor force persons aged 18–64 (unit: 100 million), under different fertility policy scenarios

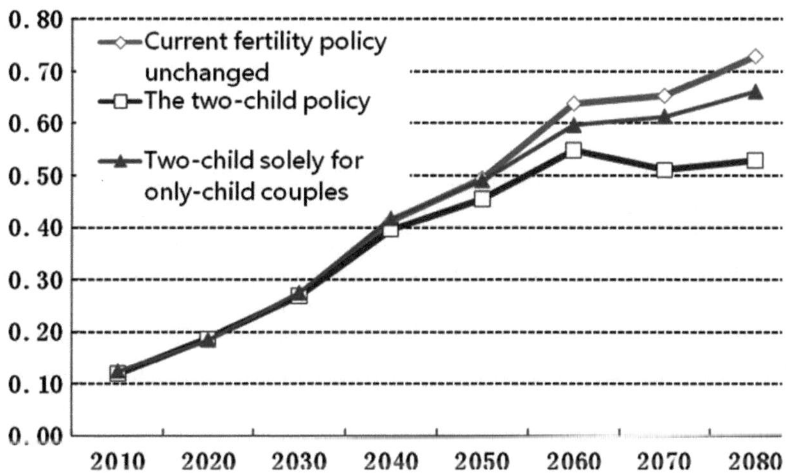

Fig. 14.6 Elderly dependency ratios, under different fertility policy scenarios

dependency ratio under the two-child solely for only-child couples scenario will be considerably worse than that under the universal two child policy scenario. More specifically, under the two-child solely for only-child couples policy scenario, the labor force will quickly shrink to 760 million in 2050 and 520 million in 2080; from 2030 to 2080, the number of persons aged 15–64 will be reduced by about 77 million every 10 years. The number of working-age persons per elder will also dramatically decrease from 8.0 in 2010 to 3.5, 2.04 and 1.5 in 2030, 2050 and 2080, respectively; the elderly dependency ratio in 2030, 2050 and 2080 will be 2.2, 3.9 and 5.3 times as high as that in 2010.

Table 14.2 Projected ratios of home-based care costs for disabled elders per working-age person in future years to that in 2010, under the fertility policy scenarios of two-child with adequate spacing and the current policy unchanged

	The two-child policy					Current fertility policy unchanged				
	2010	2020	2030	2040	2050	2010	2020	2030	2040	2050
Ratios of average numbers of disabled elders per working-age person to that in 2010	1.0	1.51	2.31	3.63	5.05	1.0	1.51	2.36	3.84	5.61
Ratios of home-based care costs in cash per working-age person to that in 2010[a]	1.0	3.18	8.29	19.14	35.12	1.0	3.18	8.44	20.24	38.99
Ratios of non-cash home-based care workdays for disabled elderly per working-age person to that in 2010	1.0	1.51	2.33	3.73	5.33	1.0	1.51	2.37	3.94	5.92

[a]The home-based care service wages are assumed to grow at the same rate as GDP growth in 2010–2050

Williamson (1997) estimated that a favorable age structure with rich labor supply contributed about one-third to one-half of the excessive growth in GDP in the four small dragons of South Korea, Singapore, Taiwan, and Hong Kong in the 1970s and 1980s. Cai and Wang (1999) estimated that about 27 % of the remarkable quick increase in GDP in China during the period 1982–2000 was due to the high proportion of the labor force among the total population. However, keeping the current fertility policy unchanged for the long-run will make China's labor force shrink quickly and thus lose its demographic resources and comparative advantage for economic growth (Lin 2006).

14.4.4 Resources of Care Providers for Disabled Elderly

Table 14.2 presents projected ratios of home-based care costs for disabled elders per working-age person in future years to that in 2010, under the two-child and current policy unchanged scenarios, based on application of the extended ProFamy model to project elderly disability status and home-based care costs, as described in detail in Chap. 5. The results show that if the current fertility policy remains unchanged, the burden of home-based care for disabled older adults per caregiver of working ages will be increasingly aggravated and substantially more serious than that under the two-child policy. The projection results demonstrate the lagging effect of fertility policy transition due to the fact that it takes about 20 years for the new births to join the labor force, but the policy transition substantially alleviates the serious challenge of population aging and caregiving in China after 2035.

Our projection results also show that, even under the two-child policy, home-based care needs and costs for disabled elders per working-age person will increase dramatically. Besides the adjustment of current fertility policy, other relevant socioeconomic countermeasures should also be taken.

14.4.5 Sex Ratio at Birth and Marriage Squeeze

Based on census data, it was estimated that the sex ratio at birth (SRB) in the two-child policy areas was 109.0. However, the SRB in the 1.5-child policy areas (if the first child is a girl, the couple is allowed to have a second birth; otherwise, only one child is allowed) was as high as 124.7, higher than the SRB in the two-child policy areas by about 16 % points (Guo 2007).

The Chinese policy makers' initial motivation for adopting the 1.5-child policy in 1984 was to account for the difficulties (such as household labor) of peasants who have one daughter only under the one-child policy scheme. However, policy makers did not expect that allowing only those rural couples whose first child was a girl to have a second birth would increase the SRB to such a large extent. It is not difficult to understand why such unexpected side effects could occur. The 1.5-child policy implicitly tells peasants that one boy is sufficient for family welfare so there is no need to have another birth, but one girl is not sufficient so the family needs to have another baby. This implies that the value of a male baby is twice as high as that of a female baby. Such implicit psychological effects may act in addition to the traditional strong son-preference to lead peasants whose first child is a girl to conduct prenatal sex determination and sex-selective abortion, both of which are illegal in China, to have at least one boy. If all couples are free to have a second birth, the side effects of the government implicitly suggesting a girl's half-value would not exist. Consequently, there would be fewer people who take the legal risk of conducting prenatal sex determination and sex-selective abortion.

Zeng (2007) estimated that the proportion of couples whose first child was a girl and who underwent prenatal sex determination and sex-selective abortion to have a boy as a second child was about 19.1 % in 1.5-child policy areas, in contrast to 4.6 % in two-child policy areas.

Furthermore, both empirical data and general logic indicate that, for couples whose first child is a boy, the SRB of second births is normal. But these couples are not allowed to have a second birth under the 1.5-child policy. This restriction eliminates about half of the second births which would have a normal SRB, and thus causes additional structural impacts towards an abnormal overall SRB. The results of the numerical simulation presented in Zeng (2007) shows that slightly more than one quarter of the excess sex ratio at birth in the 1.5-child areas is due to the structural effects of not allowing couples whose first child is a boy to have the second birth.

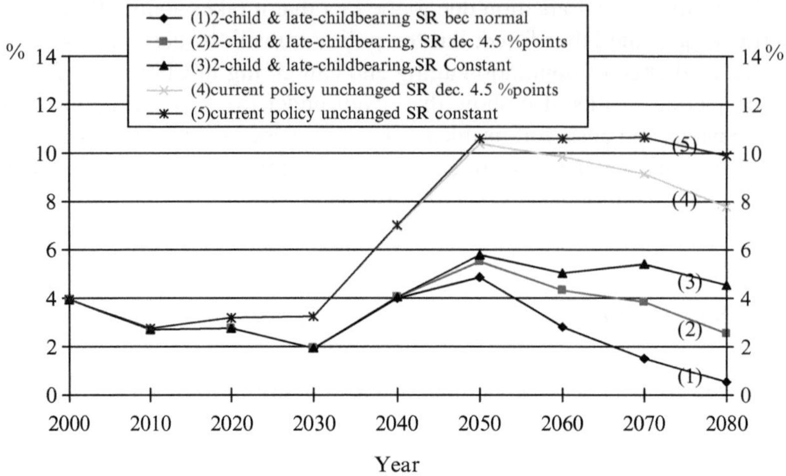

Fig. 14.7 Percentage of never-married men age 45–49 due to shortage of women, under different fertility policy scenarios. Note: "*SR*" means sex ratio at birth; "*dec*" means decline

Simulation analysis by Zeng (2007) showed that the proportion of excess men aged 20–49 as compared to women of the same age range will be higher under the current policy unchanged scenario than that under the two-child policy by 7.6 %, 46.7 %, and 102.4 % in 2030, 2050, and 2080. The excess of men aged 20–49 under the two-child policy scenario will reach a substantially lower peak value in 2030–2040 and then quickly decline and become close to normal in 2050 and continue to decline after 2050.

Zeng (2007) also applied the ProFamy extended cohort-component method to simulate the percentage of never-married middle-aged men who cannot find a wife under different fertility policies. The results show that, under the current policy unchanged scenario, the percentage of never-married middle-aged men will increase to about 10.5 % in 2050 and remain at a high level afterwards. This large percentage of middle-aged men who cannot find a wife and who are mostly less educated with low income would cause serious social problems under the current policy unchanged scenario. But, under the two-child policy scenario, even with exactly the same high SRB assumption as that under the current policy unchanged scenario, the percentage of never-married middle-aged men who cannot find a wife of any age will be much lower (see Fig. 14.7). This is demographically interpretable because the two-child policy scenario will result in larger young generations including young girls in the marriage pool in the future; thus, middle-aged never-married men could search for a younger wife. However, the current fertility policy unchanged scenario accompanied with a high SRB would result in a very low fertility level and quickly shrinking generation size, and the opportunity for future middle-aged never-married men to search for a young wife would be much smaller than under the two-child policy scenarios.

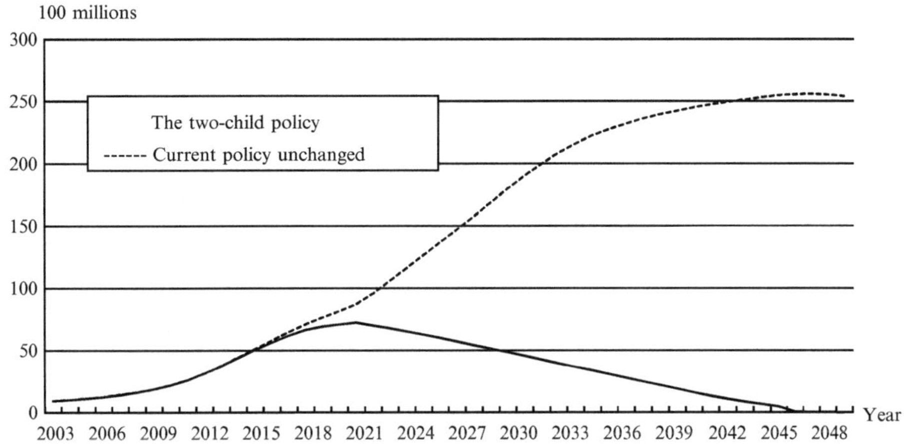

Fig. 14.8 Governmental expenditure (unit: 100 millions yuan) for subsides to be paid to rural elderly couples aged 60+ who had one-child only or two-daughters only, under different fertility policy scenarios

14.4.6 Socioeconomic Costs and Human Capital

In 2003, the Chinese government started to pay a modest compensation annually until death for those rural residents who are aged 60 years or older and have one child or two daughters only. This is a necessary and positive step forward to compensate rural people for the important contributions they made by obeying the state's fertility policy. However, the state's expenditures on this special payment will increase tremendously in the next few decades if the current policy remains unchanged. Based on forecasting by the State Population and Family Planning Commission research group, which took into consideration changes in urbanization, fertility and mortality, Zeng (2007) estimated the cost of compensation for the rural elderly who obey the current policy and have one child only or two daughters only under both the current policy unchanged and the two-child policy scenarios. It turns out that under the two-child policy scenario, the peak year of this special expenditure will be 2021 (which is 17 % lower than that under the current policy unchanged scenario in the same year), and then quickly decline and fall to almost zero in 2050. However, this annual governmental expenditure under the current policy unchanged scenario will be 3.7, 12.7, 38.7 and 4,809 times as high as that under the two-child policy scenario in 2030, 2040, 2045, and 2050, respectively (see Fig. 14.8).

If an only-child dies due to disease, accident, or natural disaster, such as earthquake or extreme flooding, and if the mother's age or health conditions do not allow a new birth, the family becomes permanently childless. When the old father or mother of the deceased only-child dies, the surviving old mother or father becomes permanently childless and widowed. The life quality of the permanently childless old widows, widowers, or couples is often miserable due to psychological damage and loneliness. According to the Chinese 2010 health statistical yearbook, there were more than one million such permanently childless families and the number is increasing by 76,000 every year. The numbers of childless old widows, widowers, and couples under the current policy unchanged scenario or the two-child solely for only-child couples scenario would be dramatically larger than that under the two-child policy. This will definitely cause social instability involving tremendous social, economic and political costs.

The policy TFR (i.e., the TFR if the fertility policy is 100 % implemented) among the ten provinces in the middle and western parts of China is around 1.5; but the actual TFR in seven of these provinces was close to or slightly over 2.0, and three of them substantially exceeded 2.0 (Wang et al. 2004). Obviously, a large number of children are born out of the "birth quota" without official registration. This large number of so-called "under-ground" children who are psychologically and socially discriminated against would be relieved into totally normal status under the two-child policy.

Both the policy TFR and the actual observed TFR after adjusting for birth underreporting in the 15 more developed and urbanized provinces and municipalities in the eastern and middle parts of China are less than 1.5 or slightly over 1.5; among them, the municipalities of Shanghai and Beijing have the lowest actual TFR, substantially lower than 1.0, and Tianjin has an actual TFR that is slightly over 1.0. Current Chinese fertility policy requires the large majority of the urban Chinese couples to have only one child, with a few exceptions such as for minority couples and couples whose first child is disabled, while more than half of the rural couples are allowed to have two children under the 1.5-child policy. On average, the overall average TFRs, after adjusting for birth underreporting in the rural and urban areas in 2010, were 2.01 and 1.24, respectively, according to our and others' analyses based on the most recent census data, as discussed earlier. Such dramatic rural–urban fertility differentials in China are obviously also due to the large differences in rural and urban fertility policies, in addition to the differences in rural/urban levels of socioeconomic development. If China adopts the two-child policy, we expect that the rural fertility will increase by about 7 %, and the urban fertility will increase by about 35 %.

As expected, there is very large gap in education between rural and urban areas in China now and in the future. We did a simple calculation to compare the overall (rural–urban combined) Chinese education level under the current fertility policy unchanged and the two-child policy scenarios, assuming the rural/urban relative education differentials remain the same as in 2010, which is very likely. It turns out that, as compared to the distributions of educational attainment under the two-child policy, the current fertility policy unchanged scenario will increase the number of people with no education and primary schooling by 8.7 %-17.5 % and 8.0 %-11.4 %; middle schooling will remain more or less stable, but high-school and college education will decrease by 3.2 %-4.3 % and 5.6 %-6.8 %, respectively, among the Chinese labor force population aged 18–59 in 2030–2050. Clearly, sustaining the current fertility policy unchanged is not good for China's human capital development because of the negative-selection effects on the overall education level of the Chinese labor force due to much better education level and much lower fertility in urban areas compared to the rural areas.

14.4.7 The "Two-Child Solely for Only-Child Couples" Is a Very Poor Policy Option

Currently, the two-child solely for only-child couples is the most popular option among many Chinese scholars and governmental officers. However, as shown in Figs. 14.1, 14.2, 14.5, and 14.6 and discussed earlier, although the two-child solely for only-child couple policy would be somewhat better than the current policy unchanged in facing the challenges of population aging and labor force shrinking, it would be substantially worse compared to the two-child policy. Because only-child couples are rather rare in rural areas, under the two-child solely for only-child couples policy the 1.5-child policy will continue to be the major component in most rural areas. Consequently, the unfortunate effects on the sex ratio at birth would continue to exist, although the effects may be smaller than under the current policy unchanged. The serious problems of high socioeconomic costs and human capital loss discussed above would continue to exist under the two-child solely for only-child couples policy scenario, because it would continue to produce many only-child high risk families, although the degree of the seriousness of the problems may be reduced to some extent.

Furthermore, the two-child solely for only-child couples policy would cause new social problems. It would create a new form of social unfairness. Couples where both or one party is an only-child are allowed to have two children, which implies that their individual total dependency ratio is 3 [=(4 old parents + 2 children)/2] or 2.5 [=(3 old parents + 2 children)/2]. Couples where neither party is an only-child are allowed to have only one child, which implies that their individual total dependency ratio is 1.5 [=(2 old parents + 1 child)/2]. This is unfair for couples with at least one party being an only-child, because their burden for caring for four or three old parents and two children is substantially higher than for couples where both parties are not only-children. Thus, only-child couples may ask the government for additional compensation based on the following rationale. Their parents made greater contributions than the non-only-child couples' parents by having only one child when the nation needed to control for excess population growth. Now they, the next generation, are again making greater contributions than others by having two children when the country needs to have more children for the future labor supply and bearing a much higher total dependency burden. How will the government answer such a request? At the same time, the non-only-child couples who are not allowed to have the second child are not happy with the governmental policy for birth restriction, so the government stands at the opposite position against all couples.

The two-child solely for only-child couples policy may cause another new social problem in marriage formulation. For example, a non-only-child may deeply love another non-only-child. Since they would not be allowed to have two children after marriage, their parents may strongly oppose the marriage if they strongly prefer to have two grandchildren. This situation may cause miserable events and social conflict. In sum, we believe that the two-child solely for only-child couples policy, which is currently favored by most policy makers and many scholars in China, is a very poor policy option.

14.5 Challenges and Opportunities Associated with Retirement Age and Rural Old Age Insurance Program

As discussed in Sect. 7.3.2 in Chap. 7, the compulsory age at retirement in China has been 60 for men and 52.2 for women, with some variation in actual age at retirement. In Sect. 14.4, when we presented and discussed the population and household aging trends in China, we used age 65 as the threshold for elderly, following the international standard. However, if we use the current Chinese compulsory retirement age 60 for men and 52.2 for women as the threshold, the proportion of those over compulsory retirement age among the total population in China in 2030 would be as high as 30.3 % or 31.6 % under the two-child policy or current policy unchanged; projections diverge to as high as 38.6 % or 43.0 % in 2050 under the two-child and the current policy unchanged scenarios, respectively. Such sky high proportions of those over the compulsory retirement age are of course not acceptable for the country's sustainable development. However, many policy makers, scholars, and ordinary people in China argue that delaying the age at retirement will reduce job opportunities for younger individuals, and thus they strongly oppose increasing the retirement age. On the other hand, the current exceptionally low retirement age offers remarkable opportunities for China to enact policy to face the serious challenges of population aging by gradually increasing the Chinese retirement age.

As presented and discussed in Sect. 7.3.3 of Chap. 7, to illustrate how a gradual increase in retirement age would offer opportunities for China to face its serious aging challenges, we applied a simple method for projecting the annual pension deficit rate based on a population projection and usually available data of a few pension program parameters developed by Zeng (2011). The analysis demonstrates that if the average age at retirement gradually and linearly increases from the current very low level to age 65 for both men and women in 2050, the annual pension deficit rate would be largely reduced or eliminated under various possible demographic regimes. With everything else being equal, the annual pension deficit rate in the scenario of medium fertility (associated with the two-child policy) would be much lower than that under the low fertility scenario (associated with the current fertility policy unchanged) after 2030 (see Fig. 7.1 in Chap. 7). This would be an additional reason, on top of the many other reasons discussed earlier, why China needs to transition towards a two-child policy as soon as possible.

Although fertility in rural areas in China is substantially higher than in urban areas, aging problems will be much more serious in rural areas because of the continuing massive rural–urban migration of young people. The social and cultural traditions and the reality of old parents relying on son(s) for financial support make it extremely important for families to have a son, especially in rural areas, where old age insurance is still much weaker compared to urban areas. The old Chinese saying "*Yang Er Fang Lao* (Having a Son for Old Age Support)" clearly explains why rural residents who have no pension strongly want to have at least one son, which is the basic cause of the high sex ratio at birth in China.

It is clear that establishment of an old age insurance program in rural China would not only be very useful for effectively responding to the serious challenges of population and household aging, but also for largely reducing the necessity of having at least one son for old age care. As consequences, the son-preference induced sex-selective abortions and the dangerous trend of an increasing sex ratio at birth may be reversed.

The Chinese rural old age insurance program, in which individuals' premium contributions were subsidized by the local collective funds and government, was first launched as an experimental project in Shandong province in the early 1990s and quickly spread to all over the country. By the end of 1995, 61.2 million peasants aged 20–60 participated in the program, and the participation rate among the population aged 20–60 was 14.2 %; there were about 80 million participants and about 890,000 peasants aged 60+ starting to receive monthly payments from the old age insurance program.[4] This encouraging program unfortunately stagnated and shrank from 1999 to 2008. By the end of 2004, there were about 53.9 million participants, a drop of 32.6 % from 1999; about 10 % of the counties completely discontinued the rural old age insurance program. The major cause of this situation was that some important policymakers and scholars argued that the pension deficit problems in urban China would be very serious due to rapid population aging and there would be no resources left to devote to the pension program in rural areas; this was wrong, as we discussed in the Sect. 7.3 of Chap. 7.

The most recent development in rural old age insurance is that in September 2009, the Chinese State Council announced that 10 % of all counties in China will launch the "New Rural Old Age Insurance (NROAI)" program before the end of 2009. Since then, the NROAI has developed rapidly and it was reported by the news media that the NROAI had almost universally covered all rural residents by the end of 2012. As compared to the previous rural old age insurance program, the governmental subsidy and back-up for the NROAI substantially increased; it is explicitly stated that the premium will be jointly paid by the individuals and local and central governments, and the state will ensure the basic and minimum income level for all

[4] Data obtained from Ministry of Civil Affairs, see Zeng (2002).

elderly who participate in the program. With this new and promising policy guidance, actions are being taken and the new rural old age insurance program is expected to further develop, as it is not yet mature.

14.6 Policy Recommendations

Our projections based on the recent census and other data have demonstrated that the Chinese population will age rapidly and on a large scale, especially the oldest-old aged 80+. The crucial question is what policy actions are optimal and feasible to deal with the challenges and utilize new opportunities given the Chinese demographic and socioeconomic context. The demographic analyses and simulations we have reported here indicate that China would be able to successfully deal with the impending challenges, if three major policy actions are taken.

14.6.1 Transfer to the Two-Child Policy with Adequate Spacing Fertility Policy as Soon as Possible

Analyses in this chapter and many other studies all indicate that the current fertility policy in China needs to be modified as soon as possible in order to avoid serious social and economic problems including too much aging in the future. To achieve this goal, we suggest that China needs to employ a series of economic and social incentives to promote voluntary late-childbearing and realize a smooth policy transition to a universal two-child policy. The programs should emphasize the following two principles:

1. Substantially increasing governmental investment in education, especially in rural areas. The aim of these programs would be to formulate social norms, contexts, and conditions in which it is in the interest of individuals, families, and society for young people to have education and technical skills first, and family formation second. Such programs will not only help China to encourage young people to postpone their marriage and childbearing for a soft-landing to the universal two-child policy, but also benefit China tremendously in increasing human capital and sustaining rapid economic development.
2. Widely disseminating scientific knowledge of the familial benefits of late childbearing. It is crucially important, for example, to inform the public that various previous studies have shown that the health and survival of children born to mothers aged 25–34 is significantly better than those who were born to mothers younger than 20 years of age, especially in developing countries (e.g., Card 1981; Koniak-Griffin and Turner-Pluta 2001; Levine et al. 2001; Nortman

1974). Media programs should also let the public know that later childbearing with longer spacing would reduce the length of time in which two children share limited family resources and help the parents to better invest for their children's education (e.g., Powell and Steelman 1995). Larger age differences between siblings are also useful so that the elder child can help the younger child (e.g., Bank and Kahn 1975).

More specifically, we propose that, while fully integrating the principles of the socioeconomic incentive programs discussed above, the Chinese local family planning offices should collect the data on:

(a) The number of women who would like to have a new baby (1st or 2nd birth) within the next few years after the two-child policy is implemented, based on simple telephone or questionnaire survey.
(b) The maximum number of 1st-year new students the local schools may recruit per year.

If (a) is significantly larger than (b), the family planning office needs to set up a starting threshold age (or age range) and encourage women whose age is below the threshold age to wait for one, two, or a few years to have a 2nd birth. All women who postpone their 2nd birth until the threshold age should be rewarded with sufficiently high economic and social incentives to avoid a new baby boom caused by too many second births by women of different ages in the same short period of time. We recommend that the government gradually deregulate the two-child family size and fertility timing policies to allow people to have more freedom to make their own choices. Such deregulation could be implemented in the most economically developed areas first, and then spread smoothly to other areas. For example, if (a) is not significantly larger than (b) in the advanced areas where most couples at reproductive ages prefer later childbearing and fewer births, no late-childbearing campaign is needed.

We expect that, after 2030, China should allow all Chinese rural and urban citizens to freely choose family size and fertility timing. We believe that such an expectation is highly possible, given the rapid socioeconomic development and changes in fertility attitudes occurring in China. The fertility level after 2030, when restrictions on the number and timing of births are all eliminated, can still be expected to remain at the relatively low level projected under the two-child policy scenario in Chinese rural and urban areas.

How feasible is the above policy recommendation? Some people may worry that the above-proposed policy transition would cause couples to thwart the limitations of two-child and adequate spacing to have three or more children without spacing. We believe that such worry is unnecessary, based on the empirical evidence presented below.

The empirical data have clearly shown that fertility preferences in China have changed remarkably since the 1980s. This is confirmed by numerous studies based on data from various national and regional surveys (e.g., Zheng 2004; Feng and Zhang 2002; Li and Zhang 2001; Li 2003; Lin 2004; Shi 2001; Xie 2000; Zhou et al. 2000). Various studies have shown that, for most areas of contemporary China, the preferred number of children per couple is two, or even one in large cities, or none among some young people. In the Eastern coastal areas, the preferred numbers of children in rural and urban areas are rather close to each other. Even in the less developed Western areas of China, the proportion of peasants who prefer three or more children is not high at all (see Zheng 2004; Feng and Zhang 2002 for details).

The policy option of relaxing or eliminating the birth spacing policy first and gradually transitioning to a two-child policy later advocated by a considerable number of Chinese policy makers and scholars is not a rational choice for two main reasons. First, it may very likely create obstacles for the transfer to the two-child policy. There are currently more than 100 million Chinese couples of reproductive age who have had only one child. A majority of these one-child couples, especially in rural areas which comprise about half of the total population, would like to have a second child as soon as possible if the "one-child" policy is relaxed to a two-child policy without a requirement of birth spacing. This is because they would likely be afraid that the policy could be reversed if the government saw many second births occurring and thus would like to catch the opportunity as soon as possible. This could actually prevent or delay the decision to move to a two-child policy, because the government does not want to see the excess population growth caused by birth heaping in the rural areas. Second, relaxing or eliminating the birth spacing policy first but preventing or delaying the decision-making of the two-child policy is unfair to the majority of the Chinese couples who are allowed to have one child only. The sacrifices among the 63 % of couples who are allowed to have only one child, such as a higher risk of induced abortion due to contraceptive failure and less support from children for parents at old ages, are much larger than that among the 37 % of the couples who are allowed to have two or more children with spacing. While the nation cannot remove the policies of birth quantum and timing simultaneously to avoid a new baby boom, it is fair to relax the limitations for those who have sacrificed the most first, i.e., allowing the one-child-only couples to have a second birth first and then giving more freedom of fertility timing to couples who are allowed to have a second birth. However, the policy option of relaxing or eliminating the birth spacing policy first and gradually transitioning to a two-child policy later on does the policy adjustment in the wrong order, and thus creates social unfairness and obstacles for timely transfer to a two-child policy. Clearly, this policy option is a poor choice.

14.6.2 Gradually Increase Age at Retirement

Currently, the average age at retirement for both men and women in China is 56 years old, while the Chinese healthy life expectancy at birth is 66 years old (WHO 2010). These statistics imply that, on average, most Chinese citizens spend about 10 years in good health after retirement. This post-retirement period with good health will continue to lengthen in the future if retirement ages remain unchanged while the trends of reaching old age in good health observed in dozens of countries (including China) in the past decades is likely to continue (Vaupel 2010). At the same time,

Chinese social security systems are facing serious challenges of pension fund deficits due to population aging. Therefore, gradually increasing the compulsory retirement age and encouraging citizens to postpone their retirement is a logical and reasonable policy action. As summarized earlier, demographic analyses have demonstrated that if the Chinese official compulsory retirement age gradually increases from 60 years old for men and 52.2 year old for women at the present time to age 65 for both men and women in 2050, the annual pension deficit would be largely reduced or totally eliminated under various possible demographic regimes (see Sect. 7.3.3.1 of Chap. 7; also ref. to Zeng 2011).

However, allowing old citizens to work more years may negatively affect the employment opportunities for younger workers. Consequently, a new policy of allowing everyone to work fewer hours per week and longer years over the whole life course would be a wise choice (Vaupel 2010). As we discussed in Chap. 7, this policy may not only play a critical role in facing the serious challenges of population aging, but also be useful in stimulating service industries of tourism and other leisure/social activities which will create more jobs and develop the economy.

14.6.3 Further Develop the Rural Old Age Insurance Program

The Chinese government also has a responsibility to promote, lead, and manage the old age insurance program for all citizens, including peasants. This is extremely important in responding to the challenges of rapid aging which will be even more serious in rural areas in the coming decades. Giving state resources (such as tax exemption) only to the urban old age insurance program and overlooking rural people is unfair, since both rural and urban residents make contributions to government revenues and state welfare. Leaving rural old age support entirely or mainly to families will not be practical in the coming decades because of the current low fertility and serious aging in many rural areas, and the continuing rural to urban migration of young people. Less developed rural old age insurance will also increase the dangerous trend of a rising sex ratio at birth, if farmers have no other way than relying on sons for old age care and thus choose prenatal sex determination and abort female fetuses to ensure having son(s).

14.7 Concluding Remarks

This chapter presents a comparative analysis of the demographic and socioeconomic implications of a few alternative options for fertility policy transitions in twenty-first century China. The results evidently show that the two-child with encouragement of adequate spacing policy option is an optimistic and feasible strategy for China to adopt to sustain socioeconomic development in the future. As compared to retaining the current fertility policy and the other options, the "two-child with encouragement

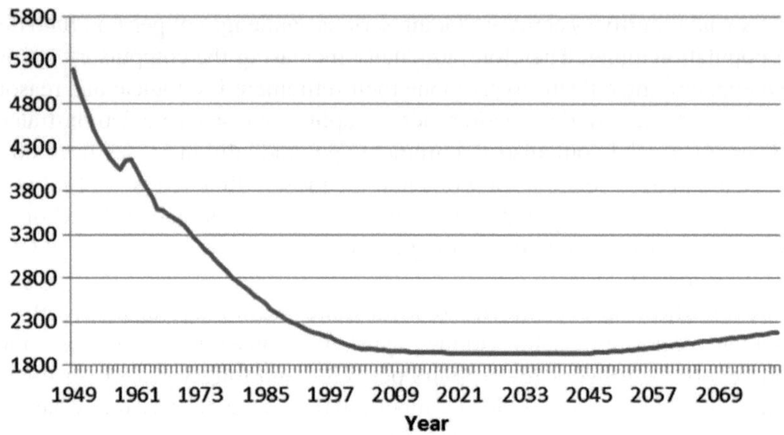

Fig. 14.9 Average water resource per capita, 1949–2080. Note: the unit of water resource (y axis) is "cubic meter" (Data source: The average water resource per capita, 1949–2033 are taken from "National Population and Development Strategies Report" (p61, published by China Population Press, 2007). We assumed that the total amount of the water resource after 2033 remains the same as that in 2033, and estimated the average water resource per capita, 2034–2080 using our projected 2034–2080 total population size under the two-child policy scenario)

of adequate spacing" soft-landing policy would create much better demographic conditions and socioeconomic implications in the future, with respect to proportions of elderly and those elderly who live alone, labor force resources, pension deficit rates, sex ratio at birth, the marriage squeeze, and socioeconomic costs. We also believe that it is highly feasible to implement the "two-child with encouragement of adequate spacing" policy without causing more un-planning births of third or higher order or new coercive events.

Note that many policy makers, scholars, and much of the public do not understand why China needs to transition to a two-child policy as soon as possible, because they are still deeply influenced by the previous propaganda that larger population size produces severe threats to the environment, natural resources, and economic growth. However, such propaganda is out-of-date in the new era today. It is true that, for example, the average water resource per capita and average arable land per capita have declined over time with population increase; in 2011 the water and land resources per capita were reduced by 62.4 % and 50.7 % compared with 1949 and reduced by 30.4 % and 35.7 % compared with 1979, and the decline substantially levels off after 2000 (see Figs. 14.9 and 14.10). However, the Chinese average GDP per capita (with compatible price in US dollars) was $5,439 in 2011, 19.7 times as high as that in 1979 ($276) and 112.4 times as high as that in 1949 ($48.4). Thus, the standard of living of Chinese people has dramatically improved compared to 30 or 60 years ago. This was mainly due to the market economic reform and opening the door to the world, although it also includes the positive effects of the family planning program in largely reducing the over-growth rate of the Chinese population observed in the 1950s–1980s. Under the two-child with encouragement of adequate spacing

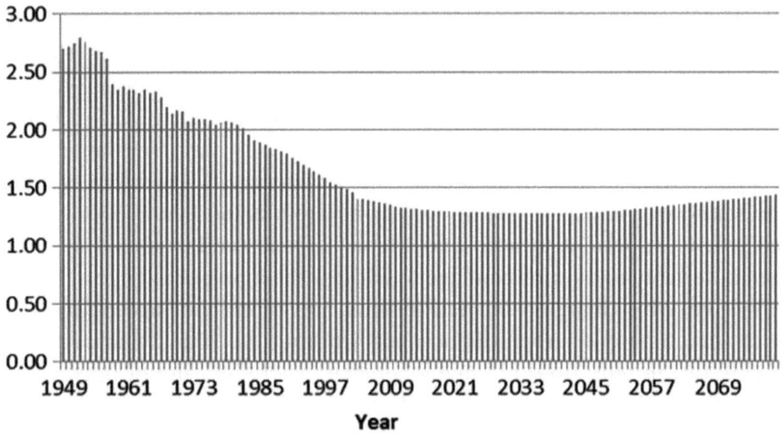

Fig. 14.10 Average arable land per capita,1949–2080. Note: the unit of water resource (y axis) is "Mu" (the Chinese unit of arable land, that is equal to 0.0667 ha) (Data source: The average arable land per capita, 1949–2033 are taken from "National Population and Development Strategies Report" (p61, published by China Population Press, 2007). We assumed that the total amount of the arable land after 2033 remains the same as that in 2033, and estimated the average arable land per capita, 2034–2080 using our projected 2034–2080 total population size under the two-child policy scenario)

policy, the Chinese average water resource per capita and average arable land per capita would reach the lowest value around 2029, reduced by about 2 % and 6 % compared to 2011, respectively, and then start to gradually increase as the total population size deceases after 2029 (see Figs. 14.9 and 14.10).

On the other hand, recent research inside and outside of China has demonstrated that the massive rural-to-urban migration in recent years has resulted in the transition of poorly cultivated land back to forestry or grassplot, which substantially improved rural areas'ecological and environmental conditions (Li et al. 2013). Therefore, a smooth transition to a two-child policy, which may result in somewhat higher population growth than under the current fertility policy unchanged scenario, will not present threatening pressures on environmental protection and natural resource utilities.

The implication of the demographic analyses reported in this chapter is that China needs to relax its one-child policy now to fully utilize the demographic dividends which are to be gained in about 15 years, up to 2025–2030. During this period, the total dependency ratio of the elderly and children will continue to gradually decline because the moderate increase in the elderly dependency ratio will be compensated for by the relatively more substantial decrease in the child dependency ratio induced by an increase in the working age population. Furthermore, the two-child policy would produce significantly more consumption and job opportunities associated with childbearing and childrearing, which would enable China to fully utilize its demographic dividend of rich labor resources in the next 20 years. These additional new babies will enter the labor force about 20 years from now and they will substantially help China to resolve the problems of accelerated population aging and potential labor shortages after the demographic dividend is gone 20 years later. However, if China postpones its transition to a two-child policy for another 5–15 years, those additional new babies will still be in childhood around the years 2025–2030; at that time, more children plus accelerated aging and a quickly shrinking labor force will seriously hurt China's future of socioeconomic development. Therefore, it is time for China to act now to revise its fertility policy before it is too late.

Finally, it is important to note that China has substantial potential, rooms for increases in fertility and retirement age to address the aging challenges. This can be done effectively by changing current fertility and retirement age policies, which largely restricted people's reproduction and working life span. Moreover, the new and quickly developing rural old age insurance programs can not only reduce the more serious problems of aging in rural areas, but also create a huge amount of capital by collecting premiums from hundreds of millions of new program participants; this is equivalent to a large collective capital accumulation and is useful for further economic development. In sum, China would be able to successfully deal with the serious challenges of population and household aging if it utilizes these opportunities with prompt and effective policy actions.

Chapter 15
Household Housing Demand Projections for Hebei Province of China

15.1 Introduction

The housing supply is one indicator of the quality of life. Real estate is an important industry that is strongly correlated with economic development. Housing is not only associated with advancements in people's living standards, but also affects the sustained and healthy development of the economy and society. Because households are the basis of residential housing demand, projections of how changes in household and population size and structure may affect future housing demand can provide an important reference in scientific decision-making for all levels of government, as well as promote the healthy development of the real estate industry and improve people's living conditions.

China currently has the largest real estate market in the world. Real estate investment in China hit $304 billion in 2012 (The Hindu 2013), and in the same year the national homeownership rate reached almost 90 % (Jiang 2012). Because there are very few studies concerning household housing demand in China (Jiang and Ren 2005; Yang and Xu 2011; Zeng, Li et al. 2013), the study described in this chapter focuses on this important but still relatively weak research field. We choose Hebei Province as our study population for this chapter, a province for which we obtained access to the 10 % sample micro-data of the 2010 census through a collaborative research project.[1] Hebei Province is in northeast China with a population of 71.85 million people in 2010 and median/representative levels of socio-economic development and urbanization (Starmass International 2009). We believe that a study in this representative province could provide important information for

[1] The micro-data for the 2010 China census, which are needed for the household housing demand projections, are currently not available for scholars and the public to use; thus, we are not able to conduct household housing demand projections for China as a whole at present time.

Y. Zeng et al., *Household and Living Arrangement Projections*,

The Springer Series on Demographic Methods and Population Analysis 36,
DOI 10.1007/978-90-481-8906-9_15,

understanding the general trends and patterns of household housing demand not only for Hebei, but also for the rest of China.

Note that this chapter explores how changes in household and population structure and size may affect future housing demand based on our most recent research (Zeng et al. 2013), rather than real estate forecasting of housing demand per se (Zeng et al. 2013). We briefly introduce the method, data sources, and input estimates in the next section. In the third section we present and describe the main results of our projections and analysis. We then discuss some related policy considerations in the fourth section. The last section concludes with a statement on the unique contributions and limitations of this study, as well as some perspectives on future research.

15.2 Method, Data, and Input Parameter Estimates

15.2.1 The Method

Various approaches have been used to project future housing demand. Some studies set up demographic and economic scenarios (e.g., Berson et al. 2006); some use econometric modeling (e.g., Green and Hendershott 1996; Meen 1998; Ng et al. 2008); some apply a cohort method (e.g., Myers et al. 2002; Pendall et al. 2012; Pitkin and Myers 1994); some rely on linear extrapolation combined with expert opinions (e.g., Forrest and Leather 1998); and some housing demand forecasts have been based on household projections (e.g., Berson et al. 2006; Nishioka et al. 2011). Previous approaches to housing consumption forecasts have integrated various economic and policy factors into the model, but have not included forecasts of household types and sizes by age, gender, and rural/urban residence. Variations in these household structure characteristics have large effects on housing consumption, especially in countries such as China where family households have been changing substantially over a relatively short period of time. Furthermore, although future socioeconomic and policy factors are included as covariates in some housing forecast models, it is extremely difficult to forecast them into future years with reasonable accuracy (Hendershott and Weicher 2002), especially for societies that are changing and developing quickly. Thus, these prior models may not be an optimal choice for our present study.

The main objective of the present study is to explore how changes in household structure may affect future housing demand, rather than to forecast real estate developments. A promising way to create more practical and reasonable forecasts is to build housing demand forecasts from household projections, because the two are closely associated (Gan 2010; Kennett and Chan 2011). In this chapter, we apply the ProFamy extended cohort-component macro model for household housing demand projections (see Chaps. 2, 3, and 4 for details).

15.2.2 Data Sources, Estimates, and Parameter Assumptions

Data on the population of Hebei Province, classified by rural/urban residence, single year of age, sex, marital status, number of co-residing children and parents, and whether living in a private versus institutional household at the projection baseline year, were extracted from the micro-data file of the 2010 census of Hebei Province. The micro-data file provided by the Hebei Provincial Statistical Bureau consists of the de-identified individual census records for seven million persons, or 10 % of the total population in Hebei.

We estimated rural/urban-single age-sex-specific mortality rates based on the micro-data files of the 2010 census, which collected detailed data on household members who died during the 12 months prior to the standard census time. Using the 2010 census micro-data file, we estimated rural/urban-single age-sex-specific occurrence/exposure (o/e) rates of first marriage and fertility by parity, age-sex-specific net migration frequencies between rural and urban areas within the province, and age-sex-specific external net migration frequencies. These estimates are straightforward based on census counts of year and month of first marriage from all adults over age 15, parity and month of births from all women aged 15–50 who gave birth within the 12 months prior to the standard census time, and residence locations at census time and 1 year and 5 years prior.

The model standard schedules of age-sex-specific o/e rates of divorce and remarriages were estimated using data from the Chinese In-depth Fertility Survey and the Chinese Longitudinal Healthy Longevity Survey, both of which included Hebei Province. The rural/urban age-sex-specific probabilities of children leaving their parental homes were estimated based on 2000 and 2010 census data in Hebei and the iterated interpolating method within cohorts proposed by Coale (1985) and further extended by Stupp (1988).

We estimated the rural/urban-specific general marriage rate and divorce rate based on age-sex-specific standard model schedules of marriage and divorce rates, 2010 census data, and the total number of marriages and divorces in 2010 published by the Bureau of Civil Affairs of Hebei Province. Based on the mortality rates collected in the 1990, 2000, and 2010 censuses and the 1 % population survey data collected in 1995 and 2005, stratified by urban and rural residence, age, and gender, we estimated the rural/urban and gender-specific average life expectancy at birth in 1990–2010 and extrapolated it to future years up to 2050 (see Table 15.1).

According to the 2010 population census data, the observed total fertility rate in 2010 was 1.3 in Hebei Province. This rate was adjusted for under-reporting as follows. Based on the 2010 census data for ages 10–19 and the "backward forecasting" method (using the estimated sex-age-specific mortality rates as described above and adjusting for the estimated migrations in and out of Hebei Province), we estimated that the average under-reporting rate for ages 0–9 was 7.6 % in Hebei in the 2000 census, which will be used as a reference for the under-reporting rate at ages 0–9 in the 2010 census. We used the 2000 population census as a source of basic data for women of reproductive age and tried various possible

Table 15.1 Estimated and projected main demographic parameters, Hebei Province, China

	Rural			Urban			Rural–urban combined		
Demographic parameters	2010	2030	2050	2010	2030	2050	2010	2030	2050
Total fertility rate	2.01	2.10	2.13	1.32	1.70	1.72	1.71	1.85	1.82
Male life expectancy at age 0 (years)	70.9	74.3	77.4	74.2	75.7	79.3	72.4	75.2	78.8
Female life expectancy at age 0 (years)	74.7	78.3	81.3	78.3	82.3	85.0	76.3	80.8	84.1
General marriage rate (per thousand)	99.0	99.0	99.0	89.8	89.8	89.8	95.0	93.2	92.1
General divorce rate (per thousand)	2.8	2.8	2.8	5.4	5.4	5.4	3.9	4.4	4.8
Mean age at first birth	25.20	26.50	26.50	26.56	27.75	27.75	25.80	27.29	27.44
% of urban population among total population							44 %	63 %	75 %

inputs of total fertility rate (TFR) in 2000–2010 in Hebei to perform "forward-forecasting" to estimate the number of children aged 0–9 in 2010. Compared to the population aged 0–9 in the 2010 census, adjusted for the under-reporting rate of ages 0–9 in 2010, we estimated that the total fertility rate in Hebei Province in 2010, adjusted for the under-reporting of births, was 1.7.[2] This estimate is highly consistent with the estimates by scholars and the Population and Family Planning Committee of Hebei Province. Considering that family planning policies will be gradually relaxed to some extent, the total fertility rate is estimated to be 1.85 in 2030. After that, the TFR is assumed to slowly decline to 1.82 in 2050 (see Table 15.1).

Following a simple trend extrapolation approach based on time series data of proportion of urban population from the censuses and annual surveys of population changes, we estimated/projected that the proportion of urban residents among the total population in Hebei will be 63 % and 75 % in the years 2030 and 2050, respectively (see Table 15.1). Based on the trend extrapolation method and data from the censuses conducted in 1990, 2000, and 2010, we estimated/assumed the gender-specific number of net-immigrants from other provinces (mainly from other poorer provinces in the middle and western parts of the country). Our estimates and projection reveal a gradual increase in net immigration, from a total of 230,410 persons in 2010 to 260,870 and 288,970 persons in 2030 and 2050, respectively, for the two sexes combined; we assumed the same age-sex distributions of net migrants in the future years as those observed in 2010.

[2] Our estimated under-reporting rate of 7.6 % was an average for ages 0–9. However, the underreporting rate of new births is much higher than other childhood ages, especially ages 6–9, when most of the under-reported children were registered for school. Our estimate of the TFR as 1.7 in 2010 in Hebei province implies an under-reporting rate of new births of 24.1 %.

From the Hebei Province 2010 census 10 % sample micro data, we estimated rural/urban-age-sex-household type/size-specific homeownership rates and home-rental rates.[3] The status-specific homeownership rate is defined as the proportion of households that own a housing unit (house or apartment) among private households with the same status. The status-specific home-rental rate is defined as the proportion of households that rent a housing unit among private households with the same status.[4] The sum of the homeownership rate and home-renter rate is equal to one. The homeownership rates and home-renter rates were classified by rural/urban, household type/size, and age of household reference person, with exactly the same categorizations as those for households. The homeownership rates and home-renter rates were further divided into three components for the three types of housing units: housing units with 1–2 rooms, housing units with 3–4 rooms, and housing units with 5 or more rooms.[5] The homeownership rates and home-renter rates were assumed to be constant in future years, and they were multiplied by the corresponding rural/urban-householder age-household type/size-specific numbers of households forecasted by the ProFamy extended cohort-component approach to yield projected future household housing demands.

Note that we employed the common approach of holding some of the current rates constant throughout the projection horizon given that scientific theories and past history do not provide a reliable basis for predicting how those rates will change (e.g., Day 1996; Smith et al. 2001; Treadway 1997). In addition, holding the homeownership and rental rates constant allows us to focus on time-varying demographic rates, which serves well the present study's purpose of exploring how changes in household structure may affect future housing demand.

---

[3] We did not include income in our estimates of homeownership rates and home-rental rates for three reasons. First, based on many others' and our own research, we do not trust the accuracy of self-reported income in the Chinese census data. Second, even assuming we might obtain reasonably accurate estimates of income categories, it would be extremely hard to forecast future changes in income for various rural and urban household types/sizes and age groups. The accuracy of the forecasts relies heavily on the validity of assumptions regarding future time paths of the covariates and parameters included in the forecasting model. Erroneous assumptions about very uncertain future years' covariates and parameters included in the model can quickly lead to forecasts that are far off the mark (e.g., Lee and Tuljapurkar 2001). Third, we have included in our projection the rural/urban dimension, which captures the income level to a considerable extent, and meets our needs in the present study.

[4] We investigate housing demand trends of private households in this chapter; housing demand for institutionalized persons is out of the scope of this chapter and therefore excluded from the present study.

[5] Unlike the U.S. census, which collects data on the number of bedrooms in housing units, the Chinese census collects data on the number of rooms in the housing units without distinguishing between bedrooms, living room, dining room, or storage rooms.

15.3 Results and Discussion

15.3.1 *A Brief Outline of the Current Household Housing Situation*

Based on the 2010 census' 10 % micro sample data of Hebei Province, the summary indices listed in Tables 15.2 and 15.3 indicate that housing units by number of rooms and ownership or rental status are closely associated with household size, household type, rural/urban residence, and age of the household reference person. For example, Table 15.3 shows that the larger the household and the more generations living in a household, the more likely residents are living in a housing unit with larger number of rooms. The average number of rooms per household increases with increasing household size and number of generations, but the average number of rooms per person decreases substantially with increasing household size and number of generations. As compared to middle-aged (aged 35–64) and young adults (aged <35), the elderly (aged 65+) have a higher homeownership rate and are more likely to have smaller housing units with 1–2 rooms (see Table 15.3). It is interesting to note that the homeownership rate among rural residents is 93.2 %, in contrast to 98.8 % among urban residents (see Table 15.3). Such rural/urban differentials may be attributed to three main factors. First, rural-to-urban migrants who most likely could not afford to buy a housing unit are likely to rent a cheap room or apartment in the villages surrounding the cities or towns, and thus be counted as rural rental housing units in the census. Second, the fast-developing township and village industries in rural China attracted many migrant workers who live in cheap rental housing units in the rural areas. Third, lower income and higher poverty rates in rural areas may result in more poor rural residents who cannot afford to build or buy their own housing.

Table 15.2 Census-observed percentage distributions of household housing units (owned- and rental combined) by number of rooms and household types/sizes in 2010, Hebei Province, China

	Household housing units by number of rooms						Average no. rooms per household	Average no. rooms per person
	1-room	2-room	3-room	4-room	5+ room	Total		
1-person	18.3	34.2	25.5	11.5	10.5	100	2.74	2.74
2-person	7.9	36.1	28.0	13.9	14.2	100	3.08	1.54
3-person	3.9	35.5	29.0	14.7	16.9	100	3.29	1.10
4-person	1.8	23.7	25.8	20.2	28.5	100	3.93	0.98
5-person	0.5	16.8	24.8	19.8	38.1	100	4.44	0.89
6+ person	0.3	4.8	14.0	17.3	63.5	100	6.36	0.79
1-generation	28.4	31.3	21.9	9.4	9.0	100	2.49	1.54
2-generation	8.9	29.8	29.3	15.0	17.0	100	3.31	0.99
3+ generation	1.3	10.1	22.1	22.2	44.3	100	3.50	0.70

Table 15.3 Census-observed percentage distributions of household owned- and rental housing units by number of rooms, rural/urban residence, age groups of the household reference persons in 2010, Hebei Province, China

	Owned-housing units				Rental housing units				
	1–2 room	3–4 room	5+ room	Sub-total	1–2 room	3–4 room	5+ room	Sub-total	Grand total
Total	34.6	42.2	18.9	95.7	3.4	0.9	0.1	4.3	100
Rural	37.8	42.0	13.4	93.2	5.4	1.3	0.1	6.8	100
Urban	30.5	42.6	25.8	98.8	0.8	0.3	0.1	1.2	100
Ages <35	36.4	40.6	16.6	93.5	5.2	1.2	0.1	6.5	100
Ages 35–64	32.7	43.1	20.5	96.3	2.8	0.8	0.1	3.7	100
Ages 65+	43.7	40.7	13.6	97.9	1.6	0.4	0.1	2.1	100

15.3.2 General Trends of Household and Population Dynamics

Our projections show that the total population in Hebei will reach 74.4 million by 2015 and will peak at 77.1 million by 2033. After that, total population will gradually decline. Over the next 40 years, the average household size in Hebei will show a declining trend associated with a soaring increase in the proportion of one-person households, and a decline in two-generation and three-generation family households.

Single-parent households with children will increase substantially for rural and urban areas combined in the next 40 years compared to 2010 (see Table 15.4). Note that we did not assume an increasing trend in divorce rates for either rural or urban areas in Hebei Province in the next 40 years (see Table 15.1). So why would single-parent with children households be expected to grow substantially? One factor may be structural changes due to rapid urbanization, given that the divorce rate in urban areas has been about twice as high as in rural areas in the recent past, and is assumed to remain so in the future. A second factor pertains to the demographic theory of family household momentum, which was first proposed and empirically verified in Zeng et al. (2006). Rural and urban cohorts who were younger in 2010 experienced and will continue to experience stabilized but higher rates of divorce than cohorts who were older in 2010 and had already completed most of their family life course. Profiles of households in 2010 represent the mixed cumulative life course experiences of younger and older cohorts over the past few decades. Although divorce rates are assumed to remain constant during the period of 2010–2050, the distributions of households will change considerably because older cohorts, who had lower divorce rates, will be replaced by younger cohorts with higher divorce rates. Such family household momentum is similar to the well-known classic population momentum (Keyfitz 1971), in which population size could continue to increase after the fertility was equal to or even below the replacement level. The work by Zeng et al. (2006) provided empirical

Table 15.4 Projected main household and population indices in 2010–2050, Hebei Province, China

Index	2010	2020	2030	2040	2050
Main household indices					
Average household size	3.34	3.02	2.82	2.70	2.66
One-person households (% of total)	9.25	12.54	14.07	15.17	16.06
Two-generation households, both parents (%)	45.72	43.91	40.29	36.55	34.70
Two-generation households, single parent (%)	4.44	6.85	10.00	12.14	13.38
Three-generation households (%)	20.54	15.29	11.22	9.50	9.13
Population and household aging					
Number of elders aged 65+ (millions)	5.92	9.71	13.74	17.15	18.95
Number of oldest-old aged 80+ (millions)	0.99	1.39	2.33	4.06	6.06
% of elderly aged 65+ among total pop	8.24	12.76	17.80	22.23	25.06
% of oldest elderly aged 80+ among total pop	1.37	1.83	3.01	5.26	8.02
% of elderly aged 65+ living alone among total pop	1.00	1.18	1.69	2.44	3.14
Households with householder is aged 65+ (%)	8.32	10.31	21.13	29.40	33.81
% of oldest-old aged 80+ living alone among total pop	0.23	0.32	0.37	0.66	1.20
% of oldest-old 80+ living with spouse only among total pop	0.25	0.31	0.52	0.96	1.68
Children dependency ratio (%)	29	33	30	29	31
Elderly dependency ratio (%)	12	19	28	37	44
Elderly and child dependency ratio (%)	41	53	58	66	75
% of 55 to 64-year-old workers among the total work force of 18–64 years old	16.65	20.73	23.09	23.31	30.09

Note: The figures presented in this table are for rural and urban areas combined

evidence to numerically illustrate family household momentum based on U.S. data, and our present study has reconfirmed it based on Chinese data.

With the acceleration of population aging, the percentages of the elderly aged 65+ and the oldest-old aged 80+ among the total population will show a rapid increase, and the elderly dependency ratio will also increase substantially. The proportion of close-to-elderly workers aged 55–64 among the working-age population aged 18–64 will also sharply increase from 16.7 % in 2010 to 30.1 % in 2050 (see Table 15.4).

15.3.3 General Trends of Owned and Rental Housing Demands

The projected results in Tables 15.5, 15.6, 15.7, and 15.8 and Figs. 15.1, 15.2, 15.3, 15.4, and 15.5 show increases in household housing unit numbers and percentages by different categories in 2015–2050 compared to 2010. These are the outcomes of our investigations of how changes in household and population size and structure may affect future housing demand, given our assumption that the status-specific homeownership rates and rental rates remain constant at the 2010 level. The projection results in Table 15.5 show that owned-housing units in Hebei Province

Table 15.5 Projected number of housing units (in millions) by number of rooms in 2010–2050, Hebei Province, China

	Owned-housing units								Rental housing units							
	Total		1–2 Room		3–4 Room		5+ room		Total		1–2 Room		3–4 Room		5+ room	
Year	# of Units	% inc. comp. 2010	# of Units	% inc. comp. 2010	# of Units	% inc. comp. 2010	# of Units	% inc. comp. 2010	# of Units	% inc. comp. 2010	# of Units	% inc. comp. 2010	# of Units	% inc. comp. 2010	# of Units	% inc. comp. 2010
2010	19.51	0.0	7.05	0.0	8.61	0.0	3.85	0.0	0.88	0.0	0.69	0.0	0.18	0.0	0.0213	0.0
2015	21.94	12.5	7.71	9.4	9.62	11.7	4.61	19.8	0.83	−6.0	0.64	−7.0	0.17	−3.8	0.0228	7.1
2020	23.50	20.4	8.17	15.9	10.27	19.2	5.06	31.4	0.80	−8.8	0.61	−10.7	0.17	−3.9	0.0233	9.4
2025	24.62	26.2	8.54	21.2	10.72	24.5	5.35	39.1	0.78	−12.2	0.59	−14.1	0.16	−6.6	0.0221	3.4
2030	25.49	30.6	8.90	26.2	11.05	28.3	5.54	43.9	0.76	−14.4	0.57	−16.3	0.16	−8.8	0.0214	0.5
2035	26.26	34.6	9.16	30.0	11.30	31.2	5.79	50.5	0.64	−27.4	0.48	−29.6	0.14	−21.3	0.0197	−7.5
2040	26.73	37.0	9.42	33.6	11.45	33.0	5.86	52.1	0.62	−29.9	0.47	−31.8	0.13	−24.6	0.0187	−12.2
2045	26.84	37.6	9.56	35.7	11.45	33.0	5.82	51.2	0.60	−31.8	0.46	−33.5	0.13	−26.9	0.0175	−17.9
2050	26.68	36.8	9.57	35.8	11.37	32.0	5.74	49.0	0.58	−34.7	0.44	−36.2	0.12	−30.3	0.0164	−23.0

Notes: (1) "% inc. comp. 2010" means "% increase compared to 2010"; (2) the estimates presented in this table are for rural and urban areas combined; (3) the unit of "# of Units" (i.e. housing units) is "million"

Table 15.6 Projected number of housing units by rural and urban residence in 2010–2050, Hebei Province, China

	Owned-housing units				Rental housing units			
	Rural		Urban		Rural		Urban	
Year	# of Units (million)	% inc. comp. 2010	# of Units (million)	% inc. comp. 2010	# of Units (million)	% inc. comp. 2010	# of Units (million)	% inc. comp. 2010
2010	10.64	0.0	8.87	0.0	0.78	0.0	0.10	0.0
2015	9.63	−9.6	12.32	38.9	0.65	−16.9	0.18	75.6
2020	9.49	−10.8	14.01	57.9	0.59	−24.2	0.21	106.5
2025	9.25	−13.1	15.37	73.3	0.56	−27.8	0.21	104.5
2030	8.87	−16.7	16.62	87.4	0.54	−30.2	0.21	103.7
2035	7.56	−28.9	18.69	110.8	0.42	−45.8	0.22	110.4
2040	7.30	−31.4	19.43	119.0	0.41	−47.3	0.21	100.9
2045	6.87	−35.5	19.97	125.2	0.40	−48.6	0.20	94.4
2050	6.36	−40.3	20.33	129.2	0.38	−50.7	0.19	85.1

Note: "% inc. comp. 2010" means "% increase compared to 2010"

will substantially increase from 19.5 million in 2010 to 25.5 million and 26.7 million in 2030 and 2050, an increase of 30.6 % and 37.0 % from 2010. Although the demand for owned-housing units with 5 or more rooms is substantially smaller than that for owned-housing units with 1–2 and 3–4 rooms, it grows at a relatively faster speed. The total number of owned-housing units in Hebei Province will peak in 2045; demand for 1–2 room housing units will continue to rise with no declines over the next 40 years, while demand for 3–4 room housing and 5 or more room housing units will peak in 2045 and 2040, respectively (see Table 15.5).

Projected results for the next four decades show a downward trend in demand for rental housing. Total rental housing will be reduced from 0.88 million units in 2010

Table 15.7 Projected number of housing units (in millions) by age of household reference person in 2010–2050, Hebei Province, China

	Owned-housing units						Rental housing units					
Ages	<35		35–64		65+		<35		35–64		65+	
Year	# of Units	% inc. comp. 2010	# of Units	% inc. comp. 2010	# of Units	% inc. comp. 2010	# of Units	% inc. comp. 2010	# of Units	% inc. comp. 2010	# of Units	% inc. comp. 2010
2010	5.13	0.0	12.72	0.0	1.66	0.0	0.356	0.0	0.491	0.0	0.035	0.0
2015	6.73	31.2	13.4	5.4	1.81	9.0	0.386	8.4	0.409	−16.7	0.034	−2.9
2020	6.27	22.2	14.76	16.0	2.46	48.2	0.352	−1.1	0.408	−16.9	0.045	28.6
2025	4.44	−13.5	16.41	29.0	3.77	127.1	0.287	−19.4	0.42	−14.5	0.069	97.1
2030	3.68	−28.3	16.36	28.6	5.45	228.3	0.272	−23.6	0.389	−20.8	0.095	171.4
2035	4.1	−20.1	15.11	18.8	7.05	324.7	0.231	−35.1	0.293	−40.3	0.117	234.3
2040	4.29	−16.4	14.52	14.2	7.91	376.5	0.239	−32.9	0.252	−48.7	0.128	265.7
2045	4.03	−21.4	14.51	14.1	8.3	400.0	0.242	−32.0	0.233	−52.6	0.128	265.7
2050	3.56	−30.6	14.03	10.3	9.09	447.6	0.229	−35.7	0.22	−55.2	0.127	262.9

Note: "% inc. comp. 2010" means "% increase compared to 2010"

Table 15.8 Projected number of owned-housing units (in millions) by household type in 2010–2050, Hebei Province, China

Ages	1-person		One-couple		1-person & others		2-parents & children		Single-parent & children		Three-generation	
Year	# of Units	% inc. comp. 2010	# of Units	% inc. comp. 2010	# of Units	% inc. comp. 2010	# of Units	% inc. comp. 2010	# of Units	% inc. comp. 2010	# of Units	% inc. comp. 2010
2010	1.76	0.0	3.62	0.0	0.36	0.0	8.81	0.0	0.87	0.0	4.09	0.0
2015	2.62	48.9	3.91	8.0	0.37	2.8	9.8	11.2	1.18	35.6	4.05	−1.0
2020	2.87	63.1	4.6	27.1	0.41	13.9	10.35	17.5	1.6	83.9	3.66	−10.5
2025	3.22	83.0	5.16	42.5	0.46	27.8	10.49	19.1	2.06	136.8	3.24	−20.8
2030	3.5	98.9	5.73	58.3	0.49	36.1	10.31	17.0	2.56	194.3	2.9	−29.1
2035	3.73	111.9	6.28	73.5	0.54	50.0	10.1	14.6	2.99	243.7	2.62	−35.9
2040	3.98	126.1	6.56	81.2	0.57	58.3	9.8	11.2	3.25	273.6	2.57	−37.2
2045	4.14	135.2	6.66	84.0	0.58	61.1	9.5	7.8	3.43	294.3	2.53	−38.1
2050	4.22	139.8	6.57	81.5	0.59	63.9	9.27	5.2	3.58	311.5	2.46	−39.9

Note: "% inc. comp. 2010" means "% increase compared to 2010"

to 0.76 million and 0.58 million units in 2030 and 2050, reductions of −14.4 % and −34.73 %, respectively (see Table 15.5). Demand for 1–2 room rental housing will show the fastest drop and that of 3–4 room housing will rank the second fastest drop (see Table 15.5). Such projected trends, which at first glance appear counter-intuitive, may be explained by the demographic and housing related factors of population urbanization and population aging. As shown in Table 15.3 and discussed/explained in Sect. 15.3.1, home rental rates were higher in rural areas than in urban areas in Hebei Province, and they mainly reflected Chinese rural–urban migrants renting a cheaper room or apartment in the villages surrounding the cities, as well as other migrant workers who live in rental housing units in villages due to the rapid development of the township and village industries in rural areas. Also, the rental rates among households with an elderly reference person are substantially lower than those among households with a younger

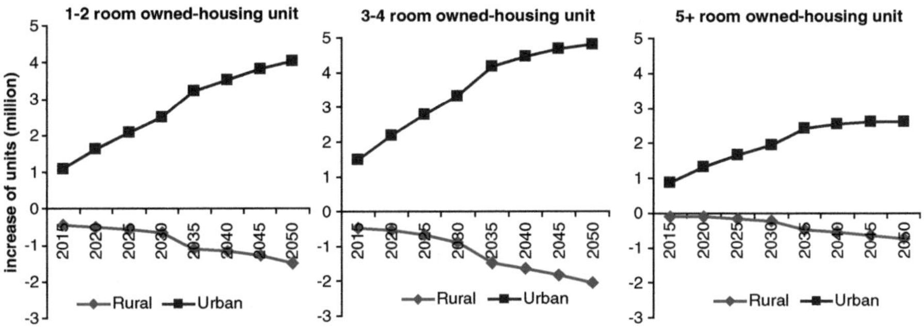

Fig. 15.1 Projected change in owned-housing units by rural and urban areas in 2015–2050 compared to 2010, Hebei Province, China

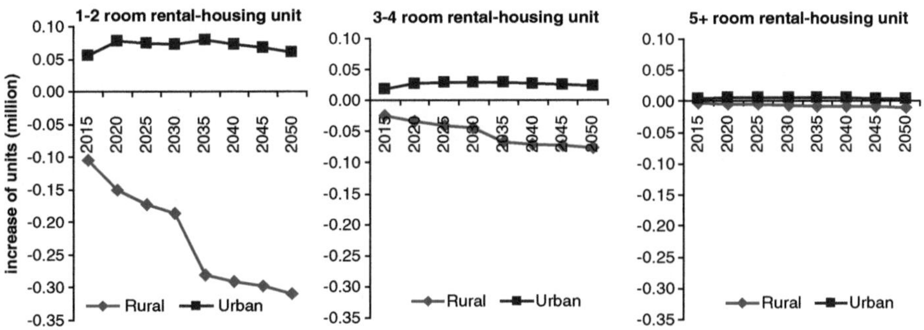

Fig. 15.2 Projected change in rental housing units by rural and urban areas in 2015–2050 compared to 2010, Hebei Province, China

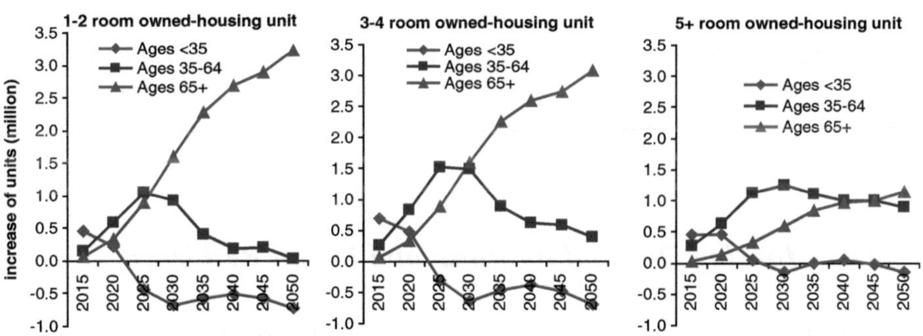

Fig. 15.3 Projected change in owned-housing units by age of the household reference person in 2015–2050 compared to 2010, Hebei Province, China

reference person. Therefore, rapid population urbanization and aging in the next 40 years will result in a general decrease in rental housing demand, especially for 1–2 room rental housing units, while owned-housing demand will continue to increase. The projected future large increase in rental housing units in urban areas

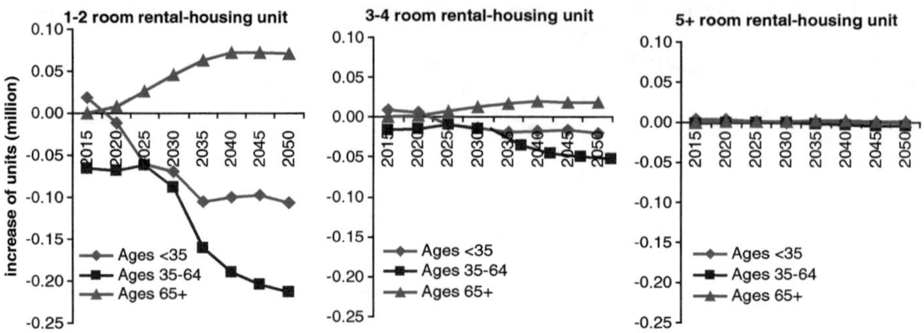

Fig. 15.4 Projected change in rental housing units by age of the household reference person in 2015–2050 compared to 2010, Hebei Province, China

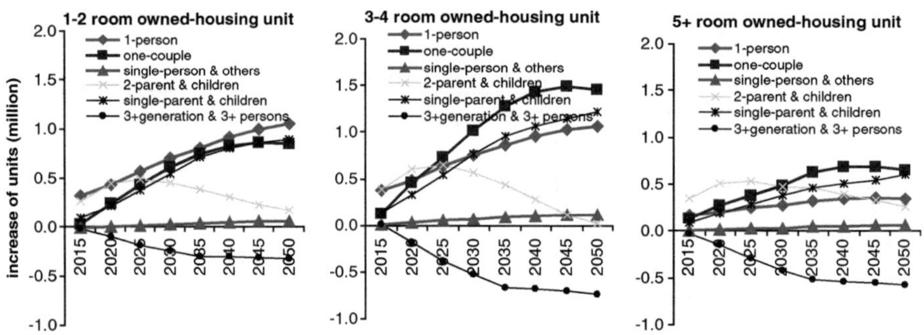

Fig. 15.5 Projected change in owned-housing units by household type in 2015–2050 compared to 2010, Hebei Province, China

and among elderly households, coupled with a substantial decrease in rental housing units in rural areas and among young people's households (to be presented/discussed in Sects. 3.4 and 3.5) reconfirm our explanations.

15.3.4 Housing Demand By Rural and Urban Residence

Since there are significant differences in urban and rural housing demand, projections of owned-housing and rental housing are made by classifying the urban and rural areas. Due to advancements in urbanization over the next 40 years, urban owned-housing demand will likely sustain continuous large growth, while the demand for rural owned-housing will be reduced substantially. More specifically, the projection results indicate that the total number of owned-housing units in urban areas of Hebei Province in 2030 and 2050 will increase by 87.4 % and 129.2 % compared to 2010 (see Table 15.6); the results also indicate an increase in 1–2 room, 3–4 room and 5+ room owned-housing in urban areas, with the largest increase in the smaller units (see Fig. 15.1). The total number of owned-housing units in rural areas in 2030 and 2050 will decrease by 16.7 % and 40.3 % respectively compared to 2010 (see Table 15.6), showing a reduction in demand for all kinds of owned-housing in rural areas (see Fig. 15.1). Compared to 2010, rental housing demand in urban areas will be 103.7 % and 85.1 % higher in 2030 and 2050, but rental housing demand in rural areas will be 30.2 % and 50.7 % lower in 2030 and 2050, respectively (see Table 15.6 and Fig. 15.2).

15.3.5 The Aging of Housing Demand

The projections show that over the next 40 years there will be a substantial increase in owned-housing demand among households whose reference persons are aged 65+. There will also be an increase followed by a decrease in owned-housing demand among households between 35 and 64 years old and a continued drop for households with a reference person aged less than 35 years old. But for households with an older adult reference person, owned-housing demand will be 228.3 % higher in 2030 and 447.6 % higher in 2050, compared to 2010 (see Table 15.7).

Due to substantial population aging and a gradual decline in the proportion of the younger population, the total demand for rental housing will continue to decrease among those households whose reference persons are younger than 35 years old and between 35 and 64 years old -- compared to 2010, they will drop by 23.6 % or 0.8 % for households with reference persons aged <35 or aged 35–64 in 2030; or drop by 35.7 % or 55.2 % for households with reference persons aged <35 or aged 35–64 in 2050. However, the elderly population aged 65+ will require a large increasing amount of rental housing – an increase of 171.4 % and 262.9 % in 2030 and 2050, respectively, compared to 2010 (see Table 15.7 and Fig. 15.4).

This phenomenon of aging in housing demand is closely linked with population aging in Hebei Province, to which the government and the real estate industry should pay serious attention.

15.3.6 Housing Demand By Household Type

Over the next 40 years, the proportion of three-generation households will drop. Therefore, demand will continue to decrease for three-generation owned-housing units – dropping by 29.1 % and 39.9 % in 2030 and 2050, respectively; but demand for owned-housing for all other household types considered here will rise to a different degree (see Table 15.8). The demand for housing units for two-parent households with children will grow slowly until 2025 and then start to decline gradually, although it will remain the largest category among the five main household types. The growth of owned-housing demand for one-person households in Hebei Province in the next 40 years is substantial – from 1.8 million units in 2010 to 3.5 million and 4.2 million units in 2030 and 2050, an increase of 98.9 % and 139.8 %, respectively. The largest relative increase in owned-housing units among the five main household types is for single-parent with children households – compared to 2010, it will show a relative increase of 194.3 % in 2030 and 311.5 % in 2050, while remaining the second smallest category of household type (see Table 15.8). As discussed in Sect. 15.3.2, rapid population urbanization and the demographic mechanism of family household momentum (Zeng et al. 2006) may explain the substantial increase in owned-housing demand from single-parent households with children.

In general, the growth in demand for owned-housing units for one-generation households will be faster compared to demand from two-generation and three-generation households. The highest increase in demand will be for 3–4 room owned-housing units among one-generation households (see Fig. 15.5). Figure 15.5 also shows that one-person households will have the largest growth in demand for 1–2 room owned-housing; one-couple households will have the largest growth in demand for 3–4 room owned-housing; and two-parent households with children will have the largest growth in demand for 5+ room owned-housing before 2030.

15.4 Summary and Relevant Policy Considerations

The household and housing demand projections for Hebei Province presented above show that household and population dynamics in the next four decades will lead to continual increases in demand for owned-housing in urban areas and decreases in rural areas; owned-housing with 3–4 rooms will show the largest increase in the urban population, while also showing the greatest decrease in rural areas. Rental housing in urban areas will present a substantial increase before 2035 and then a slow decline afterwards, while showing a declining trend in the countryside over the next 40 years.

Different types of households will manifest different housing demands. The housing demand of one-generation and two-generation households will increase quickly. One-person households will have the largest growth in demand for 1–2 room owned-housing; one-couple households will have the largest growth in demand for 3–4 room owned-housing. Since the proportion of three-generation households will decrease substantially, their demand for housing (usually with larger numbers of rooms) will go down.

The housing demand of the elderly population aged 65+ presents a remarkable fast-growing trend. With the acceleration of population aging, the demand for both owned-housing and rental housing for elderly people aged 65+ will show accelerated and dramatic growth – compared to 2010, demand will increase by 228.3 % and 447.6 % in 2030 and 2050 for owned-housing, and increase by 171.4 % and 262.9 % in 2030 and 2050 for rental housing. For those aged 35–64, owned-housing demand will at first moderately increase up to 2025 and then gradually decline; rental housing demand will decline for all time periods. Both owned and rental housing demand for younger people aged less than 35 will decline after 2020. There is no doubt that the future elderly population will become the backbone of the growth in housing demand in Hebei Province.

The analyses in this chapter should be useful for the government and real estate industries in Hebei Province and the rest of China in planning policy adjustments and housing investments, while considering future changes in household and population characteristics. For example, the most striking result of our demographic projections is that as population aging accelerates, the elderly will become the main source of growth in housing demand. Various forms of accessible housing, such as housing for the elderly with assisted living facilities, special elderly residential areas, seasonal housing, and community service centers should be developed to meet the needs of the aging population. Given the future trends of dramatic increases in housing demand for the elderly (see Table 15.7) and the substantial decrease in housing demand for three-generation households (see Table 15.8), the Chinese government and real estate enterprises may need to consider developing and investing in "dual-apartments" to encourage and facilitate adult children and their elderly parents living in independent but neighboring and directly connected apartments. These dual-apartments would allow the elderly to have their own independent lives, while receiving home-based care from their children whenever needed and also providing care for grandchildren, which may result in a mutually beneficial outcome for both older and younger generations. In addition to our present study, other recent research also supports the win-win concept and expectations of inter-generational living arrangements with dual-apartments, which facilitate direct and close connections between older and younger generations (e.g., Shen 2011; Shen et al. 2013; Zeng et al. 2013b).

To improve the housing conditions of citizens, there is a need to vigorously push forward the construction of affordable housing based on rural and urban household characteristics and the socioeconomic status of the population. Attention should be given to improving the investment structure, especially increasing the percentages of low-cost, small and medium-sized affordable housing for which demand is growing, and reducing the proportion of large housing units for which demand is shrinking. For example, the projected future trends of substantially and continually increasing small households and the gradual shrinking of three-generation households indicate that construction of extraordinarily large housing units should be reduced commensurately, while the construction of smaller housing units may need to receive more attention.

15.5 Concluding Remarks

In sum, this chapter applies the ProFamy extended cohort-component model to project future housing demands by number of rooms, rural/urban residence, age, household type and size in Hebei, a typical province with median levels of socioeconomic development and urbanization in China. Our housing demand projections are based on commonly available demographic data, including the unique and most recent 10 % micro-data sample (with a sample size of seven million persons) from the 2010 census. To our knowledge, this is the first successful attempt to integrate multistate dynamic projections of household type/size, living arrangement, and housing demand by number of rooms in China. The projections and analyses presented in this chapter quantitatively demonstrate how changes in household and population size/structure may affect future housing demand. These projections and analyses may prove useful for government and housing industry policy analyses as well as strategic plans for future public services and for private sector market potential research, although they are not real estate forecasts.

There are, however, some important limitations of the projections presented in this study that should be explored in future research. First, mainly due to space limitations, we have presented here only the general trends and patterns of owned- and rental housing demands by number of rooms and some selected major attributes of the household reference person such as rural/urban residence, large age groups, and major household types. These results are extracted from more detailed projected outcomes of the numbers of owned- and rental housing units with number of rooms by much more detailed attributes, such as rural/urban residence, 5-year age groups, marital status, and relatively detailed household type/size and living arrangement. A full presentation and discussion of the housing demand projection outcomes could be useful in more detailed academic research, governmental planning, and business market strategy analysis in future studies.

Second, as indicated in the Introduction, this study explores how changes in household and population structure may affect future housing demand, rather than providing real estate forecasts of housing market per se. The real estate market is also associated with socioeconomics, wealth, lifestyle, and governmental regulations (Cohen et al. 2003; Koklic and Vida 2009), which are not included in our multistate projection model. In this chapter, we generally follow the expert opinion approach for projecting future demographic summary measures and proportions of owned- and rental housing units among various categories of households based on rural/urban residence and different attributes of the household reference person. This serves well for the present study's purpose of demographic research for a relatively long time horizon of 40 years from 2010 to 2050.

Within our multistate model framework, future years' demographic and housing consumption summary parameters may also be forecasted using the methods of time series analysis and regression models with a focus on shorter forecasting time horizons. Furthermore, these parameters may be forecasted by regressions using the time series data of other housing related covariates of socioeconomics, wealth, lifestyle, and governmental regulations. It should always be noted, however, that the accuracy of the forecasts relies heavily on the validity of assumptions regarding future time paths of the various covariates and parameters included in the forecasting model.

Third, we present the medium projections with expected time-varying demographic rates as input without sensitivity analyses. Note that the medium projections provide a reasonable projection of future trends, because neither the direction nor the magnitude of future upward or downward changes of the parameter can be predicted accurately due to great uncertainties in the future (Smith et al. 2001:83–84). However, the high and low bounds of the sensitivity analysis may need to be conducted in the future along with more in-depth studies.

Finally, as we emphasized in other chapters of this book, projections for time horizons of less than 20 years may be used as forecasting for governmental and business strategic planning, but any results beyond that should be considered to be simulations only, due to large uncertainties in projections greater than 20 years. Thus, the projection results for the period 2010–2030 presented in this chapter may be used for housing market strategy and relevant socioeconomic analysis and planning, but results after 2030 should be regarded mainly as simulations. Such simulations are useful for addressing "what if" questions in academic and policy analyses of the effects of changes in demographics on future general trends and patterns of household housing demands in Hebei Province, but they cannot be considered to be any kind of accurate real estate forecasts.

Part IV
ProFamy: A Software for Household and Consumption Forecasting

ProFamy (Version 2.1)[1]: A SOFTWARE FOR HOUSEHOLD AND CONSUMPTION FORECASTIONS

Zhenglian Wang, Yi Zeng, Danan Gu, Wenzhao Shi[2]

USER'S GUIDE

[1] The ProFamy software (version 2.1) and User's Guide are part of the outcomes of a research project on demographic tools and databases for household forecasting supported by NIH/NIA SBIR Phase I and Phase II grants and China National Natural Science Foundation grants (71110107025 and 71233001). We also thank the Population Division of the U.S. Census Bureau, NIA, NICHD, Center for Demographic Studies of Duke University, National School of Development of Peking University, Max Planck Institute for Demographic Research, Sabre System Inc., and Digital China for supporting our applied scientific research. Contributions by Zhiheng Ding, Liquan Zhou, Shuang Liang, Ning Gao, Qing Wu, Weiheng Shi, and Changqing Gan are highly appreciated.

Buyers of the Papercover, Hardcover and eBook of this Springer book all get 50% discount for the professional version of the ProFamy software, which can be used to prepare your input data and run the household, living arrangement and population projections based on the data and parameters provided by the users. The buyers of this Springer book can purchase/download the professional version of ProFamy software with 50% discount from ProFamy website (http://www.profamy.com), simply attaching a copy of the receipt of purchasing the book.

[2] We will greatly appreciate the comments and suggestions provided by the users of ProFamy software (Contact information: Dr. Zhenglian Wang (email: wangzl6668@gmail.com; tel. 1-919-945-1940).

Chapter 16
Setting Up the Projection Model

16.1 Main menu

After you click the ProFamy icon to open the program, you will first see a colorful cover page and then you will enter the main menu (Fig. 16.1):

The sub-menu of "ProFamy File" at the top left of the screen allows you to save your work on a ProFamy file, open a ProFamy file you prepared previously, or open an example file that is provided as part of the ProFamy package for tutoring purposes. A ProFamy file has a special format with an extension of ".PRD" produced by ProFamy program. The input data, prepared by a user through ProFamy, are saved in a single ProFamy file once the user clicks "Save" or "Save as" under the "ProFamy File" sub-menu; new ProFamy files and changes in previously saved ProFamy files must be saved; otherwise the new input data will be lost. A ProFamy file can be produced only through the interface of ProFamy, and it can be transferred between the users.

Once you click either "Create New ProFamy File" or "Open Saved ProFamy File", you will see four menu trees at the left panel of the window that will guide you through model specifications:

- Menu tree ① "Specify Models and Data Types" mainly deals with specifications of model design, data type of the base population, and directories that store the input and output files of your projection.
- Menu tree ② "Data Preparation" deals with preparations of the base population and input data such as standard schedules and projected (or assumed) summary measures for future years.
- Menu tree ③ "Computation of the Projection" deals with running the ProFamy program.
- Menu tree ④ "Output and Running Information" allows you to view the output of the household projection and the aggregated information of the base population. Click "Help" at the top of the four menu trees (or the right end of the top menu) for more information to help you use ProFamy.

Y. Zeng et al., *Household and Living Arrangement Projections*,

The Springer Series on Demographic Methods and Population Analysis 36,
DOI 10.1007/978-90-481-8906-9_16,

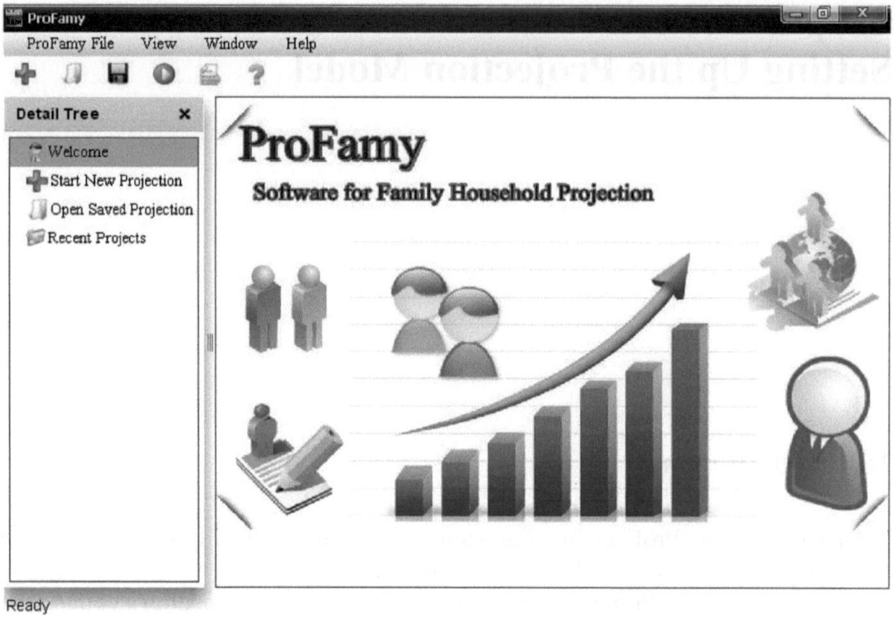

Fig. 16.1 Main menu

When working with ProFamy for the first time, you must follow the logical order of the menu trees ① → ② → ③ → ④. We begin with the "ProFamy File" menu partly because Windows menus always locate "ProFamy File" in the first position, and partly because you also need to first access the "ProFamy File" menu to open a previously completed or an incomplete file to continue the work or modify the model design or parameters. When first learning about ProFamy, you may wish to enter the "ProFamy File" menu to open an example ProFamy file provided with the software to see what the controlling parameters, standard schedules, projected summary measures, and the output of tables and graphics look like. This will help you to be familiar with how to use the ProFamy software for household and living arrangement projections. You can open one of the two example input files (one is associated with an application to the United States and another one is associated with an application to China) and click through the four menu trees in the left panel of the window to see the model design, input, and output data for this example of household and living arrangement projections.

16.2 Create, Open and Save ProFamy Files

When you click "ProFamy File", you will see the following sub-menu (Fig. 16.2).

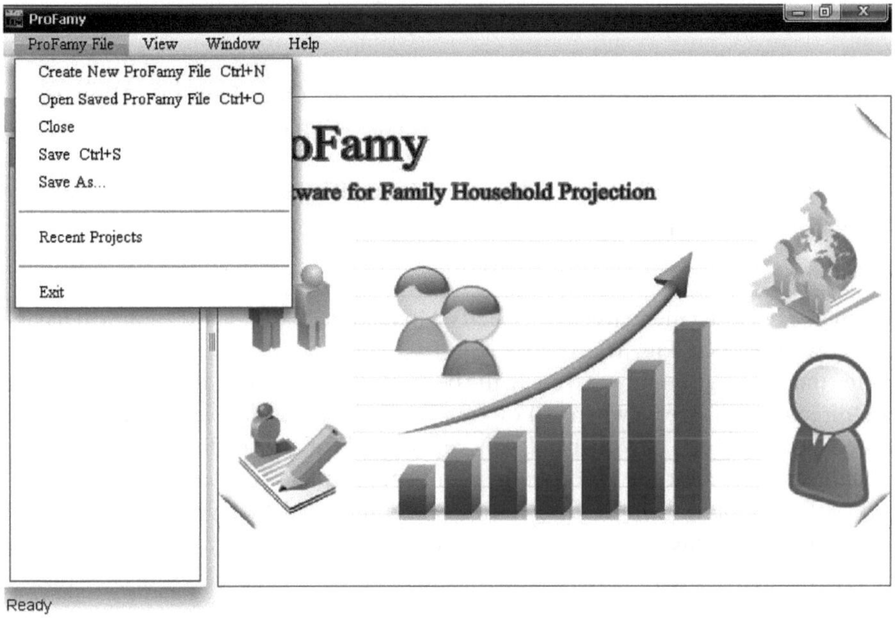

Fig. 16.2 Sub-menu of "ProFamy File"

Open saved ProFamy file

When you click "Open Saved ProFamy File", you will be asked to identify the file name of a previously prepared ProFamy file, which may be stored either in the sub-directory named "Input" under the ProFamy Directory, or any other sub-directory created by you. ProFamy will give you an error message if you try to open a file that is not produced by the ProFamy program (i.e., file extension is not ".PRD").

Save or save as

If you are just starting to prepare input data for your projection or scenario, you have no files of your own and you should go to the "Specify Models and Data Types" and "Data Preparation" menus first. When you finish the input preparation, or if you have not yet finished the input preparation but need to stop for any reason, or if you wish to save the intermediate work from time to time to prevent loss due to unexpected disruption of your computer, you must go to the "ProFamy File" menu and click "Save" (save the file in the current subdirectory) or "Save as" (save the file in a subdirectory to be identified by you) and provide a file name for the ProFamy input file to be saved. After you have completed a ProFamy input file, it is easy to open it, change some parameters or part of the data, save it, and then run the modified file.

Exit

Every time you want to quit the ProFamy program, you need to go to the "ProFamy File" menu and click "Exit". You will be asked if you want to save the work you have done.

16.3 Specify Models and Data Types

After you click either “Create New ProFamy File” or “Open Saved ProFamy File”, you will see four menu trees (①, ②, ③, ④) at the left panel of the screen as shown in Fig. 16.3.

In Fig. 16.3, you will see six sub-menu trees under the menu tree ① of “Specify Models and Data Types”. We will explain them one by one.

Directory

When you click the menu tree ① “Specify Models and Data Types”, you will have a dialog window (see Fig. 16.4).

In this dialog window, you will be asked to specify the names of the sub-directories (folders) where all of your ProFamy input data and output files are stored (see Fig. 16.3). If you install this version of ProFamy software under C:\, the default input directory is C:\ProFamy\Sample\Input; and the default output directory is C:\ProFamy\Sample\Output. It is recommended that you use the default input and output directories unless you have some special needs to have different

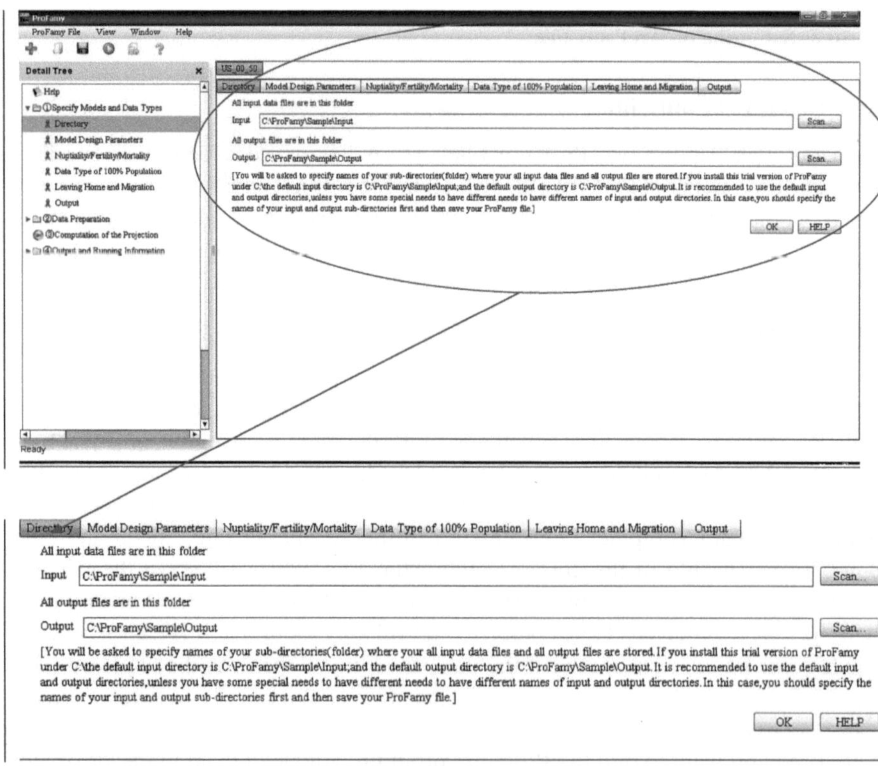

Fig. 16.3 Structure of menu tress of the ProFamy interface

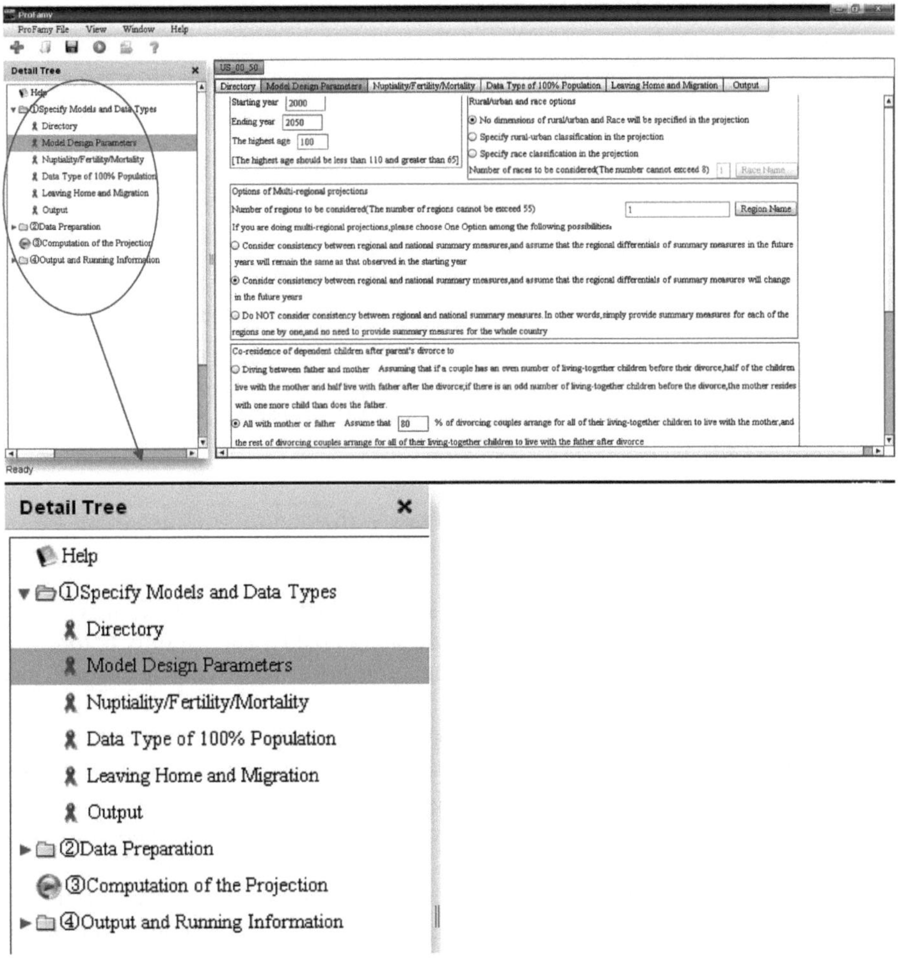

Fig. 16.4 Dialog window of "Model Design Parameters" (I)

names for them. In this case, you should specify the names of your input and output sub-directories first and then save your ProFamy files.

Model design parameters

When you click "Model Design Parameters", you will be asked to specify the starting and ending year of your projection, and the highest age considered (see Fig. 16.5). These three parameters are very important. If you have not made any ProFamy input files before on the computer you are working with, you should always specify these parameters at the very beginning of your new data preparation. Otherwise, you will not be able to continue your data preparation properly, since ProFamy will not have these model design parameters to move to the next steps.

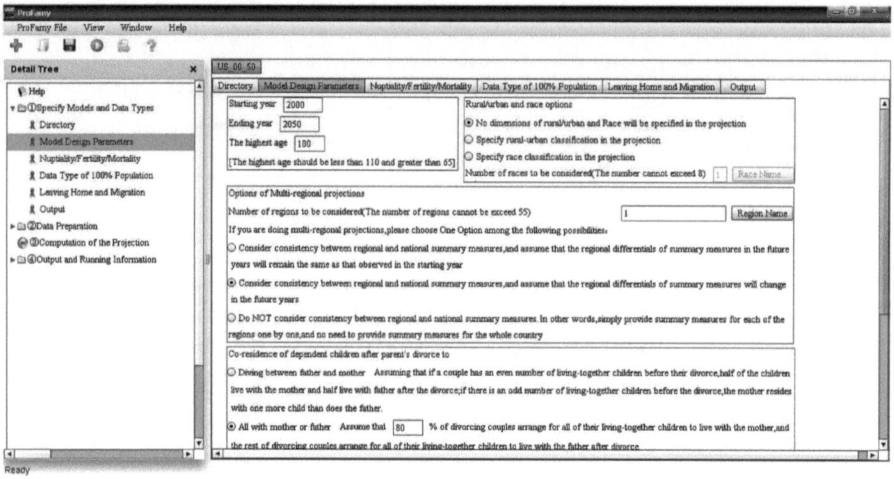

Fig. 16.5 Dialog window of "Model Design Parameters" (II)

The definitions of each of the model design parameters are given in the notes in the dialog window as shown in Fig. 16.5; you should read them carefully before you specify the model design parameters. After you have made your decisions, just move the mouse (or use the Tab key) to the proper box and click it to make the choice or key figures into the box, as appropriate. After you have made a choice, the associated box will be filled by a black point "•". There are default choices and numbers in the dialog window shown in Fig. 16.5, but these are NOT fixed and can be modified by you; you should make sure that your choices correspond to your research purpose and data availability. This principle should be followed in all other dialog windows in ProFamy. Once the values of the model design parameters are given, they will be always appear in the screen to conveniently remind you of the basic features of the current projection model design; you can modify them whenever you wish. It is very important to NOT change the parameters and click "Save" when you use an example input data file provided by the ProFamy package in order to prevent losing data of the example input file. You may click "Save as" with another new file name if you wish to save the new model design and parameters you have changed based the example input data file.

Rural/urban and race options

There are three options to deal with rural/urban and race classifications:

- No dimensions of rural/urban or race will be specified in the projection
- Specify rural–urban classification in the projection
- Specify race classification in the projection

If you click "No dimensions of rural/urban and race will be specified in the projection", you will need to provide data only for the whole population under study.

If you click "specify rural–urban classification in the projection", you will have to provide all of the data for rural and urban sectors, respectively.

If you click "Specify race classification in the projection", you will have to provide the data classified by racial group. You may not choose both rural–urban and race classifications at the same time, due to unavailability of practical data.

Options for multi-regional projections are not available for the current version as this newly added function is still under trial, but will be available in the next version.

Options for co-residence of dependent children after parents' divorce

You will need to choose how to assume the living arrangements of dependent children after their parents' divorce (i.e., whether children live with the mother or father). For societies where divorced couples do not wish their children to be separated from each other, our ProFamy model and software provide an option for the users to assume that for a certain percentage of divorced couples, all children stay with their mother after their parents' divorce; the rest of divorced couples' children stay with their father. This percentage may be estimated from survey data. Another available option is to assume that if a couple has an odd number of children living together before divorce, the mother will have one more child than the father will after divorce. If a couple has an even number of children living together before their divorce, each party would have half of the children after divorce.

Options for small area projection

Small area means an area such as a county, city, or a town where demographic summary measure data are not available. There are basically two options for small area projection: No or Yes.

If you check "No", it means that you are doing a national or sub-national projection, using available age-sex-specific standard schedules and summary parameters.

If you check "Yes", you must first do the projections for the parental nation or sub-national region where the small area is located. You also need to provide census micro data for the small area in order for ProFamy to derive the proportions of the households with various types/sizes and age groups in the small area among the corresponding households in the parental country or sub-national region, derived from the micro census data files (see Chap. 6 for details of the methodology and illustrative application).

Note that the options for small area projections are still under trial and will be available in the next version.

Option for 3-generation households

- 3-generation household is one of the common household types
- 3-generation household is NOT one of the common household types

If you check the option "3-generation household is one of the common household types", will compute households of one, two, and three generations. Note that a three-generation household is one of the common household types in Asian countries and many other developing countries, so you need to choose this option for household projections of these countries. In this case, ProFamy assumes that some children may continue to live in the parental home after first marriage or first consensual union, so that both nuclear and three-generation households will be projected. If three-generation household is not one of the common household types, as in Western countries, you need to choose the second option. In such

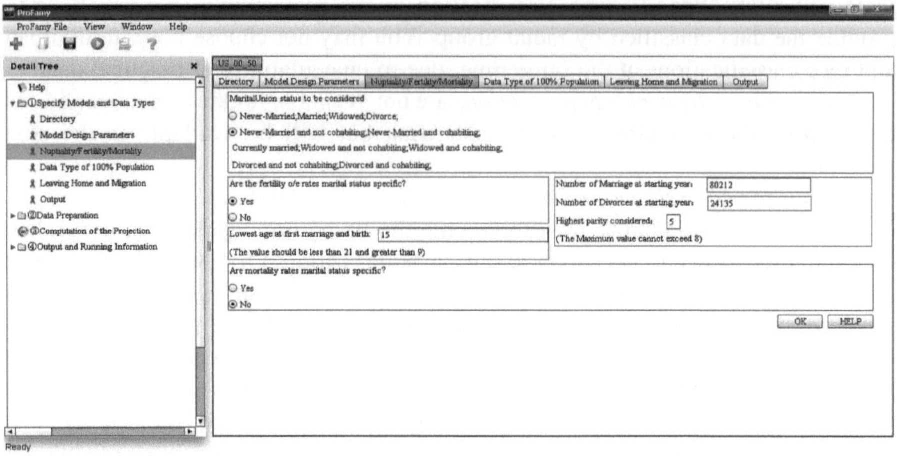

Fig. 16.6 Dialog window of "Nuptiality/Fertility/Mortality"

circumstances, ProFamy assumes that all adult children leave their parental home after first marriage or first consensual union while a few never-married and not-cohabiting adult children may stay with parents.

Are you going to project pension deficit rates?

Note that the option for projecting pension deficit rates is still under trial, and will be available in the next new version.

In the new version of ProFamy software, there will be two choices: Yes or No. If you answer "Yes", you plan to project pension deficit rates using the simple method proposed by Zeng (2011). Please also refer to Chap. 7 for the methodology and an illustrative application.

Nuptiality/Fertility/Mortality

If you click "Nuptiality/Fertility/Mortality", you will see a screen menu as shown in Fig. 16.6.

Options for marital/cohabitation statuses

1. If you click the option of **"**Never-Married; Married; Widowed; Divorced**"**, you will need to provide only 4 sets of single-year age- and sex-specific occurrence/exposure rates (abbreviated as o/e rates hereafter) of marital status transitions:

   Never-Married →Married
   Married → Divorced
   Widowed →Married
   Divorced →Married

2. If you click the option of seven marital/union statuses, you will have to provide 13 sets of single-year age and sex specific o/e rates of marital/union status transitions that also consider cohabitation:

Never-Married & not-cohabiting→Married
Never-Married & cohabiting→ Married
Widowed & cohabiting→ Married
Divorced & cohabiting→Married
Widowed & not-cohabiting→ Married
Divorced & not-cohabiting → Married
Never-Married & not-cohabiting→Never Married & cohabiting
Widowed & not-cohabiting→ Widowed & cohabiting
Divorced & not-cohabiting→ Divorced & cohabiting
Never-Married & cohabiting → Never-Married & not-cohabiting
Widowed & cohabiting→ Widowed & not-cohabiting
Divorced & cohabiting → Divorced & not-cohabiting
Married→ Divorced & not-cohabiting
Married→Widowed & not-cohabiting

Note that the age-sex specific probabilities of a transition from married to widowed and a transition from cohabiting to non-cohabiting due to death of the partner will be estimated by the ProFamy software using the age-sex-specific probabilities of death and the average difference in age between husbands and wives, usually assumed to be two years.

Option for the o/e rates of fertility

If you choose "Yes" in answering the question "Are the fertility o/e rates marital status specific", you will have to provide the estimated standard schedules of age-parity specific o/e rates of fertility for both married and non-married women.

If you choose "No" in answering the question "Are the fertility o/e rates marital status specific", this means that you assume that the births out of wedlock are negligible, and therefore, only fertility rates for married women are needed.

Number of marriages at starting year and number of divorces at starting year

You will need to provide number of marriages (including all first marriages and remarriages) and number of divorces for the population under study at the starting year of the projection. These two numbers are used to estimate the standardized general rates of marriage and divorce at the starting year (see Sect. 3.4 and Appendices 2 and 5 of Chap. 3 for details).

Highest parity considered

This specification means that the births with order higher than the "Highest parity considered" are grouped into this "highest parity" (maximum is 8).

Lowest age at first marriage and birth

Specification of the "Lowest age at first marriage and birth" implies that you assume that the events of marriages and births below this age are negligible.

Options for mortality rates

If you choose "Yes" in answering the question "Are mortality rates marital status specific", you will have to provide the age-sex-marital-status-specific life table probabilities of surviving from age 0 to age x. If you choose "No", you will use the average age-sex-specific life table probabilities of surviving for persons of all marital statuses combined.

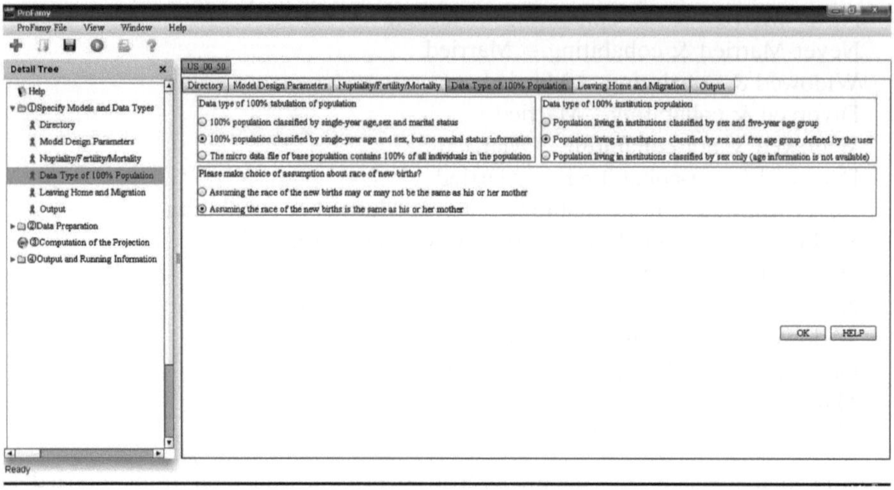

Fig. 16.7 Dialog window for "Data Type of 100 % Population"

Data Type of 100 % population

After you click "Data Type of 100 % Population", a screen menu will appear as show in Fig. 16.7:

Data type of 100 % tabulations of the baseline population

First, you need to choose one of the three options listed below:

- 100 % population classified by single-year age, sex, and marital status
- 100 % population classified by single-year age and sex, but no marital status information
- The micro data file of base population contains 100 % of all individuals in the population under study

You may very likely rely on a sample data file derived from a census to get the baseline population distributions by single year of age, sex, marital/union status, number of co-residing children and parents and living in private or institutional household, because statistical agencies usually do not publish such relatively detailed information. The weights of the sample data set will be necessary to estimate the group averages, but this may not produce the exact total population size even after the weights have been considered. Therefore, you will need the aggregated 100 % tabulations of the total population published by the statistical office in order to adjust the sample data set so as to have the correct total population size and distributions of age, sex, marital (including cohabitation if classified) status, and co-residence status between parents and children. ProFamy provides three options for the type of the data on 100 % tabulation of the total population. Because the number of persons at advanced ages derived from the sample data set may be too small to be reliable, it is recommended that you have the single-year-age-specific 100 % tabulation of the population, preferably by marital status, in order to eliminate possible sampling errors at advanced ages.

Data type of 100 % institutional population, if sample data set of base population is used

If a sample data set of base population is used, the proportion of the sample living in institutional settings may not be the same as that of those who live in private households. Therefore, 100 % tabulations of age-sex distributions of those living in institutions are needed. ProFamy provides three options of the type of the data on 100 % tabulation of the institutional population.

- Population living in institutions classified by sex and 5-year age group
- Population living in institutions classified by sex and free age group defined by the user
- Population living in institutions classified by sex only (Age information is not available)

If the micro data file of the base population contains 100 % of all individuals in the population under study, there is no need to click any of the three options listed above.

Please make choice of assumption about race of new births:

- Assuming the race of the new births may or may not be the same as his or her mother;
- Assuming the race of the new births is the same as his or her mother;

If your projection is by race and if you choose "Assuming the race of the new births may or may not be the same as his or her mother", you will need to provide the cross-tabulation data of proportions of the race of new births by the race of their mothers.

Leaving home and migration

By clicking "Leaving Home and Migration", you will see Fig. 16.8.

Net rates of leaving parental home

There are two options about net rates of leaving parental home:

- Estimating the single-year age and sex specific net rates of leaving parental home with ProFamy, based on two adjacent census data provided by you;
- Provide "age-sex-specific net rates of leaving parental home", estimated by you based on survey data.

If you check the first option, you will need to prepare single-year-age-sex-specific proportions of living with parent(s) from age 5 to 49 at two points of the census time, derived from the micro data files of the two most recent adjunct censuses. You can do so either on your own, or through preparing two BasePop input files based on micro data files of the two censuses (see Sect. 17.1.1 for the format of such input files) and running the BasePop subprogram twice. Please refer to Sect. 17.2.5 "Standard schedules of leaving the parental home" for details. With the appropriate data prepared by you, ProFamy will estimate the age-sex-specific net rates of leaving home based on a method initially proposed by Coale (1984, 1985), Coale et al. (1985), and generalized by Stupp (1988). This method was

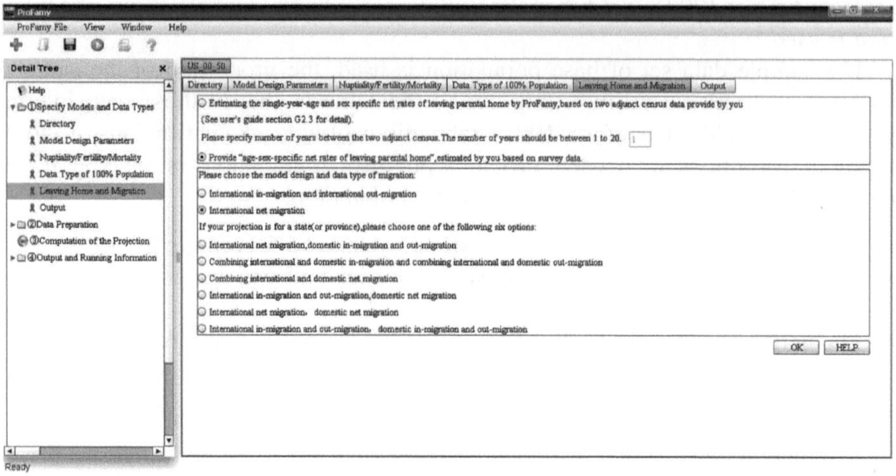

Fig. 16.8 Dialog window of "Leaving Home and Migration"

applied to estimate age-sex-specific net rates of leaving the parental home in the U.S., China, France, Sweden, Japan, and South Korea (Zeng et al. 1994).

If you click the second option, you will need to provide "age-sex-specific net rates of leaving parental home", estimated by you based on survey data. The age-specific net rate of leaving the parental home is defined as the difference between the number of persons who leave the parental home at age x and the number of persons who return to the parental home at age x divided by the total number of x-year-old persons. Note that it is less likely that an ordinary survey can provide good enough data to estimate reliable age-sex-specific (and race-specific if race classification is included) rates of leaving home.

Data type of migration

If your projection is for a country, please choose one of the following two options:

- International in-migration and international out-migration;
- International net migration;

(Note: net migration is defined as the difference of in-migration minus out-migration).

If your projection is for a sub-national region, please choose one of the following options:

- International net migration, domestic in-migration, and out-migration;
- Combining international and domestic in-migration; combining international and domestic out-migration;
- Combining international and domestic net migration;
- International in-migration and out-migration, domestic net migration;
- International net migration, domestic net migration;
- International in-migration and out-migration, domestic in-migration and out-migration.

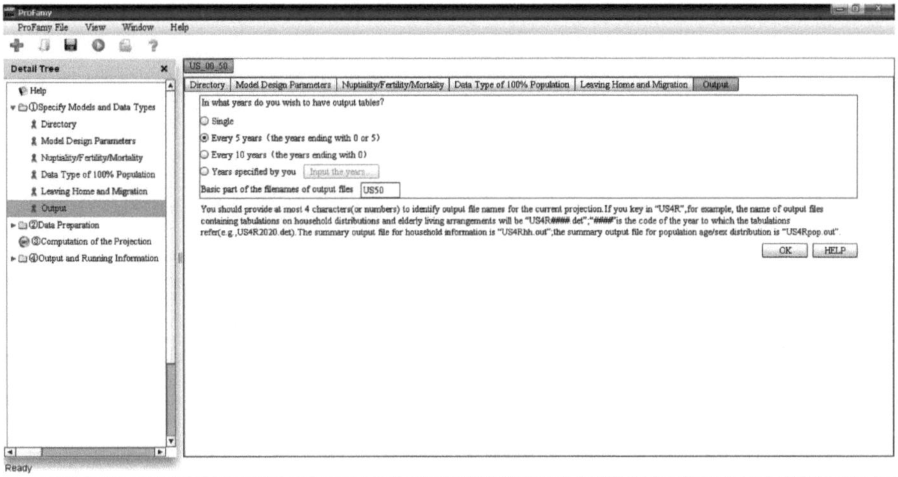

Fig. 16.9 Dialog window of "Output"

Output

When you click "Output", a screen menu as shown in Fig. 16.9 will display:

Specify the years in which you wish to have output tables

- Single year
- Every 5 years and the years ending with 0 or 5
- Every 10 years and the years ending with 0
- Years specified by you

If you choose "Years specified by you", you will need to key in the years in which you want to have output tables by clicking the "Input the years >>" button.

Basic part of the filename of output files

You should provide at most four characters (or numbers) to identify output file names for the current projection. If you key in "US4R", for example, the name of output files containing tabulations on household distributions and elderly living arrangements will be "US4R####.det"; "####" is the year to which the tabulations refer (e.g., US4R2020.det); the summary output file for household information is "US4Rhh.out"; the summary output file for population age/sex distribution is "US4Rpop.out".

Chapter 17
Preparing Input Data, Computing, and Managing Output

17.1 How to Prepare Input Data of the Base Population

In this section we discuss the contents and formats of the input data for the base population, how to prepare the base population file, and the published 100 % tabulations of the base population at the starting year of the projection.

Preparing base population is done through executing "BasePop", which is a sub-program to run the base population input file based on the census micro data file prepared by you to prepare the starting year's base population classified by the statuses recognized in the ProFamy model. When you run the BasePop for your research (rather than going through the tutor to learn how ProFamy software works), you will need to click "New Input File", and click the icon on the right side of the top line to browse/identify your input file for running the BasePop. ProFamy will automatically assign a Report file and an Output file with the same main names as the Input file but with different extensions. You can also change the file names as you wish.

Please identify the filename of the base population

You will be asked to provide the name of a file containing the data for the base population for the projection in the lower panel of the right dialog window (see Fig. 17.1). The base population in the starting year of the projection is classified by single-year of age, sex, marital/union status, parity (if any), number of co-residing children and co-residing parents, status of whether living in a private household or an institutional household, etc. You need to provide the filename of the "base population file" by directly keying in the filename or by clicking the >> button to find the file in your computer folders. The "base population file" in ProFamy has an extension of ".bpo". If you have not run "BasePop" and the file is not available at the moment, you may leave it empty. But you need to go back to this step and provide its folder path and file name before you run the calculations of the projection.

The "base population file" is derived from census data or exceptionally large survey data by running the sub-program "BasePop" provided in the ProFamy

Y. Zeng et al., *Household and Living Arrangement Projections*, The Springer Series on Demographic Methods and Population Analysis 36, DOI 10.1007/978-90-481-8906-9_17, © Springer Science+Business Media Dordrecht 2014

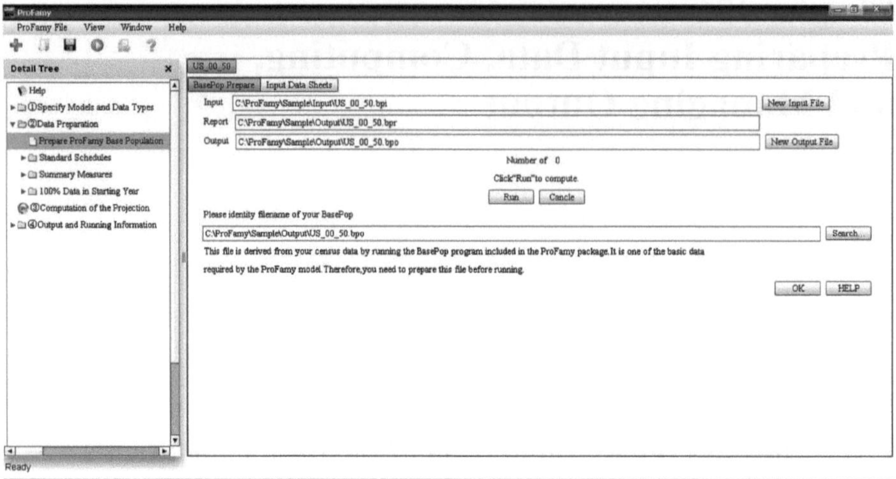

Fig. 17.1 Data preparation sub-menu

package. How to prepare the input data file for running the sub-program "BasePop" will be discussed in the next section.

You may wish to choose one of the previously prepared base population files for the current projection. For example, you may have already prepared the base population file through running the sub-program "BasePop", and already conducted household projections for population A (which may be a country or a sub-national region). You may now wish to perform an additional projection or scenario for population A with some different projected (or assumed) summary measures. In such a case, you do not need to re-run the sub-program "BasePop" again, but need only to specify the filename of the base population A prepared previously, and use the revised summary parameters to re-run the ProFamy program for the additional projection or scenario.

Choosing one of the previously prepared base population files for a new projection for the same population will also enable ProFamy to provide users with some estimates of summary measures based on the census data (base population) at the starting year (e.g., general marriage/union formation or dissolution rates), which will be shown in the corresponding screen and can be used as references when users input future years' summary measures.

17.1.1 Format and Variable Definitions for the Input File of Base Population

The base population is classified by single-year of age, sex, race (optional) marital/union status, parity (if any), number of children living at home, status of co-residence with parents, status of living in a private or institutional household, and rural–urban classification (optional).

The format of the input file is shown in the following condensed example of a sample of a population with N (e.g., N = 4,500,000) households including private and institutional households.

10392203111 ⇐ records of individuals in household 1 start from here
20371200111
30082000111
30042000111
30022000111
10351200112 ⇐ records of individuals in household 2 start from here
20302203112
30072000112
30032000112
10392403113 ⇐ records of individuals in household 3 start from here
.
.
.
10351200114500000 ⇐ records of individuals in household 4500000 start from here
20302203114500000
30072000114500000
30032000114500000

There are as many lines as the number of individuals in the sample data set of the population. The input file of the base population must be in ASCII format with an extension of ".bpi". Note that this ".bpi" contains the input data for preparing the "base population file" with an extension of ".bpo" produced by ProFamy program using the "bpi" input file.

Column number, definitions, and code of variables

Column 1	Relationship to the reference person
Column 2–4	Age
Column 5	Sex
Column 6	Marital/union status
Column 7–8	Parity
Column 9	Household type
Column 10	Residence area (rural or urban)
Column 11	Race group
Column 12 -	Household code number

The definition and code of the variables are as follows:

Column 1: relationship to the reference person

1. Reference person;
2. Spouse of the reference person;
3. Child of the reference person;

4. Grandchild of the reference person;
5. Parent of the reference person;
6. Grandparent of the reference person;
7. Other relative;
8. Non-relative

Note: If grandchildren cannot be identified and are coded as children in the original data set (as in the case of some Western countries, where three-generation households are rare), simply give them the code "3". This simplification will not affect the calculation for those countries where three-generation households are negligible. Note that ProFamy can project both nuclear and three-generation households. A reference person who lives with child(ren) but does not co-reside with parent(s) represents a two-generation household. A reference person who has both child(ren) and parent(s) living together represents a three-generation household. If, however, a grandparent in a three-generation household is reported in the census as the reference person and his or her children and grandchildren are all coded as children, this household would be recorded as a two-generation household rather than a three-generation one. Therefore, it is recommended that you distinguish between children and grandchildren when three-generation household is one of the common household types in the population under study.

Column 2–4: age

Single-year of age from 0 to the highest age reported (up to 3 digits). For ages less than 10, one needs to put the age number in the 4th column and fill either "0" or blank for columns 2 and 3; For ages greater than 9 and less than 100, one needs to put the age number at the 3rd and 4th columns and fill in either "0" or blank for column 2.

Column 5: sex

1. Male; 2. Female.

Column 6: marital/union status

As discussed earlier, you can choose either the 4-marital-status model or the 7-marital/union-status model:

(a) 4-marital-status model

1. Never-married;
2. Currently married;
3. Widowed;
4. Divorced.

(b) 7-marital/union-status model

1. Never-married & not-cohabiting;
2. Currently married;
3. Widowed & not-cohabiting;
4. Divorced & not-cohabiting;
5. Never married & cohabiting;
6. Widowed & cohabiting;
7. Divorced & cohabiting.

Column 7–8: parity

Number of children ever born (up to two digits; optional – it can be empty if no parity is identified). For parity less than 10, one needs to put the parity number at 8th column and fill either "0" or blank at column 7.

We use parity in ProFamy to consider fertility differentials of women with different parities. If there is no parity information in the original data set, we will assume that a woman's probability of giving an additional birth depends on her age, marital/union status, and the number of children who are living with her. This is not ideal because some women who give birth may have children who have already left home. We believe, however, that this kind of approximation should not cause serious errors since women with children who have left home are most likely have completed or are close to the end of their reproductive life span.

Column 9: household type

This is to specify whether this household is a private household or an institutional household. "Private household" refers to one or more persons living together who make individual or common provisions for food and other essentials for living (U.N. 1990; also see Chap. 1). "Institutional household" means a nursing home, military unit, prison, or other non-private household (i.e., group quarters).

1. Private;
2. Institution.

Column 10: residence area (rural or urban)

Code for identifying rural or urban areas (optional – it can be empty if there is no rural–urban classification in the projection).

1. Rural;
2. Urban.

Column 11: race group (optional, it can be empty if there is no race classification in the projection)

The maximum number of race groups is eight. For example, four race groups are identified as follows in the U.S. household projection included as a demonstration example of the applications in the ProFamy package:

1. White & non-Hispanic;
2. Black & non-Hispanic;
3. Hispanic;
4. Asian & others non-Hispanic.

Column 12 and the subsequent columns: household code

Household code is to identify members in the same household. It occupies column 12 and the subsequent columns. ProFamy counts individuals with exactly the same household code in consecutive lines as members in the same household. The household code can be simply 1, 2, 3, 4, 5, . . ., N, where N is the total number of households (including private and collective households) in the sample data file of the base population. In the above listed condensed example in which there are N (e.g., N = 4,500,000) households, the household codes are 1, 2, 3, 4, . . ., 4,500,000.

This simple code of Arabic digit order without geographical identification is good enough for ProFamy to identify members in the same household. The code of Arabic digit order is, however, not a necessary requirement. Users can include digits to represent province or state, county, township, etc., in the household code for their own research purposes. For example, all households in province A may have a household code starting with 01; all households in county X of the province may have a household code starting with 010001, . . ., and so on. The crucial point is that all members in the same household (either private or collective) must have exactly the same household code. The code of household type (column 9) and residence area (column 10) for those members of the same household (either private or collective) must be exactly the same, too. Unlike the data format for age and parity, there is no need to fill in the empty columns with "0" or blank for the household code; you just enter the code number with whatever length, starting with column 12.

17.1.2 How to Run BasePop

You can specify new input and output files by clicking "New Input File" and "New Output File". After you doing so, the BasePop input file window appears as shown in Fig. 17.2. You can choose drives and the path to locate your input file, specify the right file name and file type, and then click "open".

After preparation of the input data file is done, you may start to run the BasePop program. Once you click 'Run', the program starts to run BasePop and show the progress by giving the number of households that has been read and computed. After running the BasePop has completed, output tables for the base population will appear on your screen under the menu tree ④ (see Fig. 17.3). You can click any line of the tree to view the corresponding output tables and information as the result of running BasePop. The output tables of BasePop include various information (shown in the menu tree) for you to understand your data set. Comparisons between the model count and the direct count are given. The "model-counts" of the number of households by type and size are based on the characteristics of the reference persons following the ProFamy accounting system. The "direct-counts" of the numbers of households by type and size are derived following the standard census tabulation approach directly based on the codes that record household membership and relationship to the head of the household. Comparisons between the "model-counts" and the "direct-counts" reveal how accurate ProFamy is in modeling the households of your population. The other output tables are self-explanatory and thus there is no need to describe them in detail here. You can print the results by clicking 'Print'.

These outputs from running BasePop are useful for you to understand the census micro data files used to derive the starting year's population classified by the statuses recognized in the ProFamy model. If any of the output is not reasonable, you may need to check your raw data to identify and resolve the problems. We have

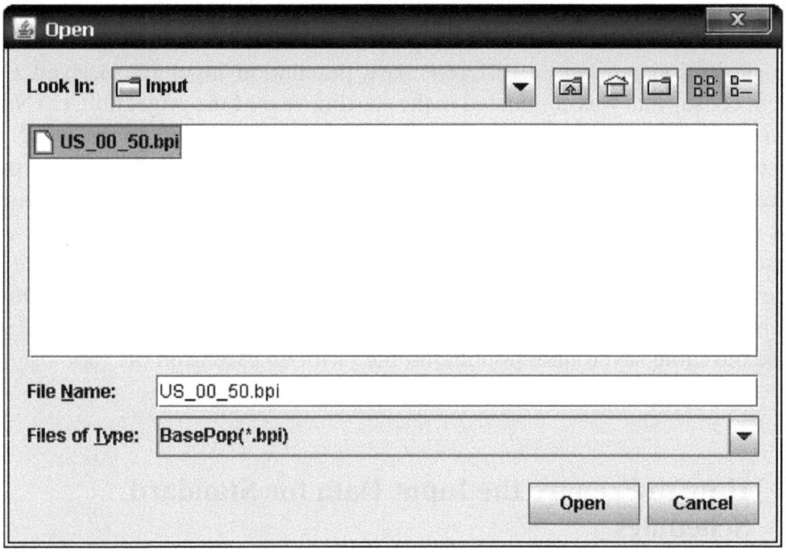

Fig. 17.2 Open input file of the BasePop

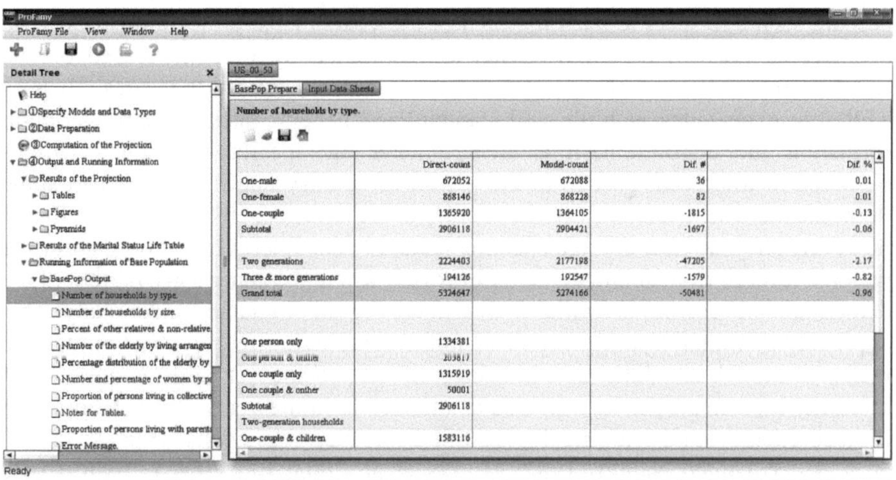

Fig. 17.3 Menu tree of the output tables of the base population

tested the BasePop subprogram using the census micro data for the whole U.S., each of the 50 U.S. states and DC, the whole China, each of the 31 provinces of China, and Germany; the results of the tests are satisfactory.

Once you have successfully executed "Run" (i.e., running BasePop) in the "Prepare ProFamy Base Population", you do not need to do it again, as long as you do not change the very basic design of the model. The very basic design consists of four aspects: (1) Starting year of the projection. Note that a different starting year of the projection requests a different base population input file derived from a different census that was conducted in the starting year of the projection. (2) Number of marital/union statuses. (3) Whether you wish to compute rural–urban differentials/dynamics in the projection. (4) Whether the family household projection is classified by race. If these four aspects of the basic design are not altered, you do not need to re-execute "Run" in the "Prepare ProFamy Base Population", since the base population file has been created and saved by the software. Each time if you do a new projection or scenario by changing the controlling parameters or summary measures or standard schedules, while keeping the basic design of the model intact, you use the same saved base population file (with an extension of ".bpo").

17.2 How to Prepare the Input Data for Standard Schedules

In the household projection using ProFamy, we need to prepare only one set of the needed *standard schedules* of age-specific demographic rates, but do not need to prepare such age-specific rates for each of the years of the projection period. The age-specific rates in the future years are estimated by ProFamy based on the standard schedules and the projected future years' summary measures, such as TFR, life expectancy at birth, and standardized general rates of marriage/union formation and dissolution. It is ideal to have age-specific standard schedules observed in the recent past from the region or country under study. When some of the standard schedule(s) from the country or region under study are not available, however, one may use the age-specific standard schedule(s) based on data at the national level or from another country or region where the general age pattern of the demographic processes is similar to that in the region or country under study. This approach is similar to the practice of jointly employing the regional model life tables as model schedules and projected life expectancy at birth as anticipated mortality level to project age-specific death rates in future years. Such an approach is theoretically and empirically justified in many previous studies (Coale et al. 1983; United Nations 1982; Murray et al. 2003; Brass 1974, 1978; Coale and Trussell 1974; Rogers 1986; Booth 1984; Paget and Timaeus 1994; Zeng et al. 1994), and reviewed and further assessed in Zeng et al. (2013a), and summarized in Sect. 2.2.4 of Chap. 2.

In the classic population projection, both single-year age-specific and the 5-year age-specific classification can be employed. This is, however, not the case for family household projection. Within a 5-year period a person may, for example, get married, leave the parental home, experience a first birth and/or second birth, and get divorced. Obviously, a model with 5-year age-specification cannot adequately handle the complicated calculations of these related events. Therefore, we

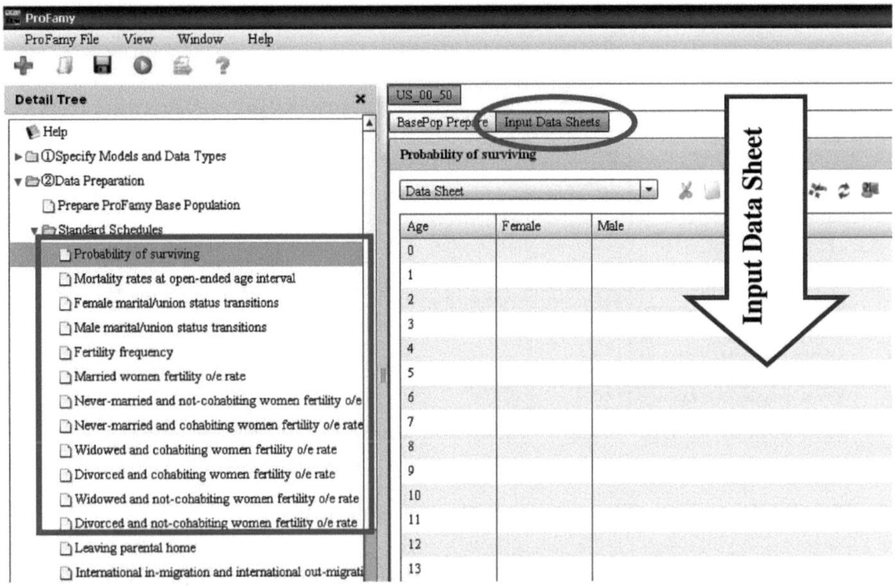

Fig. 17.4 Menu tree for standard schedules

use single-year-age classification in the ProFamy model. Due to the limitations of the sample size of survey data sets, one can first estimate the 5-year age-specific occurrence/exposure (o/e) rates, and then interpolate them into single-year age-specific o/e rates.

17.2.1 Input Data Sheets

When you click menu tree ② "Data Preparation" (see Fig. 17.1), you will also see the categories "Standard Schedules" and "Summary Measures". If you use the micro sample data file (rather than 100 % data file), there will be another category of "100 % Data in Starting Year" shown on the screen at the bottom of menu tree ② where you will provide the 100 % age-sex distributions of the population published by the statistical office.

After you click "Standard Schedules", you will see a tree of standard schedules in the left panel and "Input Data Sheet" for each schedule in the right panel of Fig. 17.4. The tree of standard schedules depends on your model design. For example, if you have chosen a model with seven marital/union statuses, the data sheet includes standard schedules for 13 sets of age-sex-specific o/e transition rates. If four marital statuses are distinguished in the model, the corresponding four sets of standard schedules of marital status transitions are shown on the list. Similarly, you will need to provide standard schedules of age-parity specific fertility for married

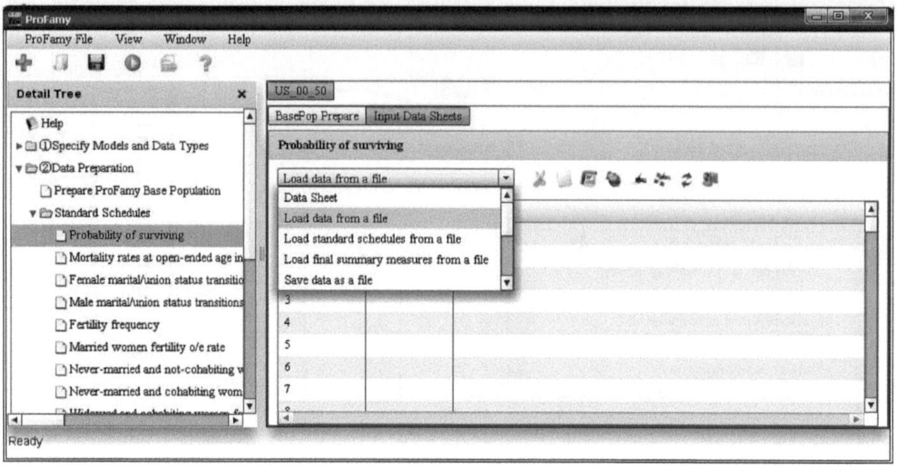

Fig. 17.5 Load data from a file under standard schedule menu tree

women only if you have chosen to assume that non-marital fertility is negligible. Otherwise, the fertility standard schedules for non-married women will need to be provided, too.

You can work on ProFamy input data sheets to prepare your standard schedules and other data by keying in the data or by reading in data from existing files. The ProFamy input data sheets are compatible with Microsoft Excel worksheets and Word so that you can copy/paste the data from Excel and Word files. To key in the data, you can simply move the cursor to the correct place, key in the figure, and then, using the arrow key, advance to the next line to key in another figure, and so on. You can highlight some lines and/or columns and then use "copy" and "paste" to copy these lines, or use "delete" to delete them. You can also copy columns and lines from another text file or Excel file or Word file and then paste them into your data sheet, and vice versa.

In addition to the option of keying in the figures using ProFamy's input data sheet, users may also choose to read in data from existing files by clicking the "Input Data Sheet" at the top line of the window and choose "Load data from a file" (see Fig. 17.5). The number of lines and the order of lines in those existing input files should be exactly the same as what are shown in the corresponding ProFamy input data sheets. The number of figures and the order of the figures in each line should also be exactly the same as what are shown in the corresponding ProFamy input data sheets. The positions of the figures, however, are flexible; they can be separated by one, two, three or more spaces or by a comma sign (","). The number of digits after the decimal point is flexible. ProFamy can read the data correctly as long as the number of figures is correct and they are in the proper order.

You can use the same methods and user-friendly data entry sheet as described above to enter and/or modify all of the standard schedules, summary measures, and 100 % data in the starting year, so we will not repeat these descriptions again.

Note that when the user clicks any of the standard schedules in the left panel menu tree of "Standard Schedules", ProFamy will display its corresponding figures in the input data sheet of the same window and highlight the selected title. When you are keying the data in the input data sheet, ProFamy will tell you what standard schedule is being keyed in by showing its title at the top of the same window.

17.2.2 Standard Schedules of Mortality

When you click "Probability of surviving" under the "Standard Schedules" tree, you get the data sheet on single-year-age- and sex-specific life table probabilities of survival from age 0 to x. These data can be easily obtained from recent life tables for males and females in the country or region under study. The ages for this data sheet to be presented below are all up to the highest age (e.g., age 100+ or age 85+) specified by the user as one of the model design parameters (see Fig. 16.5). The ages and the data shown in Fig. 17.6 and other data sheet figures printed in the text are only part of the data in the ProFamy input file, due to space limitations of the printed screen. *You can of course always view and input/edit data for other ages by scrolling (clicking the arrow ▼and ▲ or > and < at the right side or bottom of the data sheet).* This applies to all other data sheets to be presented and discussed later, and will not be repeated hereafter.

You will need to provide "Mortality rates at open-ended age interval", which is defined as the total number of deaths over the highest age of the life table divided by the total number of person-years lived over the highest age. This is totally different from the probability of surviving from age 0 to age x described above.

17.2.3 Standard Schedules of Marriage/Union Formation and Dissolution

You can prepare the female and male o/e rates of marital/union status transitions by clicking "Female marital/union status transitions" and "Male marital/union status transitions"; you will get the data sheet shown in Fig. 17.7. You simply follow the format shown on the data sheet to key in, read from a file, or copy/paste data from Excel or another kind of compatible worksheet.

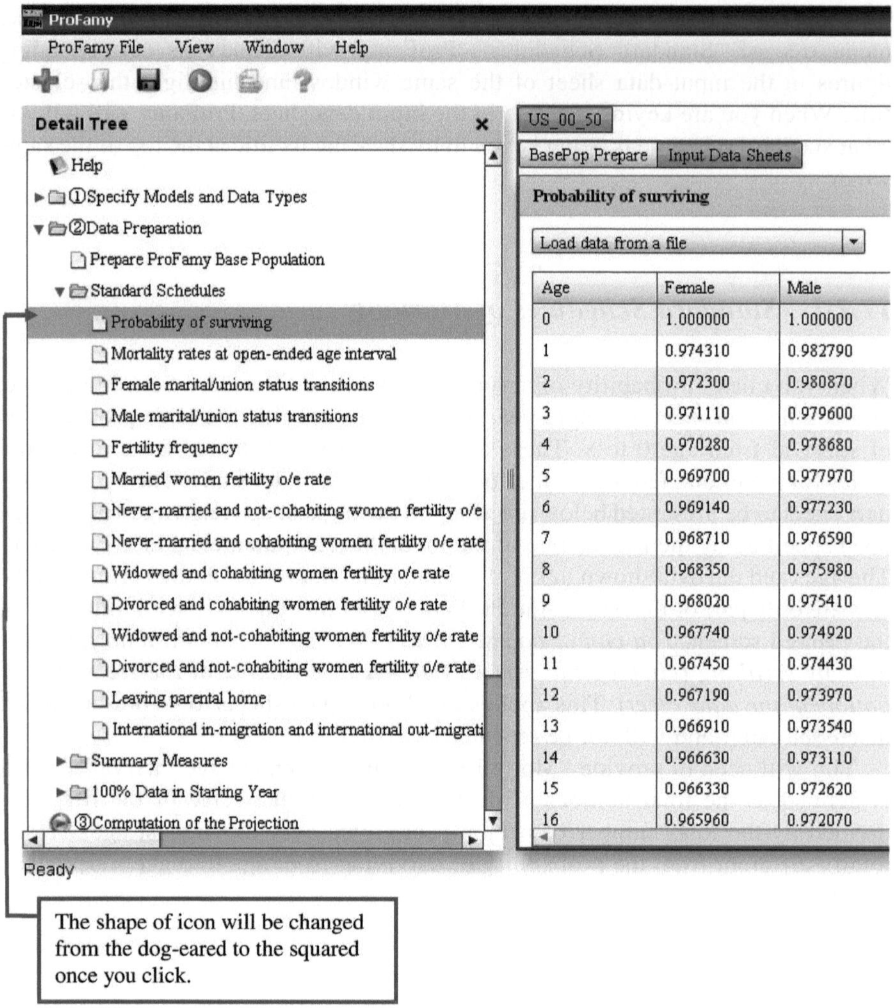

Fig. 17.6 Data sheet of age-sex-specific probabilities of surviving

17.2.4 Standard Schedules of Fertility

"Fertility frequency" is the age-specific frequencies of all birth orders combined, namely, the most commonly used age-specific frequencies, which are defined as the number of births of all orders combined divided by total number of all women aged x. Many people normally call the "age-specific frequencies" as "age-specific fertility rates."

When you click "Married women fertility o/e rate" under the standard schedule tree, you get the data sheet shown in Fig. 17.8. The columns "1Birth", "2Birth", "3Birth", "4Birth", and "5Birth" contain age-parity specific o/e rates of birth orders

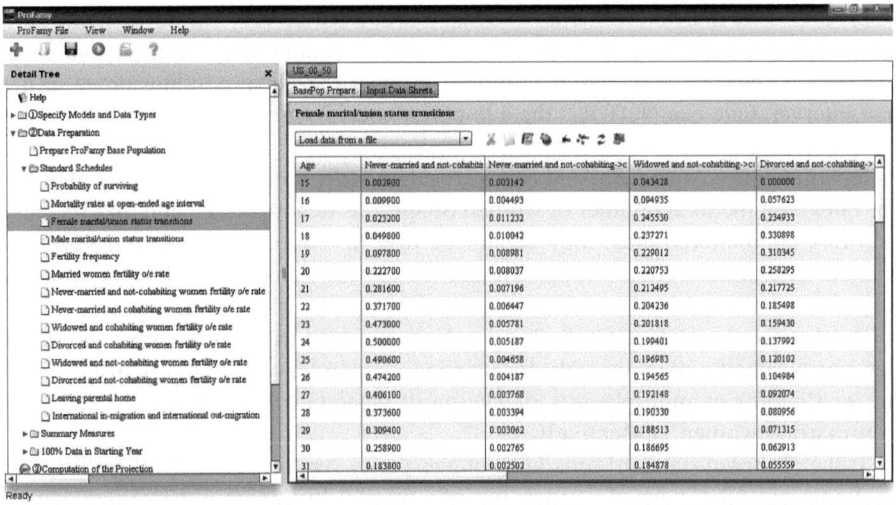

Fig. 17.7 Data sheet of age-sex-specific o/e rates of marital/union status transitions

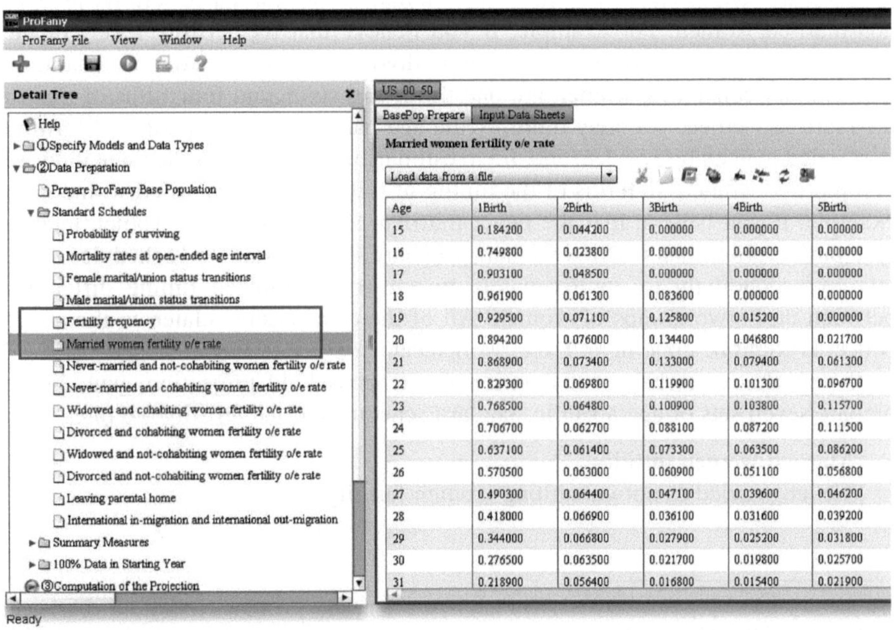

Fig. 17.8 Data sheet of fertility standard schedules

1, 2, 3, 4, 5+ by married women. Note that a user must be very careful to make sure to provide o/e rates in these columns, and the denominators of these o/e rates are person-years lived at risk of giving a birth of the corresponding order in the age interval. One can NOT use the age-parity specific frequencies of births (many people normally call the "age-specific frequencies" as "age-specific fertility rates") to substitute the o/e rates. This is because the age-parity specific frequencies of births (or rates) use the total number of women of age x as denominators and mix risk and non-risk populations; thus, they cannot be used to compute changes in status of number of children. This is important and we specially emphasize this requirement here to draw the readers' attention.

As discussed earlier in Chap. 16, you have a choice to provide either one set of age-parity specific o/e rates only for married women (assuming non-marital births are negligible) or multiple sets of age-parity specific o/e rates for married and non-married women, respectively.

If the observed standard schedules of age-parity specific o/e rates of fertility are available only for married women but non-marital births are not negligible, you may assume that the age-pattern of fertility of non-married women is the same as that of married women or assume some systematic difference in the age-pattern of fertility between married and non-married women. Of course, the non-marital fertility level differs from the marital fertility level. You may proportionally modify the standard schedules of married women to match your estimated fertility level of non-married women. For example, if you believe that the fertility levels of the never-married & not-cohabiting women, divorced or widowed not-cohabiting women, and cohabiting women are one-third, one-sixth, and four-fifths of that of the married women, you may multiply the age-parity-specific o/e rates of fertility of married women by 1/3, 1/6, and 4/5 to estimate the o/e rates for the three kinds of non-married women. In terms of the timing of fertility of non-married women, for example, if you believe that the never-married & not-cohabiting women tend to give births earlier than married women, you may shift the standard schedules to the left correspondingly to approximately match your estimated timing difference between these two groups of women. All of the calculations related to the above concerns could be done in an Excel worksheet.

If you choose to have cohabitation in the household projection, you should prepare seven sets of age-specific standard schedules of the fertility o/e rates[1]:

- "Married women fertility"
- "Never-married & not-cohabiting women fertility"

[1] In the example input data derived from the U.S. population, the fertility rates for cohabiting women with different legal marital statuses are assumed to be the same because sub-sample size problems prevented us from estimating them separately. A user can, however, provide different fertility rates for cohabiting women with different legal marital statuses if the data are available.

Note that seven marital/union statuses are distinguished in the U.S. applications. If the data for cohabitation unions are not available or cohabitation is not substantial, one may distinguish four traditional marital statuses (never-married, married, widowed, divorced) in the applications.

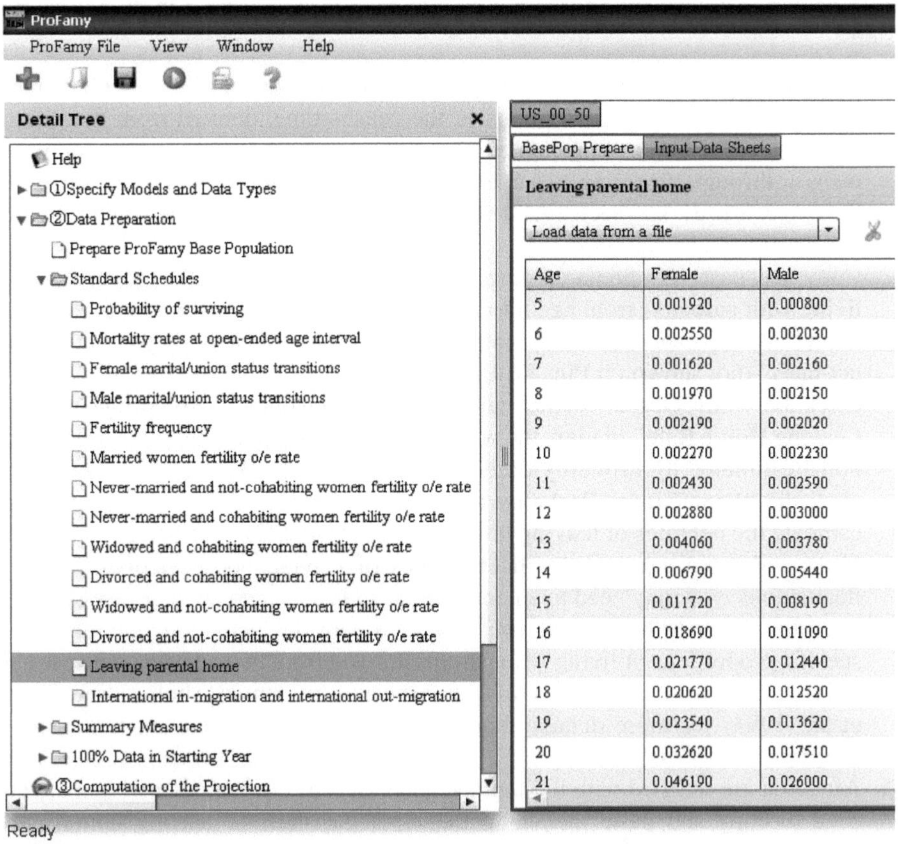

Fig. 17.9 Data sheet for standard schedules of leaving parental home

- "Never-married & cohabiting women fertility"
- "Widowed & cohabiting women fertility"
- "Divorced & cohabiting women fertility"
- "Widowed & not-cohabiting women fertility"
- "Divorced & not-cohabiting women fertility".

17.2.5 Standard Schedules of Leaving the Parental Home

When you click "Leaving parental home", you will get the data sheet shown in Fig. 17.9:

The procedure of employing the ProFamy software to estimate the age-sex-specific net rates of leaving the parental home based on two adjacent censuses data and the iterative intra-cohort interpolation method is summarized in the following steps:

1. Prepare single-year-age- and sex-specific proportions of living with parent (s) from age 5 to 49 at two points of the census time, derived from the micro data files of the two most recent adjunct censuses. You can do so either on your own, or through preparing two BasePop input files based on micro data files of the two censuses (see Sect. 17.1.1 for the format of such input files) and running the BasePop subprogram twice.
2. Copy and paste the prepared single-year-age- and sex-specific proportions of living with parent(s) from age 5 to 49 at two points of census time into the data sheet "two adjacent census data for estimating children's leaving parental home net rates" (not shown on Fig. 17.9).
3. Click the "View Output" in the main menu and then click "Estimating Children Leaving Home Rate" to view the results of age-sex-specific net rates of leaving home estimated by ProFamy based on steps (1) and (2). Note that when you click the "View" icon, ProFamy automatically performs the computation to estimate the net rates of leaving home if steps (1) and (2) have been completed.
4. If the estimated age-sex-specific net rates of leaving home involve a lot of fluctuations, you may need to smooth the rates. If the estimated age-sex-specific net rates fluctuate too much, you may also consider smoothing the age-sex-specific proportions of living with parents derived from two adjacent censuses in order to get better estimates of the net rates of leaving home (see Zeng et al. (1994) for more details about how to estimate the leaving home rates based on the census data).
5. After you are satisfied with the estimated net rates of leaving parental home, you need to copy and paste them into the input datasheet of leaving home rates (ProFamy will not do so, since the estimates need to be checked and approved by you).

17.2.6 Standard Schedules of Migration

The standard schedules of domestic migration refer to age-sex-specific (and marital status specific, if possible) frequencies of domestic in-migrants from the rest of the country and age-sex-specific o/e rates of domestic out-migration to the rest of the country.

17.2.6.1 A Note on Application of Net Migration Standard Schedules

NM(x) – number of net migrants at age x; it can be positive or negative;
n(x) – age-specific net migration frequency, it can be positive or negative.

$$n(x) = NM(x)/\sum_x NM(x); \sum_x NM(x) \text{ can be positive or negative.}$$

$$\sum_x n(x) = 1.0.$$

For U.S. international migration, we may use the sex-age-specific international net migration frequencies and the projected total numbers of international net migrants to compute future years' age-specific numbers of net migrants without problem, because the total number of the U.S. international net migrants in the past and foreseeable future years is always positive. However, such a net migration approach may not be applied to internal migration projection at the sub-national level or to international migration for countries other than the U.S. because the total number of internal or international net migrants could change from positive to negative or vice versa across years. For example, suppose that the net internal migration frequency is positive for young persons and negative for old persons, and the total number of net migrants is positive based on the census data. But the total number of net migrants may become negative in the future, due to economic problems in the population under study. Consequently, multiplying the negative total number of net migrants by the standard age-specific internal net migration frequencies based on the most recent past census data will result in negative numbers of net migrants for younger persons and positive numbers of net migrants for older persons, which are wrong.

Therefore, we cannot use the net migration approach for projection at national or sub-national level, unless one assumes that the sign of the total number of net migrants in future years is the same as that observed in the most recent census on which the sex-age-specific standard schedules of frequencies of net migration are based.

If you distinguish rural and urban areas in the projection, single-year age and sex-specific frequency distributions of rural–urban net migration will be needed. The frequency distribution is defined as the single-year age- and sex-specific number of rural–urban net migrants divided by the total number of rural–urban net migrants. The number of net migrants is the difference between the number of migrants moving from rural to urban areas and the number of migrants moving from urban to rural areas. Note that the rural–urban dynamics are internal migrations within the country or region under study. If rural–urban classification is specified, we will also need to provide, for rural and urban sectors separately, the age-sex-specific frequencies of in-migration from the outside of the country or region under study and age-sex-specific o/e rates of out-migration from the country or region under study, or the age-sex-specific frequencies of net external migration.

17.3 How to Prepare the Summary Measures

After you have prepared the standard schedules, you should click "Summary Measures". Note that the following summary measures implied by the base population and standard schedules provided by you will be computed and displayed by ProFamy at the top of the corresponding data sheets for the summary measures in grey color if the base population and standard schedules have been properly prepared.

1. Standardized general rates of marriage/union formation and dissolution at the starting year of the projection, implied by standard schedules of marital/union status transitions and age-sex-marital/union-status distributions in the starting year derived from the census data. When specifying the model and data type (described in Chap. 16), you also need to provide the total number of marriages and divorces in the starting year of the projection to facilitate the estimates of the standardized general rates of marriage and divorce.
2. Age-sex-marital/union status-specific proportions of those living in institutional households, and age-sex-marital/union-status-specific proportions of elderly living with adult children in the starting year derived from the census data.
3. The proportions of household reference persons not living with parents among persons aged 45–49 in the starting year derived from the census data.
4. Average number of other relatives or non-relatives per household classified by the number of direct family members in the starting year derived from the census data. "Direct family members" refers to a spouse or partner, parents, and children.

These summary measures implied by the standard schedules and the age, parity, marital/union status distributions at the starting year of the projection derived from the census data serve as a starting or reference point for projecting the summary measures in the future years. You can either assume that they will increase or decrease to a certain extent or remain unchanged, based on expert opinion or trend extrapolation.

You need to at least provide the projected (or assumed) summary measures at the starting year and the ending year of the projection. You can also choose some other years (or no years if you like), for which you wish to provide projected summary measures; you do so by clicking the "" under "provide summary measures in years other than starting/ending year" at the top middle panel of the right dialog window (see Fig. 17.10).

You have the following four choices:

Single Year
Year ended by 5 and 0
Year ended by 0
Key in the specific year

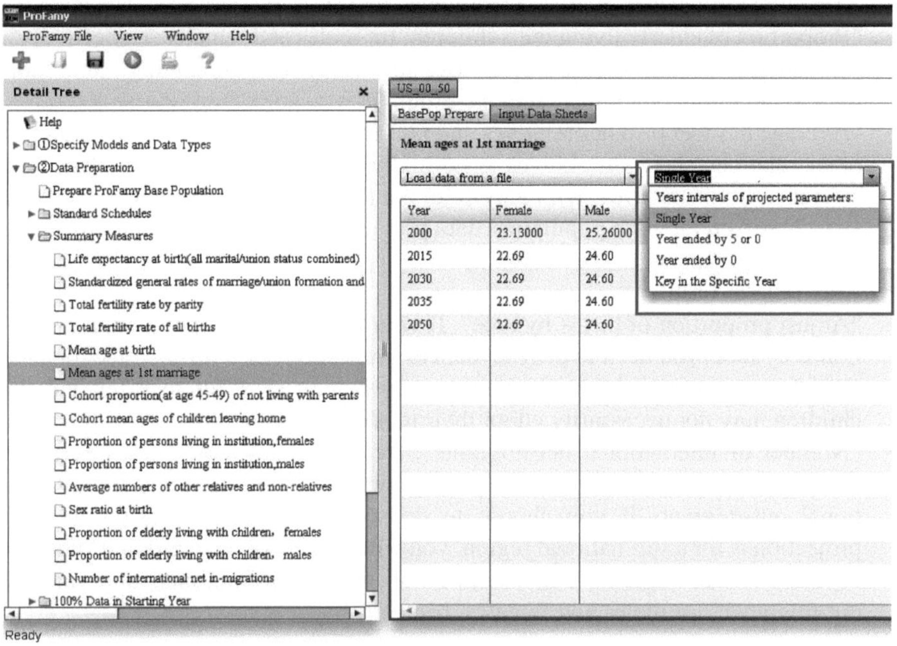

Fig. 17.10 Data sheet for summary measures

You can use the up or down arrow key or mouse to highlight your choice. If your choice is "Single year", "Year ended by 5 and 0", or "Year ended by 0", the years will automatically appear at the left column of the data sheet.

If you chose to key in the years, you have to specify the years for which you want to provide the summary measures. ProFamy will use linear interpolation between the specified years to automatically estimate the summary measures in all other years in which no projected summary measures are provided.

The following summary measures will need to be prepared:

- "Life expectancy at birth" for males and males, respectively.
- "Standardized general rates of marriage/union formation and dissolution."
- "Total fertility rate by parity" – the sum of the parity-age-specific fertility frequencies of women with all marital/union statuses combined.
- "Total fertility rate of all births" – the sum of the age-specific fertility frequencies of all births with different parities of women with all marital/union statuses combined.
- "Mean age at birth" – mean age at birth of all parities combined.
- "Mean age at first marriage" for males and females, respectively.
- "Proportion (at age 45–49) of not living with parents" – Proportion of children eventually leaving parental home, for males and females, respectively.
- "Mean age of children leaving home" – mean age of children leaving parental home, for males and females, respectively.

- "Proportion of persons living in institutions, for males and females, respectively".
- "Proportion of elderly living with children, for males and females, respectively". (the elderly are classified by categories of "married or cohabiting" and "not-married and not-cohabiting" and age groups).
- "Average number of other relatives and non-relatives" – The average number of other relatives and non-relatives is normally estimated from census data. The references in the first row with gray color are the average numbers in the starting year, based on the base population data provided by you (refer to the procedure described in Appendix 1 of Chap. 2).
- "Sex ratio at birth".
- "Adjust proportion of births by race". If your household and population projection is by race, you need to provide data of proportions of births by race for each race, because some children's parents may belong to different races and the children may not necessarily adopt their mothers' race.
- "Number of international net-migrants", for males and females, respectively. You may need to provide the numbers of international in-migrants and international out-migrants if you choose to do so in your model design. If your projection is for a sub-national region, you will also be asked to provide "number of internal net-migrants" or "number of internal in-migrants and internal out-migrants" for males and females, respectively.

Note that "standardized general rates of marriage/union formation and dissolution", "proportions of persons living in institutions", "proportions of elderly living with children" and "average number of other relatives and non-relatives" are normally estimated from census data; the numbers in the first row with gray color in this datasheet refer to the starting year, based on the base population data provided by the user.

17.4 How to Prepare Input Data for the Total Population (100 % Tabulation) By Age, Sex and Marital Status

As discussed earlier, one needs aggregated 100 % tabulations of the total population to adjust the sample data set to ensure correct total population size, age, sex, and marital status distributions at the starting year of the projection. You can choose to key in the age-specific data following ProFamy's user-friendly interface, or prepare the data files based on existing electronic files and then let ProFamy read in the age-specific data from your prepared data files. In this case, you should follow the file format described below.

The number of lines and the order of lines of the input data files must be exactly the same as in the corresponding ProFamy data sheet (e.g., Table 17.1), as shown on the screen of the ProFamy interface. The number of figures and the order of the figures in each line should also be exactly the same as in the ProFamy data sheet

Table 17.1 An example of a data file of 100 % population by age, sex, and marital status

Age	F. Never Marr	M. Never Marr	F. Married	M. Married	F. Widowed	M. Widowed	F. Divorced	M. Divorced
0	4,261,158	5,083,621	0	0	0	0	0	0
1	3,342,166	4,193,972	0	0	0	0	0	0
2	4,267,929	5,296,089	0	0	0	0	0	0
3	4,461,714	5,452,180	0	0	0	0	0	0
	.							
	.							
	.							
20	4,195,515	4,996,834	815,273	187,657	938	393	9,549	3,602
21	3,593,139	4,838,799	1,572,525	420,502	1,603	755	9,507	5,109
22	2,728,137	4,298,008	2,526,447	1,110,434	3,329	1,627	16,063	10,531
23	1,838,670	3,349,629	3,146,002	1,917,966	4,643	2,653	16,598	14,400
24	1,415,319	3,055,287	4,340,848	3,042,421	7,531	4,942	23,246	24,857
	.							
	.							
	.							
98	11	81	208	999	9,714	3,966	4	7
99	7	45	118	523	6,474	2,364	2	3
100	9	45	127	504	8,364	2,628	2	3

(e.g., Table 17.1). The positions of the figures are, however, flexible. In other words, the figures can be separated by one, two, three or more spaces or by a comma sign (","). ProFamy can read the data correctly, as long as the number of lines and their orders and the number of figures in each line and their orders are correct.

17.5 Input Data for the Total Institutional Population (100 % Tabulation) By Age and Sex

Data on the total institutional population (100 % tabulation) by age and sex will need to be provided. As shown in Fig. 16.7, they can be 5-year age specific, or they can be other age-groups specified by you (such as age 0–17, 18–34, 35–64, 65+) depending on the age classification adopted by the statistical office in their 100 % census tabulation publication. They can also simply be total numbers of institutionalized males and females if age information is not available. In the later two cases, will interpolate the figures into 5-year age specific data, based on the age and sex distribution of the institutional population derived from the micro sample data set.

An important note:

You must keep consistent between the model design parameters and data types specified (described in Chap. 16) and the input data prepared (described in this chapter). For example, if you specify that the starting and ending year of the projection are 2000 and 2050, you must provide the summary measures in

2000–2050, while it is flexible and optional to provide summary measures in years between 2000 and 2050; if you change your starting and ending year to other years in your model design, you must accordingly change the summary measures. Another example is that if you change the highest age identified from age 85 to 100, all of the relevant data (e.g., mortality, base population, etc.) must be extended to age 100. Any inconsistency between the controlling parameters of model design, the data types you specified and the input data you prepared will cause failures or errors in the computation of projections.

17.6 How to Run "Computation of the Projection"

Computation of the Projection

Up to here, you have read through and completed the tutorial parts of setting up the projection model and preparing the input data. You are now ready to enter the last part of the process – running the ProFamy program and viewing/managing the output.

Click "Computation of the Projection".

If you are doing the tutorial with the given example of ProFamy input file or if you are doing a new projection/scenario based on one of your previous projection/scenario and you changed some input data and/or model design, you will see a message on the screen saying "The output files already exist; Do you want to overwrite the output file...?" after you click the menu tree ③ "Computation of the projection". In such cases, we recommend that you click "No" and then click "Output" and provide a new name for the "basic part of the filenames of output files" to save the new output in a new file (or exit the program if you do not want to save anything). In this way, you will ensure that the example input/output files accompanied with ProFamy software will not be changed, which is important for you and others to use later. If you are doing a new projection/scenario based on one of your previous projection/scenarios and you changed some input data and/or model design, you need to keep the previous output and save your new output for comparative purposes. Only if you know the previous output is wrong and there is no need to save should you click "Yes" to overwrite the previous output file.

After you click the menu tree ③ "Computation of the projection", ProFamy will first check if all of your data preparations have been completed, and if all of the data you provided are properly within the logical ranges. For example, ProFamy will check if all of the values of age-sex-specific demographic rates of standard schedules are less than one and greater than or equal to zero, and whether other logic constraints are met. If there are no logical errors and all of your data preparations have been completed, the external DOS program will be executed. It takes about half a minute to finish the computation with a PC. When "Computation of the projection" has completed, the message "Running computation of Projection is done" will appear. After you click "OK", you will enter the output data sheets tree widow. If something is wrong and the computation does not complete, an error message "Computation of the Projection has failed" will appear on the screen.

17.7 How to View and Manage the Results

After you have successfully completed the computation of the projection, you can click the menu tree ④ "Output and Running Information" to view and print the output tables and graphics (see Fig. 17.11).

ProFamy Results of the Projection

There are three kinds of outputs for results of the projection: Tables, Figures, and Pyramids, which are indicated as sub-menus under menu tree "Results of the Projection" at the left panel of the window (see Fig. 17.11). You can switch among these three kinds of outputs by clicking the sub-menu tree.

Click "Tables" to view the projected outputs in table formats.

Within output of "Tables",

- Click "Summary of Households" to see the number and percent distributions of households by type and size in different years;
- Click "Summary of Living Arrangement" to view the projected age-sex-marital status-specific living arrangement for elderly in different years;
- Click "Summary of Population" to view the projected summary statistics and population dynamics (population size, etc.) in different years.
- Click "Detailed Tables of households" to view the projected number of households by type, size, and age, sex, and marital status of the reference person in different years.
- Click "Detailed Tables of Population" to view the projected number of individuals classified by age, sex, and marital/union status in different years.

All of the ProFamy output tables are Excel and Word compatible, and you can manage, re-tabulate, draw your own graphics, etc., through copy/paste and other

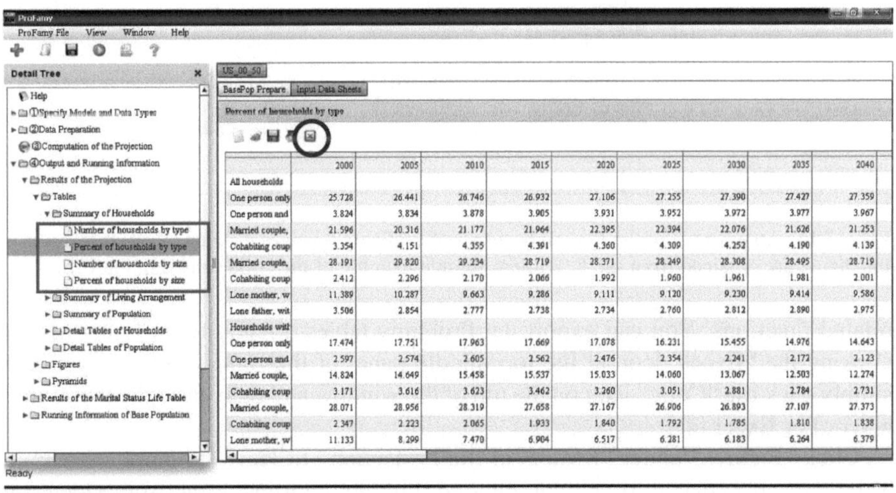

Fig. 17.11 Menu tree ④ "Output and Running Information"

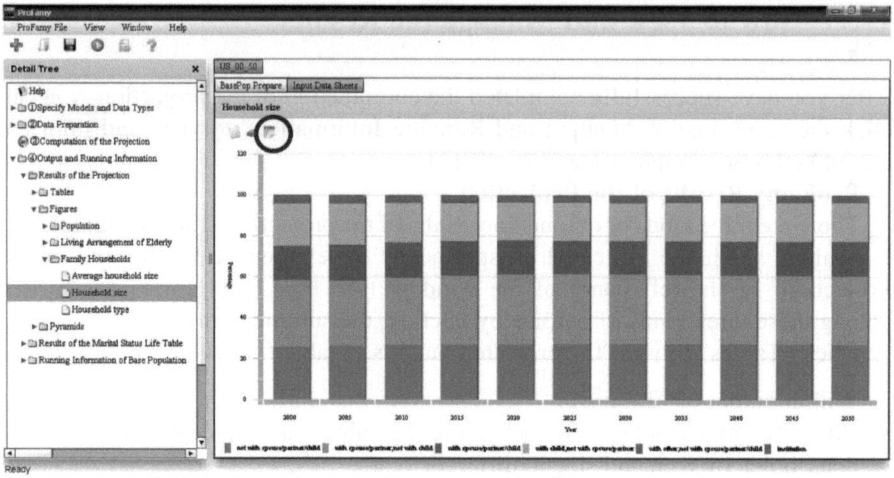

Fig. 17.12 An example of the graphics output of ProFamy

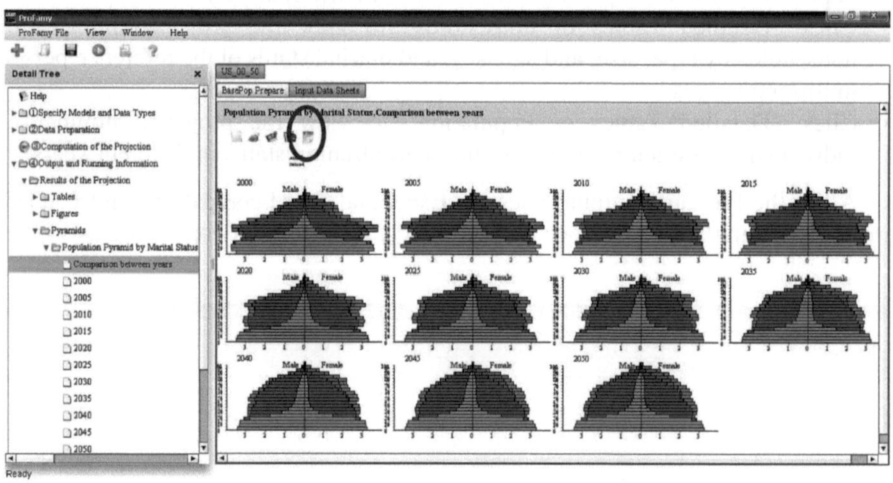

Fig. 17.13 An example of the Pyramids output of ProFamy by marital status (Note: the results in this figure are obtained from the 5 % micro sample data file of the 2000 census in the U.S.)

related functions. You can export the ProFamy tables directly to an Excel file by clicking the "Excel" icon at the top middle panel of the right screen, or you can simply click the "Copy" icon on the top of the window and then paste the tables into a Word or Excel file.

Click "Figures" under "Results of the Projection" to view the graphics output. They are self-explanatory, so you may go through them just to get some ideas on what kind of graphic outputs are available in the current version of ProFamy. Figure 17.12 shows an example of the graphs and Fig. 17.13 shows an example of the pyramids. For the graphs or pyramids, you can simply click the “Copy” icon and then paste the graphics or the pyramids into a Word or Excel file. If you want to see the data used to plot the graphs and pyramids, you can click the “data” icon at the top middle panel of the right screen. You can print the tables, graphics, and pyramids by clicking the “Print” icon.

By integrating Component One Chart version 7.0, ProFamy empowers the user to change the design of the output graphics and type of charts. By right clicking the mouse within the graphics, you can modify or refine the figures, including scale, figure title, margins of figure, 3D effects, color, and so on.

Click “Pyramids” at the bottom left corner of the ProFamy output view window to view the population pyramids by marital status. If you want to see animated dynamics of population pyramids from the base-year to the ending year of the projection, you should click “Timing Dynamics of Population Pyramids”. You can also click the “speed” icon to substantially speed-up the timing dynamics show. But please do NOT click the “Speed” icon unless you have put the cursor on one of specific pyramids under “Timing Dynamics of Population Pyramids”. Otherwise, ProFamy will exit automatically.

Chapter 18
Epilogue: Summary and Future Perspectives

18.1 Methodological Core Ideas and Empirical Assessments of the ProFamy Extended Cohort-Component Approach and Comparisons with the Classic Headship Rate Method

The methodology of the ProFamy extended cohort-component model have been presented, summarized, and justified as four core ideas (see Sect. 2.2 of Chap. 2): (1) It employs a multistate dynamic modeling framework, uses groups of individuals as the units of household projections, with demographic rates as input, and identifies households by various types and sizes based on the reference person's characteristics. (2) The ProFamy model adopts a computational strategy and accounting equations which distinguish continuously occurring from periodic demographic accounting processes and largely simplify the data requirement (Bongaarts 1987; Zeng 1991a); (3) It adopts a judicious use of stochastic independence assumptions with reasonable justifications in the model; it follows the harmonic mean methods for ensuring consistency between males and females and between parents and children; (4) it employs national model standard schedules and time-varying summary parameters at the sub-national level to specify projected demographic rates for a sub-national area in future years.

Tests of projections from an earlier census year to a later census year of 10 years apart using the ProFamy model and based on observed Chinese and U.S. demographic rates before the earlier census year show that forecast errors, measured by discrepancies between the projected values and the census observations in the later census year, are reasonably small at the national level (see Tables 4.1 and 4.2 of Chap. 4).

Applying the ProFamy extended cohort-component method and national model standard schedules of age-sex-race-specific demographic rates based on commonly available survey and census data, comprehensive and simultaneous projections

Y. Zeng et al., *Household and Living Arrangement Projections*,
The Springer Series on Demographic Methods and Population Analysis 36,
DOI 10.1007/978-90-481-8906-9_18,

of households and populations at the sub-national level requires a census micro-data file and projected (or assumed) demographic summary parameters. The test comparisons of projections of households and populations from 1990 to 2000 that used the ProFamy model with census counts in 2000 for each of the 50 U.S. states and DC show that 63.0 %, 17.4 %, 12.9 %, and 6.7 % of the absolute percentage errors are <3.0 %, 3.0–4.99 %, 5.0–9.99 % and ≧10.0 %, respectively, among 306 pairs of comparisons of the main indices of household projection between forecasts and the census observations (see Table 4.3 of Chap. 4). The test comparisons of projections of households and populations from 1990 (or 2000) to 2000 (or 2010) using the ProFamy extended cohort-component approach with census counts in 2000 for the Eastern, Middle, and Western regions and Hebei province of China also demonstrated that the forecasting errors are within a reasonable range (see Table 4.4 of Chap. 4).

These forecast evaluations provide the frame within which it can be stated that the validation test results show that the forecast errors of household and population projections using the ProFamy extended cohort-component method are within a reasonably and relatively small range. At the same time, however, we must be aware that these tests only validate the simulation properties of the ProFamy model and they do not validate the long-term projections of household structures and living arrangements, elderly home-cased care costs, and housing and household vehicle demand up to the year 2050 for the U.S and China presented in various chapters of this book, because the projected or assumed demographic summary measures in the middle- and long-term future years involve a lot of uncertainties.

Note that the classic headship rate method is still widely used for household projections (e.g. Berson et al. 2006; Nishioka et al. 2011). However, the headship rate method has been widely criticized by demographers for more than two decades. The headship rate method is not linked to demographic rates, and it projects limited household types (Bell and Cooper 1990; Mason and Racelis 1992; Murphy 1991; Spicer et al. 1992); its recent extensions project household size but with limited household types (Ediev 2007; Ediev et al. 2012). Thus, the headship rate method is insufficient for sound household consumption forecasting, especially in populations with quick and substantial familial and demographic transitions, because household consumptions are closely related to household size, composition, and other characteristics, which are heavily affected by changes in demographic rates.

In contrast to the headship rate method, the ProFamy approach projects all individuals' household statuses grouped by cohorts and specified attributes using demographic rates as input, and it projects much more detailed household types, sizes, and characteristics such as marital/union status and number of generations. These merits are particularly relevant to applications to societies which have been undergoing significant changes in household type, size, or other characteristics and demographic rates.

18.2 Extensions of the ProFamy Model to Project Elderly Disability Status, Home-Based Care Costs, and Pension Deficit Rates, with Illustrative Applications

The analyses presented in Chap. 5 project dynamic changes in disability status of older adults (measured by ADLs) and their home-based care costs, classified by age, gender, rural/urban residence, marital status, and number of co-residing children and co-residing parents. The key point of this analysis is a substantial extension of the ProFamy model by introducing and estimating changes in older adults' disability status as well as related home-based care costs. Our extended dynamic projection model includes all of the individuals in the population; it combines the projection of family structure, living arrangements, and ADL statuses for elders aged 65+ (depicted in Fig. 5.2 of Chap. 5) with the projection of family structure and living arrangements for the younger population aged 0-64 (see Fig. 2.3 of Chap. 2). The extended model in Chap. 5 projects not only ADL statuses and home-based care needs and costs for older adults, but also age-sex-specific numbers and family household structures of the working-age population, i.e., the caregivers for the disabled elderly.

The illustrative application of the extension of the ProFamy model to project elderly disability status and home-based care costs to China has resulted in several important and interesting findings: (1) Chinese disabled elders will increase much more rapidly than the total elderly population; (2) The disabled oldest-old aged 80+ will increase much faster than the disabled young-old aged 65-79; (3) Disabled elders living in empty-nest homes will increase much faster than those living with children; (4) The annual growth rate of percentage of national GDP devoted to home-based care costs for disabled elders will rise substantially faster than the growth rate of the population of disabled elderly; (5) Sensitivity analyses shown that possible changes in mortality and elderly disability status are the major direct factors affecting home-based care needs/costs; (6) Caregiver resources under two-child policy will be substantially better than that under the current fertility policy unchanged. In order to demonstrate that the ProFamy model to project elderly disability status and home-based care costs can be readily applied to other countries as well, we include these findings as an illustrative application immediately after descriptions of the extended method in Chap. 5 as part of the Part I (methodology) rather than Part III (China applications).

Chapter 7 presented a simple method associated with the ProFamy projection model and software to project the annual pension deficit rate based on (1) The elderly dependency ratio determined by demographic factors of fertility, mortality and migration; (2) The retirement age; and (3) Four (or three) pension program parameters, which can be predicted by trend extrapolation or expert opinions. These input parameters can be derived from commonly available data. The illustrative application to China demonstrates that if the average age at retirement gradually increases from the current very low level to age 65 for both men and women in

2050, the annual pension deficit rate would be largely reduced or eliminated under various possible demographic regimes up to the middle of this century. With everything else being equal, the annual pension deficit rate in the scenario of medium fertility (associated with a two-child policy) would be much lower than that under low fertility (associated with the current fertility policy unchanged) after 2030. The impact of potentially faster mortality decline is likely sizable but relatively moderate; it starts earlier than the effects of fertility change. Note that one may also use the simple method presented in this chapter to explore the magnitude and timing of impacts on future pension deficits due to alternative international migration and/or pension policies by predicting or assuming the size and age/gender structure of international migration and/or the pension program parameters through regression or expert opinion while fixing the fertility, mortality, and retirement age parameters. Again, in order to demonstrate that the simple method associated with the ProFamy projection model and software for projecting the annual pension deficit rate can be readily applied to other countries as well, we include the Chinese application as an illustrative case immediately after descriptions of the simple method in Chap. 7 as part of the Part I (methodology) rather than the Part III (China applications).

18.3 Household and Living Arrangement Projections at the Small Area Level

Chapter 6 presented and discussed the basic concepts and methodology for applying the ProFamy approach in combination with ratio methods (including the constant-share and shift-share ratio methods) to project household and living arrangement projections at the small area level. To assess the accuracy of the combined ratio method and ProFamy approach and present illustrative applications, we conducted projections from 1990 to 2000 and compared projected estimates with census-observed counts in 2000 for sets of randomly selected 25 small counties and 25 small cities which were more or less evenly distributed across the United States. The comparisons show that, in general, most forecast errors are reasonably small – mostly less than or slightly more than five percent. These results evidently illustrated the utility of the ProFamy approach in combination with ratio methods to project households and living arrangements at the small area level.

18.4 Applications to the United States

Part II (Chaps. 8, 9, 10, and 11) dealt with empirical applications to the United States. Chapter 8 projects households and living arrangements for the United States at the national level. Using data from national surveys and vital statistics, census micro files, and the ProFamy method, we presented projections of U.S. households and living arrangements from 2000 to 2050. Medium projection as well as projections based on smaller and larger family scenarios with corresponding combinations of assumptions of marriage/union formation and dissolution, fertility, mortality, and international migration were performed to analyze future trends of U.S. household structures, including their possible higher and lower bounds as well as enormous racial differentials. To our knowledge, these household projections (Zeng et al. 2006) were the first to have found empirical evidence of family household momentum and to have provided informative low and high bounds of various indices of projected future household and living arrangement distributions based on possible changes in demographic parameters in the United States.

Chapter 9 projected households and living arrangements for the five decades from 2000 to 2050 with medium, small, and large family scenarios, for each of the 50 states, DC, six counties of Southern California, and the Minneapolis-St. Paul Metropolitan Area. Among many interesting numerical outcomes of household and living arrangements projections with medium, low, and high bounds, the aging of American households over the next few decades across all states/areas is particularly striking.

Chapter 10 projected numbers of activities-of-daily-living disabled elderly and yearly payments and workdays of home-based care for them by age, gender, race, and living arrangements from 2010 to 2050 for the United States (with low, medium, and high scenarios). The chapter focused on how changes in household structure and living arrangements may affect future home-based care costs for disabled elders based on census micro datasets using the National Long Term Care Survey data and the ProFamy extended cohort-component method. The results showed a remarkable acceleration in numbers of disabled elderly aged 65+ after 2020 with a much faster increase in disabled oldest-old aged 80+, such that after 2030 they outnumber the disabled young-old aged 65–79. Increases in yearly workdays and payments of home-based care for disabled elders will dramatically accelerate after 2020, especially for the disabled oldest-old. We also discussed similarities and differentials across racial groups and genders and the policy implications of future trends in home-based care needs and costs for disabled elderly.

Chapter 11 employed the ProFamy extended cohort-component method to project household vehicle consumption from 2000 to 2025 across four regions of the United States (the Northeast, Midwest, South, and West). The results showed that the total number of household vehicles in 2025 will reach 235 million, representing a 31 % increase over the 25 years. About a half of the increase is due to the consumption of cars, while the household consumption of vans will increase at a faster rate than those of cars and trucks. Household vehicle consumption will grow more in White non-Hispanic and Hispanic households in comparison with Black non-Hispanic and Asian and other non-Hispanic households. Owners of household vehicles in the United States will be aging quickly. Among households with different sizes, the largest increase of household vehicles will come from two-person households. Across the four regions, the largest increase of household vehicle consumption will be in the South, followed by the West, Midwest, and Northeast.

18.5 Applications to China

Part III of this book (Chaps. 12, 13, 14, and 15) deals with applications to China. Using the most recent data available and the ProFamy method, Chap. 12 projected future trends of family households and elderly living arrangements in the context of rapid population aging in both rural and urban areas of China, under the medium fertility and medium mortality scenarios. Our study demonstrated that, while the population in China will be aging at a rapid speed and to a huge scale, particularly the oldest-old aged 80+, Chinese family households will continue to contract to a substantially smaller average size in the next a few decades. The proportion of elderly households with at least one person aged 65+ will increase dramatically in China in the next few decades. By the years 2030 and 2050, the proportion of the elderly aged 65+ living in empty-nest households without children among the total population will be 2.5 and 3.7 times that in 2000. The increase in percentages of the oldest-old living in empty-nest households will be even more dramatic: 4 and 11.5 times as high as in 2000 for the years 2030 and 2050. These aging population structure problems – with respect to proportion of elderly and elderly households as well as proportion of elderly living in empty-nest households – will be much more serious in rural areas than in urban areas. This strongly suggests that, to avoid serious social problems in the future China needs to change its household registration policy which restricts free movement from rural to urban areas and to adopt policies encouraging rural-to-urban family migration or family reunion after young migrants settle down in urban areas.

Chapter 13 presented the dynamics of household and living arrangements in the Eastern, Middle, and Western regions of China. The results showed that, if the current age distribution of rural-to-urban migrants with a high concentration of young people remains unchanged, the Middle region will have the most serious problems of population and household aging, followed by the Eastern region; population and household aging will be twice as severe in rural areas compared to urban areas in Middle and Eastern regions. Our multi-regional projections and analysis clearly showed that population and household aging problems under the current fertility policies unchanged would be much more serious than that under the two-child policy. Our study suggests that the inclusion of elderly parents in regional and rural to urban family migration, which implies co- or proximate-residence between old parents and adult children, would help to avoid the over-aging problems in rural areas and the Middle region; this strategy may result in a win-win outcome for both old and young generations.

Applying the ProFamy approach, Chap. 14 presented a comparative analysis among four different fertility policy transition scenarios that are currently being debated, including demographic projections of population growth and aging, elderly living alone, labor force, pension deficits, economic costs, marriage squeeze, and the socioeconomic implications. The results showed that there are three policy actions that China should consider as it faces the serious challenges of population and household aging. First, the two-child with encouragement of adequate spacing policy option is an optimistic and feasible strategy for China to sustain socioeconomic development into the future, and needs to be implemented as soon as possible. As compared to retaining the "current policy unchanged" and the other options, the "two-child with encouragement of adequate spacing" soft-landing policy would create much better demographic conditions and socioeconomic implications in the future, with respect to proportions of elderly and those elderly who live alone, labor force resources, pension deficit rates, sex ratio at birth, the marriage squeeze, and socioeconomic costs. The analyses in Chap. 14 also suggested that the "two-child with encouragement of adequate spacing" policy could be implemented without causing more out-planned births of third or higher order and new coercive events. Second, our projections and analysis led to the recommendation that the currently very low retirement age be gradually increased. A third recommendation is that rural old age insurance programs be further developed in order to reduce the more serious aging problems in rural areas compared to cities and to create a huge amount of capital for economic development by collecting premiums from hundreds of millions of new program participants. In sum, these three policy actions would enable China to successfully deal with the serious challenges of population and household aging in the coming decades.

Chapter 15 presented households and housing demand projections in rural and urban areas of Hebei, a province with 72 million residents and a median level of socioeconomic development in China, using the most recent census and other data and the ProFamy extended cohort-component model. The results showed that, due to changes in household and population structure, demand for both owned housing and rental housing for elders aged 65+ will grow dramatically, but housing demand will decline for young people aged less than 35. One-person households will have the largest growth in demand for 1–2 room owned-housing; one-couple households will have the largest growth in demand for 3–4 room owned-housing. Demand for larger housing units for three-generation households will go down. Based on our analysis, we discussed some relevant policy considerations such as "dual-apartment" housing for elderly parents neighboring directly with their children, which may result in a win-win outcome for both generations in facing the serious challenges of population and household aging.

18.6 User's Guide of the ProFamy Software for Household and Consumption Forecasting

Chapters 16 and 17 of Part IV present the User's Guide of the ProFamy software for household and consumption forecasting. More specifically, these two chapters present a tutorial with detailed explanations to help users set up the projection model, prepare the input data, compute, and manage the output. Users may use the sample input data files that accompany the software to quickly go through the main steps as a tutorial.

18.7 Limitations and Future Research Perspectives

The work reported in this book also indicates that there are substantial limitations of the household structure and living arrangement projections produced by the ProFamy extended cohort component model and that further research is needed. Some of these limitations were described at the end of each of the preceding chapters and others are described here. First, the current ProFamy model and software include and calculate the statuses of coresidence with parents and children, which is necessary to project household and living arrangements, but it cannot be used to study the availability of adult children (who may live outside the parents' household) as care providers or the survivorship of parents as care receivers. This is a limitation and further research may include the additional option of projecting numbers of surviving children and numbers of surviving parents, disregarding their co-residence status. This will be useful in studies of familial resources for old age care since an older person may very likely receive care from non-coresiding children, and an adult child may also be responsible for caring for his or her non-coresiding elderly parents. A person's status of number of surviving children and surviving parents disregarding coresidence may be considered as something between the narrowly defined "nuclear family" and more broadly defined "family kinship group" (Wachter 1987: 216). Such a further extension is highly feasible within the ProFamy multistate modeling framework, because we may simply add a new module to the model and software in which the numbers of coresiding children and coresiding parents are replaced by numbers of surviving children and surviving parents, and calculate only the survivorship or death of the children and parents by setting the rate of leaving parental home at zero.

Second, future research may add educational attainment status for all age groups and education attainment status changes for children and young adults to the ProFamy model and software. This is useful, because education level is closely related to marital/union formation and dissolution, fertility, mortality, and migration, which are determinants of households and living arrangements. Furthermore, elderly disability and home-based care costs and various household consumption behaviors are closely associated with education level. As a result of our successful addition of disability status transitions (measured by ADL) for those older adults aged 65 and older (see Chap. 5 for details), we are confident that it is possible to further extend the ProFamy multistate model by adding educational attainment status for all age groups and its status changes for children and young adults.

Third, another limitation of our present work that needs to be further investigated is application of the ProFamy extended cohort-component method to address research and policy analysis questions focused on young children and single-mothers, such as how many children will live in a two-parent or single-parent household? How many teenage and adult single mothers will have to care for their children with no spouse or partner present? Such an application is theoretically and practically feasible for the ProFamy extended cohort-component model, as it already includes children's coresidence status with two or one or no parent(s), and women's marital/union and number of coresiding children statuses. Based on reliable estimates of the age-sex-specific rates of marriage/union formation and dissolution for the young adults, different projection scenarios can be conducted to investigate differences in period and cohort proportions of children who live in a two-parent or a single-parent household and proportions of teenage and adult single mothers who have to care for one or two or more children without a spouse or partner present, under different assumptions about future propensities of marital/union formation and dissolution and marital and non-marital fertility. Such scenarios may be very useful to address policy research questions, such as: How would a reduction or elimination of teenage childbearing or reduction in divorce and union dissolution rates affect the number of single mothers? How much money would government programs subsiding single-mother families save?

Fourth, it is important to develop databases of age-sex-specific model standard schedules of demographic rates (see (2) in Table 3.1 of Chap. 3), which are similar to model life tables, for countries other than the U.S. and China for which we already have databases of model standard schedules. This is especially crucial for wide applications of the ProFamy method to household and living arrangement projections in various countries at the sub-national level, for which such detailed data are less likely to be available. Using the database of estimated age-sex-specific model standard schedules at the national level and projected (or assumed) demographic summary measures such as TFR, life expectancy at birth, standardized general rates of marriage, divorce, and cohabitation and union dissolution (if distinguished) for the population under study as input of the ProFamy software, forecasts of household structures and living arrangements can conveniently be performed at both national and sub-national levels for various countries.

Fifth, stochastic household and living arrangement projections with probabilistic distributions of the outcome and statistical confidence intervals are needed to address the uncertainties better than the medium forecasts with "low and high boundaries" presented in this book. Such stochastic household and living arrangement projections are much more complicated than stochastic population projections (Alho and Keilman 2010). The ProFamy cohort component model and software using conventional demographic input parameters can, however, provide a realistic modeling framework and tool for scientific research on stochastic household and living arrangement projections.

Finally, because household and living arrangement projections involve multiple dimensions of sex, age, fertility, marriage/union formation and dissolution, leaving the parental home, and so on, which require substantial programming and computation, wide practical application is not possible without user-friendly software, which is the focus of Part IV of this book. Further development of the ProFamy user-friendly software for household projections and living arrangement projections, including the anticipated extensions described above using demographic rates as input, is necessary for non-experts to apply the new method. We encourage more collaborative research and development work on the ProFamy software after publication of this book.

References

Ahlburg, D. A. (1992). Error measures and the choice of a forecast method. *International Journal of Forecasting, 8*, 99–111.

Ahlburg, D. A. (1995). Simple versus complex models: Evaluation, accuracy and combining. *Mathematical Population Studies, 5*, 281–290.

Alemayehu, B., & Warner, K. E. (2004). The lifetime distribution of health care costs. *Health Services Research, 39*(3), 627–642.

Alexander, J. T., Davern, M., & Stevenson, B. (2010). *Inaccurate age and sex data in the census PUMS file*. Paper presented at the 2010 PAA meeting, Dallas. Available at http://paa2010.princeton.edu/download.aspx?submissionId=100045.

Alho, J., & Keilman, N. (2010). On future household structure. *Journal of the Royal Statistical Society, 173A*(1), 117–143.

Allison, P. D. (1995). *Survival analysis using the SAS system: A practical guide*. Cary: SAS Institute Inc.

Annie E. Casey Foundation. (2012). Children in single-parent families, 2000–2011. Kids Count Data Center, Annie E. Casey Foundation. http://datacenter.kidscount.org/data/acrossstates/NationalProfile.aspx. Accessed 16 Nov 2012.

Anstey, K., Byles, J. E., Luszcz, M. A., Mitchell, P., Steel, D., Booth, H., et al. (2010). Cohort profile: The dynamic analyses to optimize ageing (DYNOPTA) project. *International Journal of Epidemiology, 39*, 44–51.

Arias, B. (2004). United States life table, 2002. In *National vital statistics reports* (Vol. 53(6)). Hyattsville: National Center for Health Statistics.

Bacon, B., & Pennec, S. (2007). *APPSIM-Modelling family formation and dissolution*. (Working Paper No. 4). National Centre for Social and Economic Modeling: University of Canberra.

Bank, S., & Kahn, M. D. (1975). Sisterhood-Brotherhood is powerful: Sibling sub-systems and family therapy. *Family Process, 14*(3), 311–337.

Bass, D. M., Looman, W. J., & Ehrlich, P. (1992). Predicting the volume of health and social services: Integrating cognitive impairment into the modified Andersen framework. *The Gerontologist, 32*, 33–43.

Becker, C. M., & Paltsev, S. V. (2001). Macro-experimental economics in the Kyrgyz republic: Social security sustainability and pension reform. *Comparative Economic Studies, 43*(4), 1–34.

Bedard, D. (1999). Stochastic pension funding: Proportional control and bilinear process. *Astin Bulletin, 29*(2), 271–293.

Bell, M., & Cooper, J. (1990). *Household forecasting: Replacing the headship rate model*. Paper presented at the Fifth National Conference, Melbourne: Australian Population Association.

Berson, D. W., Lereah, D., Merski, P., Nothaft, F., & Seiders, D. (2006). *America's home forecast: The next decade for housing and mortgage finance*. Washington, D.C.: The Homeownership Alliance.

Y. Zeng et al., *Household and Living Arrangement Projections*,
The Springer Series on Demographic Methods and Population Analysis 36,
DOI 10.1007/978-90-481-8906-9,

Bhaltacharya, C. M., Goldman, P., Hurd, D., Joyce, G. D., Lakdawalla, D. N., Panis, C. W., & Shang, B. (2004). Disability forecasts and future Medicare costs. *Frontiers in Health Policy Research, 7*, 75–94.

Bhat, C. R., & Sen, S. (2005). Household vehicle type holdings and usage: An application of the multiple discrete-continuous extreme value (MDCEV) model. *Transportation Research, 40*(B), 35–53.

Bianchi, S. M., Subaiya, L., & Kahn, J. R. (1999). The gender gap in the economic well-being of nonresident fathers and custodial mothers. *Demography, 36*(2), 195–203.

Board of Trustees of the Federal Old-Age and Survivors Insurance and Federal Disability Insurance Trust Funds. (2012). The 2012 annual report of the board of trustees of the Federal Old-Age and Survivors Insurance and Federal Disability Insurance Trust Funds. *House Document* 112–102. Washington, DC: Government Printing Office. http://www.ssa.gov/oact/tr/2012/tr2012.pdf. Accessed 6 Apr 2013.

Boaz, R. F., & Muller, C. F. (1992). Paid work and unpaid help by caregivers of the disabled and frail elders. *Medical Care, 30*(2), 149–158.

Boland, L. A. (1989). *The methodology of economic model building*. London: Routledge.

Bongaarts, J. (1987). The projection of family composition over the life course with family status life tables. In J. Bongaarts, T. K. Burch, & K. Wachter (Eds.), *Family demography: Methods and applications* (pp. 189–212). Oxford: Clarendon.

Bongaarts, J., & Feeney, G. (1998). On the quantum and tempo of fertility. *Population and Development Review, 24*, 271–291.

Bongaarts, J., & Greenhalgh, S. (1985). An alternative to the one-child policy in China. *Population and Development Review, 11*(4), 585–618.

Booth, H. (1984). Transforming Gompetrz's function for fertility analysis: The development of a standard for the relational Gompertz function. *Population Studies, 38*, 495–506.

Brace, P., Sims-Butler, K., Arceneaux, K., & Johnson, M. (2002). Public opinion in the American states: New perspectives using national survey data. *American Journal of Political Science, 46*(1), 173–189.

Bramlett, M. D., & Mosher, W. D. (2002). Cohabitation, marriage, divorce and remarriage in the United States. National Center for Health Statistics. *Vital Health Statistics Reports*, Vol. 23, No. 22. Hyattsville: National Center for Health Statistics.

Brass, W. (1974). Perspectives in population prediction: Illustrated by the statistics of England and Wales. *Journal of the Royal Statistical Society, 137*(A), 532–583.

Brass, W. (1978). *The relational Gompertz model of fertility by age of women*. London: London School of Hygiene and Tropical Medicine.

Brass, W. (1983). The formal demography of the family: An overview of the proximate determinates. In *The Family* (Occasional Paper No. 31). London: Office of Population censuses Surveys.

Breeze, E., Sloggett, A., & Fletcher, A. (1999). Socioeconomic and demographic predictors of mortality and institutional residence among middle aged and older people: Results from the longitudinal study. *Journal of Epidemiology of Community Health, 53*(12), 765–774.

Buettner, T., & Grubler, A. (1995). The birth of a 'green' generation? Generational dynamics of consumption patterns. *Technological Forecasting and Social Change, 50*, 113–134.

Bumpass, L. L., & H-H, L. (2000). Trends in cohabitation and implications for children's family contexts in the United States. *Population Studies, 54*, 29–41.

Bumpass, L. L., & Sweet, J. A. (1995). *Cohabitation, marriage, and non-marital childbearing and union stability: Preliminary findings from NSFH2* (NSFH Working Paper No. 65). Madison: University of Wisconsin, Center for Demography and Ecology.

Burch, T. K. (1999). Computer modeling of theory: Explanation for the 21st century. *PSC Discussion Papers Series* 13(4), Article 1.

Cai, F. (2006). The sustainability of economic growth of China in the 21st century. In Y. Zeng, L. Li, B. Gu, & Y. Lin (Eds.), *China's population strategy and economic development*. Beijing: Social Sciences Academic Press.

Cai, F., & Wang, D. (1999). Sustainability of economic growth and labor contribution in China. *Economic Research Journal 10*, 62–68 [in Chinese].

Cairns, A. J. G., & Parker, G. (1997). Stochastic pension fund modeling. *Insurance: Mathematics and Economics, 21*, 43–79.

California Energy Commission. (2003). *Forecasts of California transportation energy demand, 2003–2023. Prepared in support of the transportation report under the integrated energy policy report proceeding*, Docket#02-IEP-01.

Campbell, P. R. (2002). *Evaluating forecast error in state population projections using census 2000 counts* (U. S. Census Bureau Population Division Working Paper Series No. 57). http://www.census.gov/population/www/documentation/twps0057/twps0057.html. Accessed 1 Oct 2011.

Cao, X., & Mokhtarian, P. L. (2003). The future demand for alternative fuel passenger vehicles: A diffusion of innovation approach. http://aqp.engr.ucdavis.edu/Documents/AFV%20SCENARIO_June. pdf. Accessed 10 June 2010.

Card, J. (1981). Long term consequences for children of teenage parents. *Demography, 18*(2), 137–156.

Casper, L. M., & Cohen, P. N. (2000). How does POSSLQ measure up? Historical estimates of cohabitation. *Demography, 37*(2), 237–245.

Chadiha, L. A., Proctor, E. K., Morrow-Howell, N., Darkwa, O. K., & Dore, P. (1995). Posthospital home care for African-American and white elderly. *The Gerontologist, 35*, 233–239.

Chappell, N. (1991). Living arrangement and sources of caregiving. *Journal of Gerontology: Social Sciences, 46*, S1–S8.

Cherlin, A. J. (1992). *Marriage, divorce, remarriage*. Cambridge: Harvard University Press.

Cherlin, A. J. (1999). Going to extremes: Family structure, children's well-being, and social science. *Demography, 36*, 421–428.

Clark, R., & Peck, B. M. (2012). Examining the gender gap in life expectancy: A cross-national analysis, 1980–2005. *Social Science Quarterly, 93*(3), 820–837.

Clark, R. L., & Spengler, J. J. (1978). Changing demography and dependency costs: The implications of new dependency ratios and their composition. In B. Herzog (Ed.), *Aging and income* (pp. 55–89). New York: Human Sciences Press.

CMMS (Center for Medicare and Medicaid Services). (2004). Projections of national health expenditures: Methodology and model specification. Baltimore: The Centers for Medicare & Medicaid Services. http://www.cms.hhs.gov/statistics/nhe/projections-methodology/. Accessed 15 July 2006.

Coale, A. J. (1984). Life table construction on the basis of two enumerations of a closed population. *Population Index, 50*, 193–213.

Coale, A. J. (1985). An extension and simplification of a new synthesis of age structure and growth. *Asian and Pacific Forum, 12*, 5–8.

Coale, A., & Trussell, T. J. (1974). Model fertility schedules: Variations in the age structure of childbearing in human population. *Population Index, 40*, 185–258.

Coale, A. J., Demeny, P., & Vaughn, B. (1983). *Regional model life tables and stable populations*. New York: Academic.

Coale, A. J., Meredith, J., & Richards, T. (1985). Calculation of age-specific fertility schedules from tabulations of parity in two censuses. *Demography, 22*, 611–623.

Cohen, D. H., Kozak, R. A., Vidal, N., Spetic, W., & Ide, R. (2003). Performance expectations and needs of the Japanese house consumer. *Forest Products Journal, 55*(5), 37–44.

Collia, D. V., Sharp, J., & Giesbrecht, L. (2003). The 2001 national household travel survey: A look into the travel patterns of older Americans. *Journal of Safety Research, 34*, 461–470.

Congressional Budget Office (CBO). (2004). Financing long-term care for the elderly. The Congress of the United States. http://www.cbo.gov/ftpdocs/54xx/doc5400/04-26-LongTermCare. pdf. Accessed 15 May 2006.

Costa, D. L. (2000). *Long-term declines in disability among older men: Medical care, public health, and occupational change* (NBER Working Paper No. 7605). Cambridge, MA: National Bureau of Economic Research.

CRCA (China Research Center on Aging). (1994). *A data compilation of the survey on China's support systems for the elderly*. Beijing: Hua Ling Press.

Crowley, F. D. (2004). Pikes peak area council of governments: Small area estimates and projections 2000 through 2030. Southern Colorado Economic Forum, University of Colorado at Colorado Springs. http://www.ppacg.org/Trans/2030/Volume%20III/Appendix%20G%20-%20Small%20Area%20Forecasts.pdf. Accessed 2 Sept 2004.

Cutler, D. M., & Meara, E. (1999). *The concentration of medical spending: An update* (NBER Working Paper No. W7279). Issued in August. http://www.nber.org/papers/w7279. Accessed 18 July 2005.

Dalton, M., O'Neill, B., Prskawetz, A., Jiang, L., & Pitkin, J. (2008). Population aging and future carbon emissions in the United States. *Energy Economics, 30*, 642–675.

Davis, L. W. (2003). A dynamic model of demand for durable goods that consume energy. University of Wisconsin – Madison. http://www.ssc.wisc.edu/~ldavis/es.pdf. Accessed 29 Aug 2004.

Davis, L. W. (2004). The role of durable goods in household water and energy consumption: The case of front loading clothes washers. University of Wisconsin – Madison. http://rider.wharton.upenn.edu/~wred/Seminar%20Papers%2004-05/Lucas%20Davis%20cw.pdf. Accessed 3 Sept 2004.

Davis, S. C., & Diegel, S. W. (2009). Transportation energy data book, edition 24. ORNL-6973. Oak Ridge National Laboratory. Prepared for the Office of Planning, Budget Formulation and Analysis Energy Efficiency and Renewable Energy, US Department of Energy. http://cta.ornl.gov/data/index. shtml. Accessed 10 June 2010.

Day, J. C. (1996). Population projections of the United States by age, sex, race, and Hispanic origin: 1995 to 2050. US Bureau of the Census, Current Population Reports, 25–1130. Washington, DC: US Government Printing Office. http://www.census.gov/prod/1/pop/p25-1130. Accessed 10 June 2010.

Ding, J. (2003). Analysis about the impact of changes in fertility pattern on fertility level for China between 1991~2000. *Population Research, 27*(2), 55–60 [in Chinese].

Doty, P. (1986). Family care of the elderly: The role of public policy. *Milbank Memorial Fund Quarterly, 64*, 34–75.

Du, P., & Wu, C. (2006). Ability of daily life of the Chinese elderly: Status and change. *Population Research, 30*(1), 50–56 [in Chinese].

Dykstra, P. A., Kalmijn, M., Knijn, T. C. M., Komter, A. E., Liefbroer, A. C., & Mulder, C. H. (2006). *Family solidarity in the Netherlands*. Amsterdam: Dutch University Press.

Ediev, D. (2007). *On projecting the distribution of private households by size*. Vienna Institute of Demography Working paper, 4/2007.

Ediev, D., Yavuz, S., & Yüceşahin, M. (2012). Private households in Turkey: Big changes ahead. *Population Review, 51*(1), 28–49.

ESRI. (2007). Evaluating population projections—The importance of accurate forecasting. ESRI White Paper. http://www.esri.com/library/whitepapers/pdfs/evaluating-population.pdf. Accessed 1 Oct 2011.

Fancy, S. G. (1997). A new approach for analyzing bird densities from variable circular-plot counts. *Pacific Science, 51*(1), 107–114.

Federal Interagency Forum Aging Related Statistics (FIFARS). (2000). *Older Americans 2000: Key indicators of wellbeing. Federal Interagency Forum on Aging-Related Statistics*. Washington, DC: U.S. Government Printing Office, November 2000.

Federal Interagency Forum Aging Related Statistics (FIFARS). (2004). *Older Americans 2004: Key indicators of wellbeing. Federal Interagency Forum on Aging-Related Statistics*. Washington, DC: U.S. Government Printing Office, November 2004.

Federal Interagency Forum Aging Related Statistics (FIFARS). (2008). *Older Americans 2008: Key indicators of wellbeing. Federal Interagency Forum on Aging-Related Statistics*. Washington, DC: U.S. Government Printing Office, March 2008.

Federal Interagency Forum Aging Related Statistics (FIFARS). (2010). *Older Americans 2004: Key indicators of wellbeing. Federal Interagency Forum on Aging-Related Statistics*. Washington, DC: U.S. Government Printing Office, July 2010.

Federal Interagency Forum Aging Related Statistics (FIFARS). (2012). *Older Americans 2012: Key indicators of wellbeing. Federal Interagency Forum on Aging-Related Statistics*. Washington, DC: U.S. Government Printing Office, July 2012.

Feng, X., & Zhang, Q. (2002). Study on the transition of fertility desire among rural and urban residents in the past twenty years. *Market and Demographic Analysis, 8*(1), 21–31 [in Chinese].

Feng, Q., Wang, Z., Gu, D., & Zeng, Y. (2011). Household vehicle consumption forecasts in the United States, 2000 to 2025. *International Journal of Market Research, 53*(5), 593–618.

Forrest, R., & Leather, P. (1998). The aging of property owning democracy. *Ageing and Society, 18*, 35–63.

Frank, R. G. (2012). Long-term care financing in the United States: Sources and institutions. *Applied Economic Perspectives and Policy, 34*(2), 333–345.

Freedman, V. A. (1996). Family structure and the risk of nursing home admission. *Journal of Gerontology, 51B*(2), S61–S69.

Freedman, V. A., Martin, L. G., & Schoeni, R. F. (2002). Recent trends in disability and functioning among older adults in the United States. *Journal of the American Medical Association, 288*, 3137–3146.

Freedman, V. A., Crimminis, E. M., Schoeni, R. F., Spillman, B. C., Aykan, H., Kramarow, E., et al. (2004). Resolving inconsistencies in trends in old-age disability: Report from a technical working group. *Demography, 41*(3), 417–441.

Fries, J. F. (1980). Aging, natural death and the compression of morbidity. *New England Journal of Medicine, 303*, 130–135.

Gan, J. (2010). Housing wealth and consumption growth: Evidence from a large panel of households. *Review of Financial Studies, 23*(6), 2229–2267.

George, M. V. (1999). *On the use and users of demographic projections in Canada* (Working Paper No. 15). Conference of European statisticians, Joint ECE-EUROSTAT Work Session on Demographic Projections, Perugia, 3–7 May 1999.

Goldscheider, F. K. (1990). The aging of the gender revolution: What do we know and what do we need to know? *Research on Aging, 12*(4), 531–545.

Goldstein, J. R. (1999). The leveling of divorce in the United States. *Demography, 36*, 409–414.

Goldstein, J. R., & Kenney, C. T. (2001). Marriage delayed or marriage forgone? New cohort forecasts of first marriage for U.S. women. *American Sociological Review, 66*, 506–519.

Golob, T. F., & Brownstone, D. (2005). *The impact of residential density on vehicle usage and energy consumption* (Working paper UCI-ITS-WP-05-1 at Institute of Transportation Studies). Irvine: University of California.

Goodkind, D. (2009). Review on the book healthy longevity in China: Demographic, socioeconomic, and psychological dimensions. *Population Studies, 63*(3), 1–7.

Green, R., & Hendershott, P. H. (1996). Age, housing demand, and real house prices. *Regional Sciences and Urban Economics, 26*, 465–480.

Greenhalgh, S., & Bongaarts, J. (1987). Fertility policy in China: Future options. *Science, 235* (4793), 1167–1172.

Gruenberg, E. M. (1977). The failures of success. *Milbank Memorial Fund Quarterly, 55*, 3–24.

Grundy, E. (2001). Living arrangements and the health of older persons in developed countries. *United Nations Population Bulletin*, Special Issue Nos. 42/43.

Gu, D. (2008). General data assessment of the Chinese longitudinal healthy longevity survey in 2002. In Y. Zeng, D. L. Poston, D. A. Vlosky, & D. Gu (Eds.), *Healthy longevity in China demographic socioeconomic and psychological dimensions* (pp. 39–59). Dordrecht: Springer.

Gu, D., & Vlosky, D. A. (2008). Long-term care needs and related issues in China. In J. B. Garner & T. C. Christiansen (Eds.), *Social sciences in health care and medicine* (pp. 51–84). New York: Nova Publisher.

Gu, D., & Zeng, Y. (2006). Changes of disability in activities of daily living among the Chinese elderly from 1992 to 2002. *Population and Economics, 4*, 9–13 [in Chinese].

Gu, D., Wang, Z., & Zeng, Y. (2005). Time series of summary measures of elderly care needs and costs. Item 3 of Appendix of Phase I Report No.2-B. Preliminary Version of the Database to the National Institute on Aging.

Guo, Z. (2000). Lifetime fertility of Chinese women: A look at the recent period fertility behavior. *Population Research, 24*(01), 7–18 [in Chinese].

Guo, Z. (2003). *Changes in family households in China in 1990s*. Paper presented at the academic conference on the 2000 population census in China, Beijing, 28–31 Mar 2003.

Guo, Z. (2004). Study for fertility of China in the 1990s. *Population Research, 28*(2), 10–19 [in Chinese].

Guo, Z. (2007). Multi-level analysis on the sex ratio at birth in China based on the 2000 census and the regional fertility policy data. *Population Research, 5*(3), 20–31 [in Chinese].

Guo, Z., Zhang, E., Gu, B., & Wang, F. (2003). Diversity of China's fertility policy by policy fertility. *Population Research, 27*(5), 1–10 [in Chinese].

Haberman, S., & Wong, L. Y. P. (1997). Moving average rates of return and the variability of pension contributions and fund levels for a defined benefit pension scheme. *Insurance: Mathematics and Economics, 20*, 115–135.

Hamayon, S., & Legros, F. (2001). *Construction and impact of a buffer fund within the French PAYG pension scheme in a demo-economic model* (CESifo Working Paper No. 531). http://www.etla.fi/mfa/papers/fundf1.pdf. Accessed 2 Nov 2006.

Hamilton, B. E., & Ventura, S. (2012). *Birth rates for U.S. teenagers reach historic lows for all age and ethnic groups* (NCHS Data Brief. No. 89). Hyattsville: National Center for Health Statistics. http://www.cdc.gov/nchs/data/databriefs/db89.pdf. Accessed 16 Mar 2013.

Hammel, E. A. (2005). Demographic dynamics of kinship in anthropological populations. *Proceedings of the National Academy of Science, 102*(6), 2248–2253.

Hammel, E. A., Wachter, K. W., & McDaniel, C. K. (1981). The kin of the aged in A.D. 2000: The chickens come home to roost. In S. B. Kieseler, J. N. Morgan, & V. K. Oppenheimer (Eds.), *Aging: Social change* (pp. 11–39). New York: Academic.

Hammel, E. A., Mason, C., Wachter, K. W., Wang, F., & Yang, H. (1991). Rapid population change and kinship: The effects of unstable demographic changes on Chinese kinship networks, 1750–2250. In *Consequences of rapid population growth in developing countries: Proceedings of the United Nations* (pp. 243–271). New York: Taylor & Francis.

Hartman, M., Catlin, A., Lassman, D., Cylus, J., & Heffler, S. (2008). U.S. health spending by age, selected years through 2004. *Health Affairs, 27*(1), w1–w12.

He, P. (1998). Research on balance and countermeasures of pension funds. *Review of Economic Research, 9*, 20–45.

He, P. (2001). A research on the social security framework in China. *China Labor, 5*, 8–10 [in Chinese].

Heath Bureau of Shanghai. (2012). *Health status in Shanghai*, 2011. http://www.chinanews.com/df/2012/02-20/3682967.shtml. [in Chinese]. Accessed 27 Apr 2013.

Heffler, S., Smith, S., Keehan, S., Borger, C., Clemens, M. K., & Truffer, C. (2005). U.S. health spending projections for 2004–2014. *Health Affairs, 5*, 74–85.

Hendershott, P. H., & Weicher, J. C. (2002). Forecasting housing markets: Lesson learned. *Real Estate Economics, 30*(1), 1–11.

Hesketh, T., Li, L., & Zhu, W. X. (2005). The effects of China's one-child family policy after 25 years. *New England Journal of Medicine, 353*(11), 1171–1176.

Heuveline, P., & Timberlake, J. M. (2003). Toward a child-centered life course perspective on family structures: Multi-state early life tables. In E. Klijzing & M. Corijn (Eds.), *Fertility and partnership in Europe: Insights and lessons from comparative research* (Vol. II). New York/Geneva: United Nations.

Himes, C. L. (1992). Future caregivers: Projected family structures of older persons. *Journal of Gerontology: Social Sciences, 47*(1), S17–S26.

Hofferth, S. L. (1987). Recent trends in the living arrangements of children: A cohort life table analysis. In J. Bongaarts, T. Burch, & K. W. Wachter (Eds.), *Family demography: Methods and applications* (pp. 168–188). Oxford: Clarendon.

Hollmann, F. W., Mulder, T. J., & Kallan, J. E. (2000). *Methodology and assumptions for the population projections of the United States: 1999 to 2100* (Population Division Working Paper No. 38). Washington, DC: U.S. Bureau of the Census.

Houser, A., Gibson, M. J., & Redfoot, D. L. (2010). Trends in family caregiving and paid home care for older people with disabilities in the community data from the National Long-Term Care Survey. Washington, DC: AARP Public Policy Institute. assets.aarp.org/rgcenter/ppi/ltc/2010-09-caregiving.pdf. Accessed 18 Mar 2012.

Hu, P. S., & Reuscher, T. R. (2004). Summary of travel trend: 2001 national household travel survey. http://nhts.ornl.gov/2001/pub/STT. Pdf. Accessed 10 June 2010.

Hu, A., Tang, W., Wang, H., Mo, Z., Wang, B., & Chen, Z. (2003). Reasonably distribute the eldercare health service. *Population and Economics, 138*(3), 13–18.

Huang, C. (2006). Elderly health and family care burden in China. *Chinese Health Resources, 9*(5), 208–210 [in Chinese].

Ip, F., & McRae, D. (1999). Small area household projections – A parameterized approach. Population Section, Ministry of Finance and Corporate Relations, Province of British Columbia. http://www.bcstats.gov.bc.ca/data/pop/pop/methhhld.pdf. Accessed 1 Sept 2004.

Jiang, C. (2008). The analysis and prediction on the ADL care cost for the Chinese elderly. Ph.D. thesis, Beijing: Peking University, China Center for Economic Research.

Jiang, F. (2012). Report: Nearly 90 percent of Chinese families own houses. People's Daily Online. 15 May 2012. http://english.peopledaily.com.cn/90882/7817224.html. [in Chinese]. Accessed 31 Mar 2013.

Jiang, L., & Ren, Q. (2005). Study on the forecast of China's population, households and housing demand. *Market and Population Analysis, 2*, 20–29 [in Chinese].

Johnson, D. G. (1994). The shortcomings and options of China's rural population policy. *Chinese Rural Economy, 6*, 51–54 [in Chinese].

Johnson, D. G. (2000). Social security for the rural elderly in China. In J. Z. Yin, S. Lin, & D. F. Gates (Eds.), *Social security reform options for China* (pp. 229–246). London: World Scientific.

Johnson, R. W., Toohey, D., & Wiener, J. M. (2007). *Meeting the long-term care needs of the baby boomers: How changing families will affect paid helpers and institutions.* Washington, DC: The Urban Institute.

Kaye, H. S., Harington, C., & Laplante, M. P. (2010). Long-term care: Who gets It, who provides It, who pays, and how much? *Health Affairs, 29*(1), 11–21.

Keilman, N. (1985). Nuptiality models and the two-sex problem in national population forecasts. *European Journal of Population, 1*(2/3), 207–235.

Keilman, N. (1988). Dynamic household models. In N. Keilman, A. Kuijsten, & A. Vossen (Eds.), *Modelling household formation and dissolution* (pp. 123–138). Oxford: Clarendon.

Keilman, N. (2003). The threat of small households. *Nature, 421*, 489–490.

Kemper, P., Komisar, H. L., & Alecxih, L. (2005). Long term care over an uncertain future: What can current retirees expect? *Inquiry, 42*, 335–350.

Kennett, P., & Chan, K. W. (Eds.). (2011). *Women and housing: An international analysis.* New York: Routledge.

Kenney, G. M., & Dubay, L. C. (1992). Explaining area variation in the use of Medicare home health services. *Medical Care, 30*, 43–57.

Keyfitz, N. (1971). On the momentum of population growth. *Demography, 8*, 71–80.

Keyfitz, N. (1972). On future population. *Journal of the American Statistical Association, 67*, 347–363.

Khan, H. T. A., & Lutz, W. (2008). How well did past UN population projections anticipate demographic trends in six South-east Asian countries? *Asian Population Studies, 4*(1), 77–95.

Koklic, M. K., & Vida, I. (2009). A strategic household purchase: Consumer house buying behavior. *Managing Global Transitions, 7*(1), 5–96.

Koniak-Griffin, D., & Turner-Pluta, C. (2001). Health risks and psychosocial outcomes of early childbearing: A review of the literature. *Journal of Prenata l and Neonatal Nursing, 15*(2), 1–17.

Laidlaw, K., Wang, D., Coelho, C., & Power, M. (2010). Attitudes to ageing and expectations for filial piety across Chinese and British cultures: A pilot exploratory evaluation. *Aging and mental health, 14*(3), 283–292.

Lakdawalla, D., & Philipson, T. (2002). The rise in old age longevity and the market for long term care. *American Economic Review, 92*(1), 295–306.

Lakdawalla, D., Goldman, D. P., Bhattacharya, J., Hurd, M. D., Joyce, G. F., & Panis, C. W. A. (2003). Forecasting the nursing home population. *Medical Care, 41*(1), 8–20.

Land, K. C., & Rogers, A. (Eds.). (1982). *Multidimensional mathematical demography*. New York: Academic.

Land, K. C., Guralnik, J. M., & Blazer, D. G. (1994). Estimating increment-decrement life tables with multiple covariates from panel data: The case of active life expectancy. *Demography, 31*, 297–319.

Laslett, P. (1986). *Kinship and its numerical history: A newly recognized source of change in the social structure*. Presented in a conference on kinship and family, Majorca, March 1986.

Laslett, P. (1988). La parenté en chiffres, in Annales ESC, janvier-fevrier 1, pp. 5–24.

Laslett, P. (1994). *Kinship within and kinship beyond the household: Instrumental kin relations and their availability in the European past, present, and future*. Paper presented at the Murcia meeting on History of the Family. New Views on the Social Structure in Europe, December, 1994.

Lawton, M. P., Rajagonpal, D., Brody, E., & Kleban, M. (1992). The dynamics of caregiving for a demented elder among Black and White families. *Journal of Gerontology: Social Sciences, 47*, S156–S164.

Lee, R., & Miller, T. (2002). An approach to forecasting health expenditures, with application to the U.S. Medicare system. *Health Services Research, 37*(5), 1365–1386.

Lee, R., & Palloni, A. (1992). Changes in the family status of elderly women in Korea. *Demography, 29*(1), 69–92.

Lee, R. D., & Tuljapurkar, S. (2001). Population forecasting for fiscal planning: Issues and innovation. In A. J. Auerbach & R. D. Lee (Eds.), *Demographic changes and fiscal policy*. Cambridge: Cambridge University Press.

Lesthaeghe, R., & van de Kaa, D. J. (1986). Twee demografische transities? In D. J. van de Kaa & R. Lesthaeghe (Eds.), *Bevolking: Groei en Krimp* (pp. 9–24). Deventer: Van Loghum Slaterus.

Levine, J. A., Pollack, H., & Comfort, M. E. (2001). Academic and behavioral outcomes among the children of young mothers. *Journal of Marriage and Family, 63*(2), 355–369.

Li, J. (1997). Options of fertility policy and China's future population. *Population Research, 21*(1), 13–20 [in Chinese].

Li, J. (1998). Economic analysis of burdens for elderly population in China. *Population Research, 22*(6), 5–10 [in Chinese].

Li, J. (2003). An analysis of the survey of personal fertility attitudes in Beijing's one child family. *Chinese Journal of Population Science, 4*, 74–78 (in Chinese).

Li, X. (2004). On the long term strategy and countermeasures to China's population growth. *Strategy and Management, 3*, 35–47 [in Chinese].

Li, L., & Chen, Q. (2006). Impacts of population changes on the national medical and health care system. In Y. Zeng, L. Li, B. Gu, & Y. Lin (Eds.), *21st century population and economic development in China* (pp. 221–235). Beijing: Social Sciences Academic Press of China [in Chinese].

Li, H., & Zhang, C. (Eds.). (2001). *The road to solve the root problem*. Beijing: China Population Press [in Chinese].

Li, X., Xin, L., Tan, M., & Zhao, Y. (2013). Population urbanization may reduce the ecological and environmental pressures. In Y. Zeng, G. Gu, J. Liang, & Z. Guo (Eds.), *Fertility policy adjustment and development in China*. Beijing: Social Science Literature Press [in Chinese].

Liang, Z. (1979). Suggestions concerning strategies of population and development in the next a few decades. Paper presented at the second national symposium on population theories/ Science, December, 1979, Beijing, and published in Liang Zhongtang *Strategies of Population and Development*, Shanxi People's Press, 1985, pp. 12–16. [in Chinese].

Liang, H. (1999). Studies on self-supporting abilities of rural elderly in China. *Population and Economics, 115*(4), 21–25 [in Chinese].

Lin, B. (2004). An analysis for the fertility preference and determinants for childbearing age of women. In P. Guiyu (Ed.), *Collection of 2001 national family planning and reproductive health survey research papers* (pp. 57–65). Beijing: China Population Press [in Chinese].

Lin, J. Y. (2006). Development strategy, demography and population policy. In Y. Zeng, L. Li, B. Gu, & Y. Lin (Eds.), *China's population strategy and economic development* (pp. 3–9). Beijing: Social Sciences Academic Press [in Chinese].

Liu, Z. (1984). *Population theory issues*. Beijing: China Social Sciences Press (in Chinese).

Liu, K., Manton, K. G., & Aragon, C. (2000). Changes in home care use by disabled elderly persons: 1982–1994. *The Journals of Gerontology: Social Sciences, 55B*(4), S245–S253.

Liu, J., Daily, G. C., Ehrlich, P. R., & Luck, G. E. (2003). Effects of household dynamics on resource consumption and biodiversity. *Nature, 421*, 530–533.

Lutz, W., & Prinz, C. (1994). The population module. In W. Lutz (Ed.), *Population – development – environment: Understanding their interactions in Mauritius* (pp. 221–231). Berlin: Springer.

Ma, Y., & Zhang, X. (1984). Some issues in population control and population policy. Policy report submitted to Chinese central government in 1984, and published. In Ma Yingtong (Ed.), *Practice and thoughts on population control*.1993. Lan Zhou: Gan shu People Press.

Mackellar, F. L., Lutz, W., Prinz, C., & Goujon, A. (1995). Population, households, and CO_2 emissions. *Population and Development Review, 21*, 849–866.

Manton, K. G. (1982). Changing concepts of morbidity and mortality in the elderly population. *Milbank Memorial Fund Quarterly, 60*, 183–244.

Manton, K. G., Singer, B. H., & Suzman, R. M. (1993). *Forecasting the health of elderly populations*. New York: Springer.

Manton, K. G., Gu, X., & Lamb, V. L. (2006). Changes in chronic disability from 1982 to 2004/2005 as measured by long-term changes in function and health in the U.S. elderly population. *Proceedings of the National Academy of Sciences, 103*(48), 18374–18379.

Martin, L. G., Schoeni, R. F., & Andreski, P. (2010). Trends in health of older adults in the United States: Past, present, future. *Demography, 47*(s), S17–S40.

Mason, A., & Racelis, R. (1992). A comparison of four methods for projecting households. *International Journal of Forecasting, 8*, 509–527.

Mayer, G. C., Torrey, B. B., & Kineslla, K. G. (1992). The paradox of the oldest old in the United States: An international comparison. In R. M. Suzman, D. P. Willis, & K. G. Manton (Eds.), *The oldest old* (pp. 58–85). Oxford: Oxford University Press.

Mayhew, L. (2000). *Health and elderly care expenditure in an aging world*. Laxenburg: IIASA.

Meen, G. (1998). Modeling sustainable home-ownership: Demographics or economics? *Urban Studies, 35*(1), 1919–1934.

Minicuci, N., Noale, M., Bardage, C., Blumstein, T., Deeg, D. J., Gindin, J., et al. (2003). Cross-national determinants of quality of life from six longitudinal studies on aging: The CLESA project. *Aging and Clinical Experimental Research, 15*, 187–202.

Moffitt, R. (2000). Demographic change and public assistance expenditures. In A. J. Auerbach & R. D. Lee (Eds.), *Demographic change and public policy* (pp. 391–425). Cambridge: Cambridge University Press.

Morgan, S. P. (2004). Interstate differentials in demographic rates are mostly caused by differences in racial compositions. Personal e-mail communication.

Morgan, P. P., Botev, K., Chen, R., & Huang, J. (1999). White and Non-white trends in first birth timing: Comparisons using vital registration and current population surveys. *Population Research and Policy Review, 18*, 339–356.

Morris, R., Caro, F. G., & Hansan, J. E. (1998). *Personal assistance: The future of home care*. Baltimore: The Johns Hopkins University Press.

MSNBC. (2008). Gas prices put Detroit big three in crisis mode. http://www.msnbc.msn.com/id/24896359/. Accessed 10 June 2010.

Mui, A. C., & Bumette, D. (1994). Long-term care services used by frail elders: Is ethnicity a factor? *The Gerontologist, 34*, 190–198.
Murphy, M. (1991). Modelling households: A synthesis. In M. J. Murphy & J. Hobcraft (Eds.), *Population research in Britain* (A supplement to population studies, Vol. 45, pp. 157–176). London: Population Investigation Committee, London School of Economics.
Murphy, M. (2004). Tracing very long-term kinship networks using SOCSIM. *Demographic Research, 10*(7), 171–196.
Murphy, M. (2011). Long-term effects of the demographic transition on family and kinship networks in Britain. *Population and Development Review, 37*(s1), 55–80.
Murray, C. J. L., Ferguson, B. D., Lopez, A. D., Guillot, M., Salomon, J. A., & Ahmad, O. (2003). Modified logit life table system: Principles, empirical validation, and application. *Population Studies, 57*, 165–182.
Myers, D., Pitkin, J., & Park, J. (2002). Estimation of housing needs amid population growth and change. *Housing Policy Debate, 13*(3), 567–596.
National Center for Health Statistics (NCHS). (1998). Declines in teenage birth rates, 1991–97: National and State patterns. In S. J. Ventura, T. J. Mathews, & S. C. Curtin (Eds.), National Vital Statistics Report 47, No. 12. Hyattsville: National Center for Health Statistics.
National Center for Health Statistics (NCHS). (2012). Health, United States, 2011: List of trend tables. http://www.cdc.gov/nchs/hus/contents2011.htm. Accessed 16 Nov 2012.
National Research Council. (2001). *Preparing for an aging world: The case for cross-national research. Panel on a Research Agenda and New Data for an Aging World, Committee on Population and Committee on National Statistics, Division of Behavioral and Social Sciences and Education*. Washington, DC: National Academy Press.
Natural Resources of Canada. (2004). Light-duty vehicle fuel efficiency scenario: model year 1990 to 2010. http://atlas.gc.ca/site/english/maps/climatechange/humanactivitiesemissions/methodologyprojection.html. Accessed 10 June 2010.
Ng, S. T., Skitmore, M., & Wong, K. F. (2008). Using genetic algorithms and linear regression analysis for private housing demand forecast. *Building and Environment, 43*(6), 1171–1184.
Nishioka, H., Koyama, Y., Suzuki, T., Yamauchi, M., & Suga, K. (2011). Household projections by prefecture in Japan: 2005–2030: Outline of results and methods. *The Japanese Journal of Population, 9*(1), 78–133.
Nortman, D. (1974). *Parental age as a factor in pregnancy outcome and child development* (Reports on Population Family Planning, No. 16). New York: The Population Council.
O'Neill, B. C., & Chen, B. S. (2002). Demographic determinants of household energy use in the United States. *Population and Development Review, 28*, 53–88.
OECD. (2013). OECD Health Data 2012: Frequently requested data. http://www.oecd.org/els/health-systems/oecdhealthdata2012-frequentlyrequesteddata.htm
OECD Statistics. (2007). *Pensions at a glance 2007: Highlights*. Paris: OECD.
OICA (International Organization of Motor Vehicle Manufacturers) (2010). http://www.oica.net/. Accessed 10 Feb 2011.
Olshansky, S. J., Rudberg, M. A., Carnes, B. A., Cassel, C. K., & Brody, J. A. (1991). Trading off longer life for worsening health: The expansion of morbidity hypothesis. *Aging Health, 3*, 194–216.
Orsini, C. (2010). Changing the way the elderly live: Evidence from the home health care market in the United States. *Journal of Public Economics, 94*(1–2), 142–152.
Paget, W. J., & Timaeus, I. M. (1994). A relational Gompertz model of male fertility. Development and assessment. *Population Studies, 48*, 333–340.
Pendall, R., Freiman, L., Myers, D., & Hepp, S. (2012). *Demographic challenges and opportunities for U.S. housing markets*. Report of Economic Policy Program, Housing Commission, prepared for the Bipartisan Policy Center, March 2012.
Pinquart, M., & Sörensen, S. (2005). Ethnic differences in stressors, resources, and psychological outcomes of family caregiving: A meta-analysis. *The Gerontologist, 45*(1), 90–106.

Pitkin, J., & Myers, D. (1994). The specification of demographic effects on housing demand: Avoiding the age-cohort fallacy. *Journal of Housing Economics, 3*, 240–250.

Pollard, J. H. (1977). The continuing attempt to incorporate both sexes into marriage analysis. In Volume 1 of the papers of *the General Conference of the International Union for the Scientific Study of Population*. Mexico City.

Poston, D. L., & Duan, C. C. (2000). The current and projected distribution of the elderly and eldercare in the People's Republic of China. *Journal of Family Issues, 21*, 714–732.

Powell, B., & Steelman, L. C. (1995). Feeling the pinch: Child spacing and constraints on parental economic investments in children. *Social Forces, 73*(4), 1465–1486.

Preston, S. H. (1984). Children and the elderly: Divergent paths for America's dependents. Population Association of America (PAA) presidential address. *Demography, 21*(4), 435–457.

Preston, S. H., Heuveline, P., & Guillot, M. (2001). *Demography: Measuring and modeling population processes*. London: Blackwell Publishers.

Prskawetz, A., Jiang, L., & O'Neill, B. (2004). Demographic composition and projections of car use in Austria. In *Vienna yearbook of population research* (pp. 274–326). Vienna: Austrian Academy of Sciences Press.

Pudney, S. E., & Sutherland, H. (1994). How reliable are micro simulation results? An analysis of the role of sampling error in a UK tax-benefit model. *Journal of Public Economics, 53*, 327–365.

Qiao, X. (2001). *Aging issues and policy choices in rural China*. Paper presented at the 24th international union of scientific studies of population. Salvador, 20th–24th Aug 2001.

Rao, J. N. K. (2003). *Small area estimation*. New York: Wiley.

Rix, S. E., & Fisher, P. (1982). *Retirement-age policy: An international perspective*. New York: Pergamon Press.

Robinson, K. (2007). *Trends in health status and health care use among older women* (Aging trends, Vol. 7). Hyattsville: National Center for Health Statistics.

Rogers, A. (1975). *Introduction to multiregional mathematical demography*. New York: Wiley.

Rogers, A. (1986). Parameterized multistate population dynamics and projections. *Journal of American Statistical Association, 81*, 48–61.

Ruggles, S. (1987). *Prolonged connections: The rise of the extended family in nineteenth century England and America*. Madison: University of Wisconsin Press.

Ruggles, S. (1993). Confessions of a microsimulator: Problems in modeling the demography of kinship. *Historical Methods, 26*(4), 161–169.

Ryder, N. B. (1987). Reconsideration of a model of family demography. In J. Bongaarts, T. K. Burch, & K. Watcher (Eds.), *Family demography: Methods and their applications* (pp. 102–122). Oxford: Clarendon.

Schenker, N., & Raghunathan, T. E. (2007). Combining information from multiple surveys to enhance estimation of measures of health. *Statistics in Medicine, 26*, 1802–1811.

Schoen, R. (1981). The harmonic mean as the basis of a realistic two-sex marriage model. *Demography, 18*, 201–216.

Schoen, R. (1988). *Modeling multigroup populations*. New York: Plenum Press.

Schoen, R., & Standish, N. (2001). The retrenchment of marriage: Results from marital status life tables for the United States, 1995. *Population and Development Review, 27*(3), 553–563.

Serup-Hansen, N., Wickstrøm, J., & Kristiansen, I. S. (2002). Future health care costs-do health care costs during the last year of life matter? *Health Policy, 62*, 161–172.

SFPC (State Family Planning Commission) Research Group. (2000). Future population development and fertility policy in China. *Population Research, 21*(01), 18–34 [in Chinese].

Shekell, P. G., Ortiz, E., Newberry, S. J., Rich, M. W., Rhodes, S. L., Brook, R. H., & Goldman, D. P. (2005). Identifying potential health care innovations for the future elderly. *Health Affairs, 24*(2s), W5-R67–W5-R76.

Shen, K. (2011). *Comprehensive analyses of the living arrangement among Chinese elderly—its influential factors and the effects on well-being*. Ph.D. dissertation, Peking University, Supervisor: Professor Yi Zeng. [in Chinese].

Shen, K., Yan, P., & Zeng, Y. (2013). The impact of co-residence with elderly parents on female labor force participation in China. Manuscript submitted to a peer-reviewed journal for consideration of publication.

Shi, Q. (2001). Study on the marriage, fertility behavior and attitudes among rural households in Zhejiang Province. *Chinese Journal of Population Science, 4*, 44–54 [in Chinese].

Shi, W. (2012). *A micro–macro simulation modeling system for population projection and aging analysis*. Paper presented at International Conference on "Advances in Methodology and Applications: Biodemography and Multistate Event History Analysis on Healthy Aging", 15–18 Oct 2012, Hangzhou.

Shu, R. (2005), Why did the nine provinces eliminated or relaxed the birth spacing policy, internal policy explanation note from General Office of State Population and Family Planning Commission. [in Chinese]

Sin, Y. (2005). *Pension liabilities and reform options for old age insurance* (Working Paper Serial on China. 2005–1). Washington, DC: The World Bank.

Smith, D. (1992). *Formal demography*. New York: Plenum Press.

Smith, S. K. (1997). Further thoughts on simplicity and complexity in population projection models. *International Journal of Forecasting, 13*(4), 557–565.

Smith, S. K. (2003). Small area analysis. In P. Demeny & G. McNicoll (Eds.), *Encyclopedia of population* (pp. 898–901). Farmington Hills: Macmillan Reference.

Smith, S. K., & Morrison, P. A. (2005). Small area and business demography. In D. L. Poston & M. Micklin (Eds.), *Handbook of population* (pp. 761–786). New York: Kluwer/Plenum.

Smith, S. K., & Sincich, T. (1992). Forecasting state and household populations: Evaluating the forecast accuracy and bias of alternative population projects for states. *International Journal of Forecasting, 8*, 495–508.

Smith, S. K., Tayman, J., & Swanson, D. A. (2001). *State and local population projections: Methodology and analysis*. New York: Kluwer/Plenum.

Smith, S. K., Rayer, S., & Smith, E. A. (2008). Aging and disability: Implications for the housing industry and housing policy in the United States. *Journal of the American Planning Association, 74*(3), 289–306.

Smith, S. K., Rayer, S., Smith, E. A., Wang, Z., & Zeng, Y. (2012). Population aging, disability and housing accessibility: Implications for sub-national areas in the United States. *Housing Studies, 27*(2), 252–266.

Smock, P. J. (2000). Cohabitation in the United States. *Annual Review of Sociology, 26*, 1–20.

Soldo, B. J., Wolf, D. A., & Agree, E. M. (1990). Family, households, and care arrangements of the frail elderly: A structural analysis. *Journal of Gerontology, 45*, S238–S249.

Spain, D. (1997). *Societal trends: The aging baby boom and women's increased independence*. Report prepared for the US Department of Transportation, DTFH 61-97-P-00314.

Spicer, K., Diamond, I., & Bhrolcham, M. N. (1992). Into the twenty-first century with British households. *International Journal of Forecasting, 8*, 529–539.

SSB (State Statistical Bureau). (2001). *Statistical year book of China, 2000*. Beijing: China Statistical Press.

Stallard, E. (2000). *Retirement and health: Estimates and projections of acute and long-term care needs and expenditures of the U.S. elderly population. Retirement Needs Framework* (SoA Monograph M-RS00-1) (pp. 159–207). Schaumburg: Society of Actuaries

Starmass International. (2009). Hebei demographic analysis and economy overview. http://www.starmass.com/china-review/provincial-overview/hebei-demographic-economy.htm. Accessed 20 Feb 2013.

Strow, C. W., & Strow, B. K. (2006). A history of divorce and remarriage in the United States. *Humanomics, 22*(4), 239–257.

Stupp, P. W. (1988). A general procedure for estimating intercensal age schedules. *Population Index, 54*, 209–234.

Suthers, K., Kim, J. K., & Crimmins, E. M. (2003). Life expectancy with cognitive impairment in the older population of the United States. *Journal of Gerontology: Social Sciences, 58B*(3), S179–S186.

Swanson, D. A. (2010). *The methods and materials used to generate two key elements of the housing unit method of population estimation: Vacancy rates and persons per household.* Paper presented at the 2010 PAA meeting, Dallas. Available at http://paa2010.princeton.edu/download.aspx?submissionId=100049.

Swanson, D. A., & Pol, L. G. (2009). Applied demography: Its business and public sector Components. In Y. Zeng (Ed.), *Demography volume of the Encyclopedia of Life Support Systems (EOLSS)* (www.eolss.net), coordinated by the UNESCO-EOLSS Committee. Oxford: EOLSS Publishers Co. Ltd.

The Hindu. (2013). India among world's top 20 real estate investment markets. The Hindu. http://www.thehindu.com/todays-paper/tp-business/india-among-worlds-top-20-real-estate-investment-markets/article4549070.ece. Accessed 31 Mar 2013.

The Research Group. (2004). Research on Chinese fertility policy in the 21st century. Report presented by Baochang Gu at the "Conference on Population Projection and Analysis" held in Heibei University in June 2004 and organized by China Population Association.

Thornton, A., Axinn, W. G., & Xie, Y. (2007). *Marriage and cohabitation.* Chicago: University of Chicago Press.

Treadway, R. (1997). *Population projections for the state and counties of Illinois.* Springfield: State of Illinois.

U. N. (United Nations). (1973). *The determinants and consequences of population trend.* New York: United Nations.

U. N. (United Nations). (1982). Model life tables for developing Countries. ST/SEA/SER, A/77, Sales No. E. 81. XIII, 7, New York: United Nations.

U. N. (United Nations). (2008). *Principles and recommendations for population and housing censuses.* (Revision 2. Statistical Papers Series M No. 67/Rev.2). New York: United Nations.

U.N. (United Nations). (2013). *World population prospects: The 2012 revision.* New York: United Nations.

UNICEF, WHO, The World Bank and UN Population Division. (2007). *Levels and trends of child mortality in 2006: Estimates developed by the inter-agency group for child mortality estimation.* New York: United Nations.

U.S. Census Bureau. (1996). *Projections of the number of households and families in the United States: 1995 to 2010.* U.S. Department of Commerce Economics and Statistics Administration, Current Population Reports (P25–1129). Washington, DC: U.S. Government Printing Office.

U.S. Census Bureau. (1998). *Statistical abstract of the United States: 1998.* Washington, DC: GPO.

U.S. Census Bureau. (2001). U.S. adults postponing marriage. Census Bureau Reports. (http://www.census.gov/Press-Release/www/2001/cb01-113.html). Accessed 20 Mar 2008.

U.S. Census Bureau. (2008). Interim projections of the U.S. population by age, sex, race, and Hispanic origin. http://www.census.gov/population/www/projections/usinterimproj/idbsummeth.pdf. Accessed 27 Sept 2011.

U.S. Census Bureau. (2011). The 2011 American community survey. See Table B12007 in American FactFinder for the median age at first marriage. http://factfinder2.census.gov. Accessed 16 Nov 2012.

U.S. Census Bureau. (2012). The 2012 statistical abstract. Table 1337. Single-parent households. http://www.census.gov/compendia/statab/cats/international_statistics/population_households.html. Accessed 16 Nov 2012.

U.S. Department of Transportation. (2000–2009). Highway statistics 2000–2009. http://www.fhwa.dot.gov/policyinformation/statistics.cfm. Accessed 24 Apr 2011.

Van de Kaa, D. J. (2008). Demographic transition. In Y. Zeng (Ed.), *Demography: Encyclopedia of Life Support Systems (EOLSS), Developed under the Auspices of the UNESCO.* Oxford, UK: EOLSS Publishers. http://www.eolss.net. Retrieved 2 Sept 2010.

Van Gaalen, R., & van Poppel, F. (2009). Long-term changes in the living arrangements of children in the Netherlands. *Journal of Family Issues, 30*(5), 653–669.

Van Imhoff, E. (1999). *Modelling life histories: Macro robustness* versus *micro substance*. Paper presented at the international workshop on synthetic biographics: State of the art and developments, San Miniato, 6–9 June 1999.

Van Imhoff, E., & Keilman, N. (1992). *LIPRO 2.0: An application of a dynamic demographic projection model to household structure in the Netherlands*. Amsterdam: Swets & Zeithinger.

Van Imhoff, E., & Post, W. (1998). Microsimulation methods for population projection. New methodological approaches in the social sciences. *Population: An English Selection, 10*(1), 97–138.

Van Imhoff, E., Kuijsten, A., Hooimeiger, P., & van Wissen, L. (1995). Epilogue. In E. Imhoff, A. Kuijsten, P. Hooimeiger, & L. van Wissen (Eds.), *Household demography and household modeling* (pp. 345–351). New York: Plenum.

Vance, C., & Buchheim, S. (2004). *Household demographic composition, community design and travel behavior: An analysis of motor vehicle use in Germany*. Paper prepared for presentation at the Research on Women's Issues in Transportation Conference, Chicago.

Vaupel, J. W. (2010). Biodemography of human ageing. *Nature, 464*, 536–542.

Vaupel, J. W., & Zeng, Y. (1991). Population tradeoffs in China. *Policy Sciences, 24*, 389–406.

Vincent, G. K., & Velkoff, V. A. (2010). *The next four decades: The older population in the United States: 2010 to 2050*. Current Population Reports, Population Estimates and Projections. Washington, DC: U.S. Census Bureau.

Vlasic, B., & Bunkley, N. (2008). Hazardous conditions for the auto industry. New York Times. http://www.nytimes.com/2008/10/02/business/02sales.html?_r=1&partner=rssnyt&emc=rss. Accessed 10 June 2010.

Wachter, K. W. (1987). Microsimulation of household cycles. In E. Bongaarts, T. K. Burch, & K. W. Wachter (Eds.), *Family demography: Methods and applications* (pp. 215–227). Oxford: Clarendon.

Wang, F. (2005). *Can China afford to continue its one-child policy?* (Asian Pacific issues, Vol. 77). Honolulu: East–West Center.

Wang, Z. (2009a). *Households and living arrangements forecasting and the associated database development for Southern California six counties and the whole Region, 2000–2040* (In response to RFP No. 09-043-IN of SCAG: Household Projection Model Development: Simulation Approach) (Technical report No. 09–02 of Households and Consumption Forecasting Inc.).

Wang, Z. (2009b). *Household projection and living arrangements projections for Minneapolis-St Paul region, 2000–2050* (Technical report No. 09–03 of Households and Consumption Forecasting Inc.).

Wang, Z. (2011a). *The 2011 updating of households and living arrangements forecasting for Southern California six counties and the whole region, 2000–2040*. (Technical report No. 11–01 of Households and Consumption Forecasting Inc.).

Wang, Z. (2011b). *The 2011 updating of household projection and living arrangements projections for Minneapolis-St Paul region, 2000–2050*. (*Technical report* No. 11–02 of Households and Consumption Forecasting Inc.).

Wang, F., & Mason, A. (2006). Demographic dividend and prospects of economic development in China. In Y. Zeng, L. Li, B. Gu, & Y. J. Lin (Eds.), *Population and economic development in the 21st century in China* (in Chinese). China Social Science Academic Press. English version of this paper was presented at the U.N experts group meeting on "Social and economic implications of changing population age structure", Mexico City, 31 Aug–2 Sept 2005.

Wang, J., He, Y., Wang, Z., & Duan, C. (2004). Province-level total fertility rate in China in 2000. *Population Research, 2*, 20–28 [In Chinese].

Watkins, S. C., Menken, J. A., & Bongaarts, J. (1987). Demographic foundations of family change. *American Sociological Review, 52*(3), 346–358.

Weiss, C. O., Gonzalez, H. M., Kabeto, M. U., & Langa, K. M. (2005). Differences in amount of informal care received by non-Hispanic Whites and Latinos in a nationally representative sample of older Americans. *Journal of the American Geriatric Society, 53*, 146–151.

West, L. A. (1999). Pension reform in China: Preparing for the future. *Journal of Development Studies, 35*(3), 153–183.

West, L. A. (2000). Demographic underpinnings of China's old-age pension reform. In J. Z. Yin, S. Lin, & D. F. Gates (Eds.), *Social security reform options for China* (pp. 117–131). London: World Scientific.

Whiteford, P., & Whitehouse, E. (2006). Pension challenges and pension reforms in OECD countries. *Oxford Review of Economic Policy, 22*(1), 78–94.

WHO. (2010). World health statistics 2010. Geneva: World Health Organization. http://www.who.int/whosis/whostat/EN_WHS10_Full.pdf. Accessed 16 May 2012.

Willekens, F. J., Shah, I., Shah, J. M., & Ramachandran, P. (1982). Multistate analysis of marital status life tables: Theory and application. *Population Studies, 36*(1), 129–144.

Williamson, J. (1997). *Growth, distribution and demography: Some lessons from history*. NBER Working Paper Series, (Working Paper 6244).

Wolf, A. P. (1986). The preeminent role of government intervention in China's family revolution. *Population and Development Review, 12*, 101–116.

Wolf, D. A. (1988). Kinship and family support in aging societies. In *Economic and social implications of population aging* (pp. 305–330). Population Division, United Nations Department of International Economic and Social Affairs. New York: United Nations.

Wolf, D. A. (1990). *Coresidence with an aged parent: Lifetime patterns and sensitivity to demographic changes*. Paper presented at the United Nations Conference on Population Aging in the Context of Family, Kitakyushu.

Wolf, D. A. (1994). The elderly and their kin: patterns of availability and access. In L. G. Martin & S. H. Preston (Eds.), *Demography of aging* (pp. 146–194). Washington, DC: National Academy Press.

Wooldridge, J. (2003). *Introductory econometrics—A modern approach*. Mason: Thomson South-Western Press.

World Bank. (1985). *The problems and solutions for long term development in China*. Beijing: China Financial and Economic Publishing House.

World Bank. (2003). Modeling pension reform: The World Bank's pension peform Options Simulation Toolkit. Washington, DC: The World Bank. http://info.worldbank.org/etools/docs/library/76907/february2003/readings/prost.pdf. Accessed 12 Dec 2005.

World Bank and the Development Research Center of State Council of China. (2013). *China 2030: Building a modern, harmonious, and creative society*. Washington, DC: The World Bank.

Xie, Z. (Ed.). (2000). *Study on family planning and women's status*. Beijing: China Population Press.

Xinhuanet. (2003). Investigation report: Ten characteristics of reproductive desire among Shanghainese. Nov 18. http://www.china.org.cn/chinese/renkou/444418.htm

Xinhuanet. (2004). Doing the DINK. May 25. http://news.xinhuanet.com/english/2004-05. Accessed 21 June 2005.

Yang, X., & Xu, D. (2011). Study on the population structure change and the demand for housing in the city development. *Research and development, 2*, 84–87 [in Chinese].

Ye, Z. (2009). Establish a green cover for people – The 60-years of the social security program in China. *People's Daily* (oversee edition), 29 Sept 2009.

Yelowitz, A. S. (1998). Will extending Medicaid to two-parent families encourage marriage? *Journal of Human Resources, 33*, 833–865.

Yin, J. Z., Lin, S., & Gates, D. F. (Eds.). (2000). *Social security reform options for China*. London: World Scientific.

Yu, X. (1992). *Data on family expenditure for children and elderly*. Beijing: China Population Information and Research Center [in Chinese].

Zagheni, E. (2011). The impact of the HIV/AIDS epidemic on kinship resources for orphans in Zimbabwe. *Population and Development Review, 37*(4), 761–783.

Zeng, Y. (1986). Changes in family structure in China: A simulation study. *Population and Development Review, 12*, 675–703.

Zeng, Y. (1988). Changing demographic characteristics and the family status of Chinese women. *Population Studies, 42*, 183–203.

Zeng, Y. (1990). *User's manual and software of "FAMY" – A PC computer program for family status life table analysis*. Groningen: iec ProGAMMA (Inter-university Expert Center for Computer Software Development and Distribution).

Zeng, Y. (1991a). *Family dynamics in China: A life table analysis*. Wisconsin: The University of Wisconsin Press.

Zeng, Y. (1991b). Impact of the increase of age at childbearing on population development in China. *Population and Economics, 2*, 3–11 [in Chinese].

Zeng, Y. (2002). Old Age insurance and sustainable development in rural China. In G. Kochendörfer-Lucius & B. Pleskovic (Eds.), *Gudrun sustainable development with a dynamic economy*. Washington, DC: World Bank.

Zeng, Y. (2005a). Smooth transition to the two-child policy using late-childbearing as a lever. *Population and Economics, 149*(02), 7–14 [in Chinese].

Zeng, Y. (2005b). Population aging, pension deficits and old age insurance program in China. *Economic Quarterly, 3*, 1043–1066 [in Chinese].

Zeng, Y. (2006). A Soft-landing with the two-children-late-birth policy: need and feasibility. *Social Sciences of China, 58*(2), 93–109 [in Chinese].

Zeng, Y. (2007). Options of fertility policy transition in China. *Population and Development Review, 33*(2), 215–246 (Heading article).

Zeng, Y. (2011). Effects of demographic and retirement-age policies on future pension deficits, with an application to China. *Population and Development Review, 37*(3), 553–569.

Zeng, Y. (2012). Suggestions about implementing as soon as possible the new policy of universally allowing two-children with adequate late childbearing – new lessons learned from data analysis of the 2010 population census. CCER Working Paper (No. C2012004). http://www.nsd.edu.cn/cn/article.asp?articleid=16237. Accessed 3 Mar 2013.

Zeng, Y., & George, L. K. (2010). Population ageing and old-age insurance in China. In D. Dannefer & C. Phillipson (Eds.), *The SAGE handbook of social gerontology* (pp. 420–430). London: Sage.

Zeng, Y., & Land, K. C. (2001). A sensitivity analysis of the Bongaarts-Feeney method for adjusting bias in observed period total fertility rates. *Demography, 38*(1), 17–28.

Zeng, Y., & Land, K. C. (2002). Adjusting period tempo changes–with an extension of Rdyer's basic translation equation. *Demography, 39*(2), 269–285.

Zeng, Y., & Vaupel, J. W. (1989). Impact of urbanization and delayed childbearing on population growth and aging in China. *Population and Development Review, 15*, 425–445.

Zeng, Y., & Wang, Z. (2003). Dynamics of family and elderly living arrangements in China: New lessons learned from the 2000 Census. *The China Review, 3*(2), 95–119.

Zeng, Y., & Wu, D. (2000). A regional analysis of divorce in China since 1980. *Demography, 37*(2), 215–219.

Zeng, Yi. (2009). Chinese fertility urgently needs soft-landing transition. *Social Science Weekly*, issues of Jan. 15 and Jan. 22, 2009 (continuously published in two issues).

Zeng, Y., Coale, A., Choe, M. K., Liang, Z., & Liu, L. (1994). Leaving parental home: Census based estimates for China, Japan, South Korea, the United States, France, and Sweden. *Population Studies, 48*, 65–80.

Zeng, Y, Chen, H., Wang Z., & Land, K. C. (2014a). Implications of changes in households and living arrangements for future home-based care needs and costs of disabled elders in China. *Journal of Aging and Health*.

Zeng, Y., Gu, D., Wang, Z., & Land, K. C. (2014b). Effects of changes in household structure and living arrangements on future home-based care costs for disabled elders in the United States. Manuscript re-submitted to a peer-reviewed journal based on reviewers' comments, for consideration of publication.

Zeng, Y., Vaupel, J. W., & Wang, Z. (1997). A multidimensional model for projecting family households—With an illustrative numerical application. *Mathematical Population Studies, 6*(3), 187–216.

Zeng, Y., Vaupel, J. W., & Wang, Z. (1998). Household projection using conventional demographic data. *Population and Development Review supplementary issue: Frontiers of Population Forecasting, 24*, 59–87.

Zeng, Y., Wang, Z., Ma, Z., & Chen, C. (2000). A simple method for estimating α and β: An extension of brass relational Gompertz fertility model. *Population Research and Policy Review, 19*(6), 525–549.

Zeng, Y., Stallard, E., & Wang, Z. (2004). Computing time-varying sex-age-specific rates of marriage/union formation and dissolution in family household projection or simulation. *Demographic Research, 11*, 264–299.

Zeng, Y., Land, K. C., Wang, Z., & Gu, D. (2006). U.S. family household momentum and dynamics–Extension of ProFamy method and application. *Population Research and Policy Review, 25*(1), 1–41.

Zeng, Y., Land, K. C., Wang, Z., & Gu, D. (2008a). *Households forecasting at the state and sub-state levels.* Invited paper presented at the conference of "Federal-State Cooperative Program for Population Projections, New Orleans, 16 Apr 2008.

Zeng, Y., Wang, Z., Jiang, L., & Gu, D. (2008b). Future trend of family households and elderly living arrangement in China. *GENUS-An International Journal of Demography, LXIV*(1–2), 9–36.

Zeng, Y., Chen, H., & Wang, Z. (2012a). Analysis on trends of future home-based care needs and costs for elderly in China. *Economic Research Journal, 47*(1), 134–149.

Zeng, Y., Morgan, P., Wang, Z., Gu, D., & Yang, C. (2012b). A multistate life table analysis of union regimes in the United States– Trends and racial differentials, 1970–2002. *Population Research and Policy Review, 31*, 207–234.

Zeng, Y., Land, K. C., Wang, Z., & Gu, D. (2013a). Household and living arrangements projections at the sub-national level: An extended cohort-component approach. *Demography, 50*, 827–852.

Zeng, Y., Li, L., Wang, Z., Huang, H., & Norris, J. (2013). Effects of changes in households and cohorts on future housing demand in Hebei Province, China. *GENUS, LXIX*(2), 85–111.

Zeng, Y., Shen, K., & Wang, Z. (2013b). Multistate projections and policy analysis of population and household aging in Eastern, Middle, and Western Regions of China in the 1st half of the 21st century. Manuscript submitted to a peer-reviewed journal for consideration of publication.

Zeng, Yi., Chen, H., Wang, Z., & Land, K. C. (2013c). Implications of changes in households and living arrangements for future home-based care needs and costs of disabled elders in China. Manuscript revised and resubmitted to *Journal of Aging and Health* based on reviewers' comments.

Zhang, Z. (2004). *Impact of intergenerational support on healthy longevity.* Ph.D. thesis, Beijing: Peking University. Supervisor: Professor Yi Zeng.

Zhang, J. (2007). Research on individual account contribution rate of the Chinese pension system based on the average life expectancy. *Population and Economics, 5*, 61–66 [in Chinese].

Zhang, G., & Zhao, Z. (2006). Reexamining China's fertility puzzle: Data collection and quality over the last two decades. *Population and Development Review, 32*(2), 293–321.

Zheng, Z. (2004). Fertility desire of married women in China. *Chinese Journal of Population Science, 5*, 73–78 [in Chinese].

Zhou, C., Zhang, Z., & Tao, B. (2000). Changing views on childbearing preference and motivation of rural daughter-only families. *Journal of Nanjing College of Population Programs Management, 16*(04), 34–38.

Index

Y. Zeng et al., *Household and Living Arrangement Projections*, The Springer Series on Demographic Methods and Population Analysis 36, DOI 10.1007/978-90-481-8906-9, © Springer Science+Business Media Dordrecht 2014

MIX
Papier aus verantwortungsvollen Quellen
Paper from responsible sources
FSC® C105338

If you have any concerns about our products,
you can contact us on
ProductSafety@springernature.com

In case Publisher is established outside the EU,
the EU authorized representative is:
Springer Nature Customer Service Center GmbH
Europaplatz 3, 69115 Heidelberg, Germany

Printed by Libri Plureos GmbH
in Hamburg, Germany